PHOTOGRAPHIC REGIONAL
ATLAS OF BONE DISEASE

Third Edition

PHOTOGRAPHIC REGIONAL ATLAS OF BONE DISEASE

A Guide to Pathologic and Normal Variation in the Human Skeleton

By

ROBERT W. MANN, PH.D., D-ABFA, FCPP

and

DAVID R. HUNT, PH.D., D-ABFA

With a Foreword by
O'Brian C. Smith

With an Introduction by
Donald Ortner

CHARLES C THOMAS • PUBLISHER, LTD.
Springfield • Illinois • U.S.A.

Published and Distributed Throughout the World by

CHARLES C THOMAS • PUBLISHER, LTD.
2600 South First Street
Springfield, Illinois 62704

2012 by CHARLES C THOMAS • PUBLISHER, LTD.

ISBN 978-0-398-08826-2 (hard)
ISBN 978-0-398-08827-9 (ebook)

Library of Congress Catalog Card Number: 2012016071

First Edition, 1990
Second Edition, 2005
Third Edition, 2012

With THOMAS BOOKS *careful attention is given to all details of manufacturing
and design. It is the Publisher's desire to present books that are satisfactory as to their
physical qualities and artistic possibilities and appropriate for their particular use.*
THOMAS BOOKS *will be true to those laws of quality that assure a good name
and good will.*

Printed in the United States of America
UB-R-3

Library of Congress Cataloging-in-Publication Data

Mann, Robert W
    Photographic regional atlas of bone disease : a guide to pathologic and normal variation
in the human skeleton / by Robert W. Mann and David R. Hunt ; foreword by O'Brian C.
Smith ; introduction by Donald Ortner. – 3rd ed.
        p. cm.
    Includes bibliographical references and index.
    ISBN 978-0-398-08826-2 (hard) – ISBN 978-0-398-08827-9 (ebook) 1. Bones–Diseases–
Atlases. 2. Paleopathology–Atlases. I. Hunt, David R. II. Title.

RC930.4.M35 2012
616.7'100222–dc22
                                                                                                          201201671

Photo: Chip Clark

Dedicated to the memory of
Donald J. Ortner...visionary,
luminary and friend.

FOREWORD

Art is I, Science is We. – Claude Bernard

Enthusiasm. The most motivating force in a student is enthusiasm. Many bring it with them, already on fire for their particular area of interest. Most though are infected with it by their instructors and fellow students as a sense of discovery, for advancement and competency develops. Enthusiasm dwarfs things petty to science; egos, attitudes, personal agendas, and the like. It creates an aura of academic purity, an environment without fear where "we" is paramount, and "I" becomes a measure of capacity, not importance. It is a time where we can be smart together and we can be dumb together without pride or fear.

Cultivating enthusiasm is one of the hardest tasks for an educator, especially in students just entering an area of study. Many disciplines have their own language, because it requires precisely defined concepts to advance the field. The introductory student needs to acquire some of this to be facile in developing his/her knowledge and thinking, but too much can intimidate and dampen enthusiasm. The educator, well versed in terminology, needs to introduce his/her topic in the language of the layman in order to assure communication. This latter is not an easy task because precision of concept suffers.

It is a bold step then for any introductory text to be written especially for the entering student. Colleagues who have already achieved their knowledge-base can always be critical of the authors' license and charge oversimplification; and in part will always be right. My reply is that I've rarely found reference books to have a well-thumbed appearance. If I have to choose between precision and enthusiasm for the new student, it will always be enthusiasm! If the fire gets stoked, the opportunity for full potential is achieved.

Let the above be enough to explain this book to my colleagues. I hope too, that they will learn some things from the authors, because I did. For you, the most important reader, the newest generation, I welcome you as colleagues and invite you to these pages. Read! Enjoy! Discover! Think!

O'BRIAN C. SMITH, M.D.
Professor of Pathology;
University of Tennessee School of Medicine,
    Knoxville
Chief Medical Examiner;
State of Tennessee

INTRODUCTION

Careful description and classification are basic methodological tools in all categories of science. This is particularly the case in biomedical research where substantial resources are invested in a continuous process of refining diagnostic criteria (classification) for various diseases. The field of paleopathology has been slow in confronting some long-standing problems in description and classification and this has limited its development. The result is a substantial proportion of the existing literature that is of minimal value in clarifying many of the broader questions that must be addressed if paleopathology is to reach its full potential as a legitimate biomedical discipline.

For example, it would be very helpful to have a database that provides information on the antiquity, geographical distribution and evolutionary trends of disease. We also need data that will help to clarify the evolution of complex relationships that exist between the many factors that affect the human response to disease, including: (1) the pathogenic agent, (2) environmental factors (e.g., air pollution) that affect health, (3) nutrition and (4) the immune response of a patient to disease. However, without a clearly defined and generally accepted descriptive terminology and classificatory system it is difficult, if not impossible, to compare the research of one scientist with that of another in building a relevant base of data.

An important contribution to the study of skeletal paleopathology would be the development of a rigorous method to describe the abnormal conditions encountered in archeological human skeletons. It is both surprising and frustrating that after 150 years of research in paleopathology there is still much to do in creating a careful and comprehensive descriptive terminology, as well as a general classification of the abnormal conditions encountered in skeletal specimens. Much of how we describe pathological conditions in archeological skeletons is derivative of medical and particularly orthopedic nomenclature and classificatory systems. These systems continue to develop and staying conversant with current usage is a challenging exercise. The major problem, however, is not one of semantics. Rather it is that many of the lesions and their distribution patterns in archeological dry bone specimens bear minimal relationship to descriptive and classificatory features that are central in clinical orthopedic practice. What is crucial for paleopathology is a nomenclature and classificatory system that integrates all of the pathological information that is apparent in skeletal paleopathological specimens. Such a system would necessarily include orthopedic terms and classification where the features were closely related to those used in a clinical setting. There are, however, occasional conditions in paleopathological cases that are not well known in clinical orthopedic practice and a precise classificatory system might demonstrate relationships that previously had not been understood.

In working with both professional colleagues and graduate students I have, for many years, emphasized the importance of first describing carefully what one sees in cases of skeletal paleopathology. Careful description is timeless and, if done well, forever gives future readers of reports the option of reinterpreting your conclusions (i.e., diagnoses). Demographic data, including age and sex, are important factors in interpreting descriptive information. However, the most important element in paleopathological research is the basic description of abnormal bone including the type and distribution pattern within the skeleton. There are four basic abnormalities of bone: (1) abnormal size, (2) abnormal shape, (3) abnormal bone formation and, (4) abnormal bone destruction. There are additional features associated with these general abnormalities that provide helpful supplemental information. For example, is the abnormal bone formation poorly organized (this typically means rapid growth) or well organized (usually slow growth)? Do destructive lesions have well-defined margins with evidence of well-organized bony repair (circumscribed and generally less aggressive) or poorly-defined margins (permeative and generally more aggressive)? These and other features are all critical elements in any interpretation of a paleopathological case of skeletal disease.

The location of lesions within the skeleton provides an important link with clinical experience but one needs to be cautious in making such associations. In dry-bone paleopathological cases one often sees lesions that would not be apparent in clinical radiographs and are thus not well documented in the medical descriptive and classificatory systems. Indeed the added information on skeletal lesions is one of the significant potential contributions that careful study of paleopathological cases can make to a more complete understanding of the skeletal manifestations in orthopedic pathology. A pathology based on dry-bone conditions also means that some distribution patterns of abnormal tissue within a pathological skeleton will vary from patterns established on the basis of radiology in living patients.

Careful description is not easy and I do not wish to underestimate the difficulty of the process. Nevertheless, most people can, with discipline, learn to recognize the essential features of bone reaction to disease. The first step is, of course, a thorough knowledge of normal gross anatomy of bone at all ages from fetal through old age. Archeological skeletal samples are a wonderful source of anatomical knowledge since the entire age spectrum is usually represented. Classification or diagnosis is a much more complicated matter and for many cases encountered by the researcher investigating paleopathology, years of experience and a comprehensive knowledge of orthopedic pathology may be necessary and, even so, may not be possible.

For those conducting research on skeletal paleopathology great attention needs to be paid to using a well defined and widely recognized terminology in describing pathological skeletal lesions. Excellent reference materials in radiology exist such as Resnick's five-volume work (2002). The second edition of my book on human skeletal paleopathology (Ortner 2003) may also be helpful in highlighting both the terminology and the diagnostic options for some cases of skeletal disease.

I am optimistic that further advances will be made in developing descriptive and classificatory methodology for paleopathology in the near future. In the meantime it is important to use descriptive terms and modifiers that are part of the general lexicon we all share. Bone addition, bone destruction, porous bone, and destructive lesions are examples of terms that are descriptive and have wide recognition in many disciplines and I encourage their use. Jargon, is one of the biggest barriers to effective communication that exists and should be eliminated or, at the very least, kept to a minimum. At some point, however, you will need to acquire a working knowledge of medical terminology if only to understand and interpret the existing literature on paleopathology and communicate with medically trained colleagues.

The third edition of the *Regional Atlas of Bone Disease* is a new attempt to assist the beginning skeletal paleopathologist to recognize some of the more common pathological conditions that may be encountered in dry-bone specimens. The authors have provided new examples and cases to illustrate their points but continue to insist that their endeavor be viewed as an initial step in any classificatory process. This is wise counsel, given the complexity of classification. One of the fundamental problems for any classificatory system is that the bone reaction to disease is limited. In view of this it is not surprising that a given pathological condition (i.e., osseous response) may be the result of any one of several pathological processes.

The reader should also be aware of the strengths and weaknesses of a regional approach to skeletal paleopathology. Archeological skeletal samples often do not have complete skeletons. This is particularly true of older museum collections where only the skull and mandible may have been recovered. However, even where an attempt was made to excavate the entire skeleton the result is usually only partially successful. In this context a regional review of pathological conditions may be the only one possible and is certainly helpful. It is also true that many pathological conditions occur in a single location in the skeleton (solitary or unifocal conditions). A regional focus is generally adequate for such lesions.

However, a regional approach is less helpful in multifocal pathological conditions. In this type of skeletal paleopathology, the distribution pattern of abnormal bone is a critical element in classification and the user of a regional approach will need to reconstruct the overall pattern by carefully reviewing the information for each region of the skeleton. A review of the distribution pattern of abnormal bone is important for classification but also contributes to the general understanding of pathogenesis in orthopedic disorders.

Despite this cautionary note, the beginning skeletal paleopathologist should find the new edition of the Regional Atlas a helpful starting point when he or she encounters a skeletal abnormality in archeological burials. Remember, however, first provide a careful and detailed description of the abnormalities you see including the nature of the abnormalities and their location in the skeleton. An attempt at diagnosis can then be made with the assurance that others will at least have the option of reaching a different diagnostic conclusion on the basis of the descriptive information you have provided should that be appropriate. The authors' counsel to seek advice on diagnosis from specialists in skeletal disease is wise. Keep in mind, however, that very few medical specialists have experience with dry-bone specimens and are often as baffled by a pathological case as is the osteologist. The orthopedist does, however, have the advantage of knowing what most of the diagnostic options are and this is a very useful point of departure.

DONALD ORTNER, Ph.D.
Department of Anthropology
National Museum of Natural History Smithsonian
Institution

ACKNOWLEDGMENTS

The authors would like to extend a special debt of gratitude to Dr. Donald Ortner, Dr. Douglas W. Owsley, Mr. Paul S. Sledzik, and Mr. Sean P. Murphy to this enterprise. Each of these scientists played a significant role not only in the development of this book, but its contents. Dr. O'Brian C. Smith contributed substantially to writing the Foreword and Chapter IV. The authors would also like to express their gratitude to Drs. J. Lawrence Angel, Ethne Barnes, William M. Bass, Hugh E. Berryman, Bruce Bradtmiller, Ms. Kari Sandness Bruwelheide, Mr. Henry W. Case, Drs. Leslie E. Eisenberg, Eugene Giles, Thomas D. Holland, Lee Meadows Jantz, Richard L. Jantz, Ms. Erica Bubniak Jones, Drs. Marc A. Kelley, Linda Klepinger, Ms. Marilyn R. London, Drs. Keith A. Manchester, Marc S. Micozzi, Bruce Ragsdale, Charlotte A. Roberts, Jeno I. Sebes, T. Dale Stewart, Douglas H. Ubelaker, and P. Willey. It was through their friendship, teaching, and professional guidance that this book came to fruition. All illustrations were drawn by Robert W. Mann, except Figure 124 (Elizabeth C. Lockett) and Figure 93b (Neil Fallon). Drawings and most of the photographs were based on anatomical specimens at the Smithsonian Institution and Khon Kaen University Department of Anatomy, Khon Kaen, Thailand. Unless stated otherwise, all photographs were taken by the authors. Mr. Chip Clark, of the National Museum of Natural History, Smithsonian Institution, Mr. Hugh E. Tuller and Dr. Joseph T. Hefner of the Central Identification Laboratory graciously allowed us to reproduce several of their photographs. Dr. Panya Tuamsuk, Dr. Kamoltip Brown and Dr. Kowit Chaisiwamongkol provided unrestricted access to the Osteological Collection and records at Khon Kaen University, Thailand. Ms Anna Dhody and Dr. Robert Hicks of the Mütter Museum in Philadelphia, Pennsylvania facilitated access to their unique osteological and anatomical collections and provided helpful insights on many of the specimens. Ms. Evi Numen, Exhibits Manager of the Mütter Museum graciously provided us with the photographs of Mr. Harry Eastlack that grace the front and back covers of this book. Dr. Janet Monge and her assistant Mr. Paul Mitchell of the University of Pennsylvania Museum of Archaeology and Anthropology in Philadelphia allowed access to the Morton Collection and provided assistance and historical information on many of the specimens. The opinions expressed in the Regional Atlas are the sole responsibility of the authors. Last but certainly not least we thank our wives, Vara J. Mann and Kim Wells, and our parents Adele D. Mann and Arlys Roehm Hunt and James C. Hunt for their love and support.

CONTENTS

PHOTOGRAPHIC REGIONAL
ATLAS OF BONE DISEASE

Chapter I

USING THE PHOTOGRAPHIC REGIONAL ATLAS

The information contained in the *Photographic Regional Atlas of Bone Disease* – herein referred to as the Regional Atlas – is based on paleopathological examination of more than 10,000 complete or nearly complete skeletons from archaeological sites and forensic investigations throughout the world. The majority of these studies are from prehistoric collections from North America, particularly from the Great Plains, Pacific Coastal regions and the Northeastern United States; historic cemeteries and burials from Canada, Louisiana, Maryland, Nevada, Virginia, Washington, D.C. (including War of 1812, Civil War soldiers and iron coffin burials); as well as over two hundred forensic cases including Americans missing in action (MIA). Skeletal collections from Africa, Australia, Central Asia, and South America have also been investigated by the authors. Contemporary skeletal samples have been studied by the authors from the Hamann-Todd (Cleveland, Ohio), Robert J. Terry Anatomical and George S. Huntington Anatomical (Smithsonian Institution) collections, and the Osteology Collection in the Department of Anatomy, Faculty of Medicine at Khon Kaen University, Khon Kaen, Thailand.

The *Regional Atlas* approaches the recognition of disease according to the bone affected. The format of this handbook begins with a description of how to use the *Regional Atlas* (Chapter I), followed by a brief history of paleopathology (Chapter II). Chapter III gives step-by-step instructions on how the authors conduct a paleopathological analysis. Chapter IV briefly covers the mechanics of bone remodeling. The bulk of the *Regional Atlas* is Chapter V. This chapter deals with specific diseases affecting each bone in the body, beginning with the skull and progressing down the skeleton. Accompanying some lesion descriptions is a statement of the relative frequency (e.g., uncommon to rare finding) or percentage that one might expect to find in most archaeological skeletal samples, in most cases for Native American groups since the majority of the author's studies encompassed these populations.

References cited within a sentence indicate that the information was derived from these sources. References at the end of a paragraph (following the period or "cf.") were included as additional sources for the reader seeking additional information. Many of these references are the seminal reports of research for these pathological conditions or anomalies or extensively cover the condition. It is not necessary to reiterate the information published and available in these volumes.

The reader will find that many of the references used in this book were culled from the medical, clinical, anatomical, and radiological literature rather than the anthropological literature. The reason is multiple. First, clinical studies and case reports provide information based on findings, often accompanied by patient histories, known age, race, and sex of the individuals in living groups. Anthropological studies, in comparison, tend to focus on samples of unknown age, race and sex individuals in recent or ancient groups. Second, most diseases can be specifically identified in modern clinical studies, but not necessarily in ancient remains.

Chapter VI provides information on fungal infections. The treponematoses (i.e., syphilis and allied conditions) are summarized in Chapter VII. Chapter VIII briefly discusses tumors, perhaps the most difficult skeletal condition to be diagnosed. Chapter IX discusses perimortem fractures and postmortem breakage. Chapters VI through IX are designed to only briefly present the effects of these pathological skeletal conditions on the human skeleton – the references cited in those chapters much more extensively cover these particular diseases and should be referred to by the reader for more in-depth research.

Chapter X, "Entheseal Change," covers a much discussed and disputed topic variously identified as muscle markers of stress, enthesophytes, and activity-induced stress markers, to name but a few. Chapter XI presents a case report reflecting the advanced stages of a disease rarely encountered in osteological collections, reflecting the body's extraordinary capacity to deal with a disease condition that at first blush appears to be incompatible with life. As an anatomical overview, dorsal and ventral views of the human skeleton are provided in Chapter XII. The major muscle attachments are illustrated in Chapter XIII. Chapter XIV consists of many "larger than life" color photographs (plates) of common, uncommon and even rare diseases, lesions, and a few non-metric traits that are sometimes confused with lesions that one might encounter in a skeletal sample. These color photographs expand our knowledge of disease and allow the reader to see lesions with sufficient clarity and detail that might otherwise be missed in black and white or smaller photographs. The authors have learned firsthand how frustrating it is to try and find a lesion or subtle skeletal feature referred to in a figure caption, but not highlighted in the image or photo.

This book was rewritten after being "field tested" for more than two decades by physicians, students, and paleopathologists around the world. The evolution of this book, built upon the shoulders of such luminaries as Thomas Dwight, Aleš Hrdlička, J. Lawrence Angel and T. Dale Stewart, was the result of trial and error, countless trips to medical and other libraries, web searches, and a learning process that is ongoing. Truly, the more we the authors study the human skeleton, the more we are humbled by how much there is still to learn. Having worked at the Smithsonian Institution and afforded the opportunity to peruse the vast skeletal collection and Smithsonian libraries over the years, we are in awe of the endless possibilities for the subtle and not-so-subtle variation present in the human skeleton, This awe, of course, focuses not only on the extent of skeletal variation, but its etiology, as well as geographical and temporal distribution.

In order to stay current, updated references and findings in the field of archaeology, paleopathology, anatomy, and medicine have been incorporated into the *Regional Atlas* since the 2005 edition. To provide a more historical perspective that helps us "trace our roots" in paleopathology and anatomy, the authors have combed the early literature in search of the first reported examples of some pathological lesions and conditions bearing the names (eponyms) of their "discoverers." These groundbreaking anatomists and physicians include Jean-Martin Charcot (1825–1893; Charcot's joint), James Paget (1814–1899; Paget's disease), Bartolomeo Eustachi (c. 1524–1574; Eustachian tube), Antonio Pacchioni (1665–1726; Pacchionian depressions), and Sir Percival Pott (1714–1788; Pott's disease of the spine), to name but a few. This book also reflects the authors' own experiences examining more than 10,000 human skeletons from around the world since the *Regional Atlas* was first published in 1990. Diseases, lesions, and skeletal anomalies too rare to be expected to be encountered in a routine skeletal analysis have been removed and replaced with those that might be expected to be encountered in most skeletal collections around the world.

It should be remembered that no text could fully or even adequately cover or explain the etiology or geographical distribution of every disease, anomaly, or normal anatomical variant present in the human skeleton; the present text is no exception. While some topics in the *Regional Atlas* are discussed in great detail, others are conspicuously brief owing to their extreme difficulty in differential diagnosis, or rarity in most skeletal collections (e.g., tumors). One goal of the *Regional Atlas* was to include the findings and hypotheses of contemporary clinical practitioners (e.g., paleopathologists, radiologists, etc.) to supply the reader with a number of interpretations from which to choose. Such an approach also serves to inform the reader of the complexity and controversy surrounding the identification, classification, and etiology of many bone diseases, as well as the vast array of information that is available in the literature.

It is hoped that the experiences of the authors will make it possible for anyone with a sound knowledge of human osteology and skeletal morphology to conduct a basic **descriptive** paleopathological analysis of one or many skeletons. It should be noted, however, that the field of paleopathology is filled with ambiguities and subtleties. Committing this atlas to memory doesn't make one a paleopathologist; only knowledge, training and, above all, experience will qualify you for such a title. The *Regional Atlas* will, however, enable you to conduct your own analysis and, in questionable cases, alert you to seek the advice of an experienced paleopathologist, radiologist, orthopaedist, or other specialist. The importance of a thorough descriptive analysis and utilization of an accepted vernacular for paleopathology, however, cannot be overemphasized.

To use the *Regional Atlas*, first become familiar with what and where lesions, conditions, and anomalies might be expected in the skeleton, locate and identify them in the text, and then refer to the excellent paleopathology, developmental and clinical texts by Allison and Gerszten (1982), Aufderheide and Rodriguez-Martin (1998), Barnes (1994), Beighton (1978), Brothwell and Sandison (1967), Cockburn and Cockburn (1980), Currey (2006), Dieppe et al. (1986), Greenfield (1975), Hauser and DeStefano (1989), Jarcho (1966), Keats (1988), Manchester (1983), McCarty (1989), Morse (1969), Moskowitz et al. (1984), Ortner (2003), Ortner and Aufderheide (1991), Ortner and Putschar (1981, 1985, 1988), Ortner (2002, 2003, 2008), Pinhasi and Mays (2008), Resnick (2002), Resnick and Niwayama (1981, 1988), Robbins (1968), Rogers and Waldron (1995), Steinbock (1976), Thijn and Steensma (1990), Tyson and Dyer (1980), Waldron (2008), Webb (1995), Wells (1964), Zimmerman and Kelley (1982), or other references in the text, especially the *Journal of Bone and Joint Surgery* that deals primarily with the skeleton, the *Journal of Anatomy and Physiology* and *Spine* (as well as numerous radiology journals) that cover a vast array of clinical (patient-based), osteological, imaging, and anatomical research topics. Refer also to paleopathology bibiliographies compiled by Crain (1971) and by Elerich and Tyson (1997). While some of these texts may seem outdated, they continue to serve the scientific and medical community as some of the most relevant and useable texts in circulation to date. It is hoped that the *Regional Atlas* will serve as a valuable companion to the existing paleopathology literature.

The illustrations in this atlas are predominately specimens from the National Museum of Natural History (Smithsonian Institution), the National Museum of Health and Medicine (Armed Forces Institute of Pathology), Washington, D.C., Khon Kaen University, Thailand, Mütter Museum of the College of Physicians of Philadelphia and the University of Pennsylvania Museum of Archaeology and Anthropology. Catalog numbers of the particular specimens are included where appropriate. Other specimens not from these particular institutions are indicated as to their origin. To avoid continuous lengthy location and collection's names, the abbreviations below will be used for many of the specimens:

AFIP	National Museum of Health and Medicine, Armed Forces Institute of Pathology, Washington, D.C.
CSC	CIL Study Collection, JPAC
CUJ	Chiba University Japan
FSA	Forensic Science Academy
KKU	Khon Kaen University, Khon Kaen, Thailand
MM	Mütter Museum
NMNH	National Museum of Natural History, Smithsonian Insitution, Washington, D.C.
NMNH-H	George Huntington Collection
NMNH-T	Robert J. Terry Collection
UHWO	University of Hawaii, West Oahu
UPENN	University of Pennsylvania
UTK	University of Tennessee (Knoxville), Forensic Anthropology Center, William M. Bass Collection

Chapter II

A BRIEF HISTORY OF PALEOPATHOLOGY

Paleopathology is the study of disease in pre-modern anthropological and paleontological specimens as well as historic and anatomical skeletal collections. Research on this subject dates back nearly two hundred years and has evolved from an area devoted to the study of medical curiosities and exploration of the human body to an integrated discipline of medicine. Johann Esper, who, in 1774, reported on a pathological femur of a cave bear from France, made one of the earliest studies on bone diseases in dry specimens. Since this earliest study, some scholars have drawn arbitrary temporal divisions marking periods where the primary focus of paleopathology has shifted. For example, before 1900 the focus was on traumatic lesions and syphilis, and between 1900 and 1930 was on infectious disease. Today, the focus is comparative and multidisciplinary with ecology as a major component (Buikstra and Beck 2006; Larsen 1997; Ubelaker 1982).

Some of the more notable early researchers in paleopathology include A. Hrdlička, D. Brothwell, C. J. Hackett, E. A. Hooton, S. Jarcho, J. Jones, V. Møller-Christensen, R. L. Moodie, M. A. Ruffer, G. E. Smith, R. Virchow, F. Wood Jones, and C. Wells to name a few. Each of these researchers made considerable contributions to the early study of bone disease in humans and are well known in many scientific circles.

Although some early researchers studied large skeletal populations (e.g., Hooton 1930; Smith 1910), the population approach to disease frequencies was used by only a few, Wyman (1868) and Hrdlička (1914) being notable examples. Because of the seemingly unlimited field of opportunity and discovery in paleopathology, much of the focus of bone disease was on the more exciting and rare diseases and trauma such as tuberculosis, leprosy, syphilis, trephination, rickets, and the like. Skeletons exhibiting syphilitic-like lesions served to fuel many heated debates (ongoing) and spawned a number of hypotheses concerning its diagnosis in dry bone, as well as its time and place of origin and geographical dispersion.

While the "exciting" diseases were being debated by some of the world's leading scholars, other researchers around the turn of the twentieth century focused on the less exotic skeletal indicators of health such as cribra orbitalia (Welcker 1888) and symmetric osteoporosis (Hrdlička 1914). All of the early researchers had to rely on gross examination of the bones because radiographs, refinements in biochemistry, cytology (cell technology), and soft tissue pathology lay before them. Macroscopic examination and a descriptive analysis was the primary method of diagnosing bone disease when the "patient" was an archaeological skeleton with no clinical history. However, this did not stop some paleopathologists from offering their opinions and giving the disease a name (differential diagnosis), including ones that were incorrect. Although the cornerstone of contemporary paleopathological analysis continues to be the descriptive analysis, researchers now have at their disposal extremely sophisticated radiologic, immunologic, and microscopic techniques and imaging such as computer assisted tomography (CAT), computerized tomography (CT), and magnetic resonance imaging (MRI) to aid in diagnosing skeletal lesions.

Contemporary paleopathology has carried on many of the traditions established by its forefathers. For example, much debate still continues on whether syphilis-like lesions in archaeological bones represent venereal syphilis or one of the other treponematoses (e.g., yaws, pinta, endemic syphilis); the age-old debate of whether the Europeans spread syphilis to the New World also has yet to be resolved, although recent phylogenetic DNA studies by Harper and colleagues (2008) report evidence supporting the "Columbian theory." Research by Pineda and colleagues (2009), however, has yielded numerous skeletons with skeletal manifestations of treponematosis in Mexico. One of the bones was carbon dated to AD 1100 to AD 1300, suggesting treponematosis was present in this group prior to European contact. Research by Marden and Ortner (2011) of the skeleton of a middle-aged female from Chaco Canyon (dated to between AD 950 and AD 1150), Pueblo Bonito, New Mexico, revealed lesions most consistent with acquired syphilis, although they might also reflect any of the other three forms of *t. pallidum* (treponematosis); this case provides strong evidence of pre-Columbian treponematosis in Chaco Canyon. Cf. Akrawi (1949), Baker and Armelagos (1988), Bloch (1908), Brothwell (1970), Cockburn (1961), Dutour et al. (1994), Hackett (1967), Holcomb (1930, 1935), Kaye (2008), Livingstone (1991), Ortner (2003), Powell and Cook (2005), Pusey (1915), Rothchild (2005), Rothchild and Rothchild (1995), and Williams (1932).

While the exotic diseases are still of keen interest to paleopathologists, an added emphasis has emerged focusing on the subtle bony changes that reflect physical stresses and activities of everyday life (e.g., mild indicators of osteoarthritis), nutrition, and diet (Angel 1966, 1976, 1981; Angel et al. 1987; Cook 1979; Cook and Buikstra 1979; Dale 1994; İşcan and Kennedy 1989; Mann et al. 1987; Owsley et al. 1987; Schoeninger 1979; Sillen 1981).

Paleopathologists, through the cumulative studies of disease in the human skeleton, now have a better understanding of how many diseases developed and spread both geographically and temporally. The "mundane" skeletal indicators of physical stress, nutrition, and diet are now studied with the same enthusiasm as the exotic diseases. Only by including the full range of bone disease and testing hypotheses are we able to make valid biocultural conclusions on the general and specific health status of skeletal populations. As the late Dr. Larry Angel (1981:513) once wrote, "We are still all amateurs in paleopathology looking toward a bright cooperative future." For a more thorough history of paleopathology refer to Aufderheide and Rodriguez-Martin (1998:1–10), Jarcho (1966), Ortner (2003:8–10), and Ubelaker (1982).

Chapter III

PREPARING FOR A BASIC PALEOPATHOLOGICAL ANALYSIS

In preparing to examine a skeletal series it is important to first determine the focus of the study. It is simply not possible to gather every bit of information on every skeleton. While one researcher might think it imperative to take bone core samples for lead and other trace mineral analyses, another may restrict his or her research to nondestructive techniques. Prior written permission should always be obtained before performing any destructive or invasive bone studies.

Before beginning the skeletal analysis you might first like to review, or at least have at hand, some of the more comprehensive texts on human osteology, including Bass (1995), Brothwell (1981), Krogman and İşcan (1986), Shipman et al. (1985), Steele and Bramblett (1988), Stewart (1979), Ubelaker (1999), and White and Folkens (1991). As with any project, planning and organization are the keys to a successful skeletal study.

Before going further, we would like to clarify the differences between "pathology" and lesion or pathological condition. A "pathology" is not synonymous with a lesion but, rather, the *study* of disease; a lesion is the response to disease or a wound (Thomas 1985). Although many in the scientific community, including learned paleopathologists and clinicians, continue to use the words synonymously, the accurate term for a disease state in bone is lesion, wound, injury, or pathological condition.

The following suggestions are offered as guidelines on how to prepare for a paleopathological analysis. The technique can be applied to one or one thousand skeletons and has proven to be comprehensive enough to satisfy the needs of most researchers. Much of this technique was developed by Doctors Douglas Owsley and Bruce Bradtmiller in the 1980s and has since been successfully implemented on perhaps more than 10,000 skeletons.

First, lay out each skeleton on a separate tray in the correct anatomical position and, if at all possible, have a real bone skeleton at hand for comparison. For example, place the ribs in proper order according to number and side (e.g., left 1st, 2nd, 3rd, etc.) (see Mann 1993). The vertebrae should also be laid out beginning with the first cervical and ending with the fifth lumbar (possibly using a "spine tray" when being photographed; see Mann 2009). The bones of the hands and feet should be separated by side; with practice the hand and wrist bones, with the possible exception of the distal phalanges, can be identified, numbered, and sided (Case and Heilman 2006) (which is important in identifying defense wounds of the hands). This method allows comparison of bones from different trays (or archaeological features) for evaluation, comparison and pair-matching (antimeres) without commingling the bones.

When working with commingled skeletons (i.e., two or more skeletons mixed together) one method is to cut standard 3 × 5 index cards in half, punch a hole in one corner, affix a rubber band, and attach a card to each bone with a burial or feature number. This method allows bones to be pulled from various trays or archaeological features and compared for matches without commingling the remains or writing on them. Each element can later be put back on its original tray if no matches are made. Tagging the bones also allows them to be put aside for photography and radiography. After fragmentary skeletal elements are reassociated, reconstruction of the elements with glue and/or masking tape can be performed and then inventory and measurements, and a biological profile including the individual's age at death, ancestry (race), sex, and stature can be compiled. If you are working a skeletal case for the police, medical examiner or coroner, check to see if you can or should number each bone fragment as evidence. Each element in archaeological, historical and donated skeletons, however, usually doesn't have to be numbered unless it is necessary to track the elements for excavation location and provenience.

Regarding the topic of reconstructing bones, brief deviation concerning the use of particular adhesives on skeletal remains is appropriate. The choice of adhesive will depend on a number of factors, including whether the skeleton will be subject to additional forensic investigations where the introduction of adhesive chemicals may be detrimental to other tests, such as DNA, trace element and isotopic analysis. In archeological contexts, questions remain of whether the skeletons will be reburied, how long they will be available for study, will the bone be dated by C-14, and is the reconstruction planned for long-term curation (Greta Hansen, personal communication 1989). In the latter questions, the stability of the adhesive is of interest. Although Duco™ cement has been widely used for reconstructing bones; in time it becomes dry, yellow, brittle, and separates at the glued "joints." The inherent instability of Duco cement renders it unsuitable for long-term use. If the reconstruction is meant to be permanent, a better choice is Acryloid B-72 or B-48N (available from Rohm and Haas) or solvented polyvinyl acetate (PVA) (available from Union Carbide Corporation). For more information on reconstructing, preserving, and consolidating bone, consult an objects conservator or refer to Johansson (1987) or Sease (1987).

The next step in conducting a skeletal examination is to examine each bone for pathological conditions and lesions. The most important technique in the analysis is examining each bone closely, using a standard fluorescent lamp. Hold the bone approximately 3–5 inches from the lamp and carefully inspect the surface inch by inch. Do not pick up a bone and quickly examine it for obvious pathological changes since many subtle lesions and, possibly, cut marks from scalping (Neumann 1940; Ortner 2003; Steinbock 1976) or defleshing (Ubelaker 1999) can easily be overlooked. It's a good idea to examine each bone by holding it in different positions to change the angle of the light as it strikes the bone. This, too, may reveal a small lesion or trauma that was formerly hidden by shadows or it may reveal a cut, lesion or other defect by casting its own shadow. It should be stressed that the use of a hand magnifying lens or an illuminating magnifying lamp is necessary to observe subtle features on or in the lesion or defect. If you find yourself staring at a questionable lesion, put the bone aside and return to it later. You might use this time to reference other paleopathology texts to help clarify your questions. Being consistent with your scoring criteria (e.g., mild, severe), for judgment is critical.

If, after examining ten or fifteen skeletons, you find that you have been scoring a lesion as moderate in severity and feel that it should only be scored as slight, then you should modify the criteria for the trait or lesion in question before going further. Obviously, it's easier to go back and change data sheets for ten or twenty skeletons than to realize this after you've analyzed one hundred or more. At such a late point in your examination, you might find that you've run out of time or are too frustrated to reexamine the skeletons. Be observant early in your analysis and note those conditions that are ambiguous in either severity or presence/absence. After the skeletons have been examined for pathological conditions, bones needing to be radiographed and/or photographed should be clearly labeled and set aside.

Note at the top of each inventory sheet and notes (separate pages for each individual) the site information, burial number, age, sex, and any unusual features present in the bone, or associated with the particular burial (e.g., "Projectile point in 3rd lumbar. Photo/x-ray"). This information will later save you time. Also ensure that you put the examiner's name and date of the research on each page for future researchers to reference.

Finally, using a laptop computer will save you many hours of deciphering, reexamining and rewriting laboratory or field notes for each burial. This technique is especially important in conducting a thorough **descriptive** paleopathology analysis as it is sometimes easier to type than write by hand. If you use the computer as a source of recording your methods, findings, and thoughts, a large part of the write-up will already be done; this "rough draft" can then be edited and will produce a very detailed report. (You'd be surprised how much information is forgotten after the analysis is completed.) The following is an example of how a skeleton might be described, although many detailed analytical, scoring sheets and diagrams are available to the reader:

Skeleton 1. Present is the poorly preserved skeleton of a young adult male, perhaps 20–25 years of age. Present are the cranium, mandible (missing the right ramus), both femora, tibiae, left innominate, right foot, and left clavicle.

Age is based on examination of the pubic symphysis (billowy), auricular surface of the ilium (no microporosity), dental development and eruption, cranial and maxillary suture closure (the incisive is nearly obliterated), and medial epiphysis of the clavicle (early stage of formation).

Sex is assessed as male based on the morphology of the innominate (very narrow sciatic notch and triangular pubic bones), robust skull (well-developed nuchal crest, large mastoids, pronounced browridge, and a sloping frontal bone), and a left femoral head measuring 54 mm in diameter.

Pathological Conditions and Trauma. There is an ovoid, healed depressed fracture measuring 23 mm × 38 mm in the frontal bone immediately above the right orbit. Fracture lines extend from the center of the injury and radiate into the superior orbital plate. There is no evidence of healing, which suggests that this injury may have occurred at or near the time of death (perimortem). The inner vault of the skull was not affected, as there are no fractures or depressed bone visible endocranially. The shape of the injury suggests that an oblong object struck this individual above the right eye traveling from the decedent's left to right side, from front to back, and from above to below (these three directions are easy to understand and provide the reader with enough information to recreate the event). Examination of the remainder of the skeleton reveals no further evidence of trauma (it should be noted, however, that many bones are missing). Cf. Browner et al. (2003), Galloway (1999), Kimmerle and Baraybar (2008).

The distal right femur exhibits a small area of healed periostosis along the lateral surface. There is slight porosity (osteoarthritis or OA) in both temporal fossae (temporomandibular joints) and the right mandibular condyle is slightly flattened. The teeth show moderate wear, two periapical abscesses of the right mandibular first and second molars, and slight linear enamel hypoplasia.

Comments and Observations. The distal left radius exhibits a small area of green copper salts staining (about the size of a quarter) on its posterior surface (need to photo). The location of the stain suggests that a bracelet or other metallic artifact was in contact with the left wrist at the time of burial. The field notes do not, however, report any artifacts recovered in association with the burial.

Chapter IV

FUNDAMENTALS OF BONE
FORMATION AND REMODELING

The purpose of this chapter is to give the reader a brief account of normal and pathological features of remodeled bone in the human skeleton. With a core knowledge of normal bone growth and remodeling, abnormal responses of bone can be appreciated by identifying any reactive changes in gross appearance. The full developmental sequence of the skeleton is left to the domain of the embryologist and pediatrician. Here basic facts are presented, sufficient to impart an understanding of bone formation and remodeling. The reader can also refer to an excellent overview on bone biology, including cell proliferation, types of bone cells and chemistry, growth and development and remodeling by Ortner and Turner-Walker (2003), Enlow (1963), Bronner et al. (2010) and Scheuer and Black (2000:18–47). See also bone biology and articular joint physiology overviews in Shipman et al. (1985:18–77) and Steinbock (1976:3–16).

THE NECESSITY OF KNOWING

Recognizing a bone as abnormal raises questions about the cause of the event (etiology) resulting in remodeling. These findings may be of forensic significance for the individual or reflect the culture of the population. Certain fracture patterns (e.g., parry fractures) may tell us about interpersonal violence of the people, or indicate the introduction of the horse (e.g., femoral/pelvic fractures; distal fibular fracture). The ability to recognize infection (e.g., tuberculosis) helps us to understand the epidemiology of certain diseases and reflects hygiene and social activities or public health practices. Other changes may be characteristic of nutritional deficits (e.g., rickets) (Mankin 1974b–c, 1990), genetic predisposition to deformity (congenital) or tumorous lesions. But before any ability in recognizing disease can be acquired, a confident practical knowledge of what is normal is a must.

Functionally the skeleton supports the body, protects internal organs, and serves as attachment for the muscles and soft tissues. Bone is a living tissue consisting of 92 percent mineral or solids and 8 percent water. The solid matter is chiefly collagen matrix hardened by impregnation with calcium salts (Thomas 1985). Bones develop either from small cartilage models (anlage) in the eight-week-old embryo or from condensed embryonic tissue (mesenchyme) that forms a dense membrane (Arey 1966). Facial bones and portions of calvaria, mandibles, and clavicles derive from the latter and are called "membranous," all other bones form in areas occupied by cartilage, which they gradually replace and are therefore "cartilaginous."

Bone forms along the paths of invading blood vessels. Chondrocytes enlarge (hypertrophy) and proliferate (hyperplasia) about the blood vessels, becoming osteocytes as the cartilaginous matrix becomes mineralized. Endochondral ossification is a well-ordered sequential process of converting the cartilaginous model into bone. It is present under the perichondrium, a blood vessel rich layer of cells outside the model, and in active centers that develop within the bone. The fewer bones of membranous origin form along the many blood vessels and develop within the membrane. Both of the processes are referred to as *modeling* and any subsequent changes requiring resorption of preexistent bone followed by deposition of new bone is remodeling. Thus modeling is an early process, while *remodeling* occurs during normal growth and continues until death.

All bone is remodeled along a blood vessel advancing through cartilage or bone. Just ahead of this blood vessel is a cutting core of osteoclastic activity dissolving the bone. New bone is formed by the osteoblasts trailing beside the vessel. This is often referred to as a bone-remodeling unit (Rockwood 1989).

Osteoblasts form the bone matrix osteocytes, serve to maintain it, and osteoclasts remove it. The ratio of osteoblasts to osteoclasts determines whether bone is deposited or resorbed. Since a single osteoclast destroys in 36 hours the same amount of bone that ten osteoblasts produce in ten days, it is obvious that if microscopic examination reveals an abundance of osteoclasts, bone resorption must be occurring (Snapper 1957).

Seven basic categories of disease may affect any sort of tissue or organ and are best recalled by use of the mnemonics KITTENS:

K = Congenital/Genetic
I = Inflammatory/Infections/Idiopathic/ Iatrogenic
T = Traumatic
T = Toxin
E = Endocrine/Metabolic
N = Neoplastic/Neuro-mechanical
S = Systemic

Congenital may refer to structural anomalies at birth, or genetic defects appearing later such as osteogenesis imperfecta. Inflammation of any sort, especially chronic infection, frequently causes remodeling. These changes may also be due to unknown causes or a poorly understood disease (idiopathic osteoarthritis) or caused by medical intervention (iatrogenic). Trauma is

usually followed by the osteoblastic reparative process after initial resorption at the site of hemorraghe. Toxins may interfere with bone growth (lead poisoning) or become incorporated in the bone (e.g., lead, arsenic, tetracycline). Endocrine or metabolic effects vary markedly, from excessive growth of immature bones (gigantism) or the mature skeleton (acromegaly). Metabolic effects are most pronounced in nutritional deficiency states – vitamins (rickets, scurvy) and calcium deficiency (osteoporosis). Neoplasia may present as areas of resorption, with or without reactive deposition to produce a sclerotic margin. Neuromechanical defects such as tabes doralis from tertiary syphilis or the loss of sensation from diabetes mellitus destroy the control of smooth joint action producing traumatized points from the stressful gait. Systemic diseases like rheumatoid arthritis and lupus produce inflamed joints with recognizable patterns of destruction. See Lawrence (1970a), Mankin (1974a–c, 1990), and Rogers and Waldron (1995:55–63) for a further discussion.

Osteoarthritis, however, is an example of a widespread inflammatory disease that is poorly understood and can be thought of as systemic (degeneration accompanying increased age) or traumatic in origin (Mankin 1968, 1982). The important point is that any classification system is not rigid but serves as a basis for organizing thought. Another factor to keep in mind as one holds the dry bone in hand is that extraordinary changes may not be due to a primary disease of bone, but the bone has been affected secondarily in the natural course of a disease, as in general inflammatory responses, necrotic results of effected blood flow, as in sickle cell and hemophilia. Refer to Raisz (1999) for discussion on the metabolic changes in bone remodeling and the pathological changes which produce osteoporosis and bone loss.

Description of the lesion is important in developing a list of differential diagnosis. Normal bone has a continual balance of resorption and formation and is subject to gradual remodeling of trabeculae to meet changes in stress loading. When stimulated, this balance is upset, but bone can react in only three ways (resorb, deposit or both). Recognizing the *predominant* process is the key to description. Once the pattern has been described by location, gross, and radiographic appearance, the number of disease possibilities can be narrowed.

For example, a single resorptive (lytic) lesion of a vertebral body could represent infection (e.g., tuberculosis), a primary cancer (plasmacytoma), or a benign lesion (Schmorl's node). If the lytic margin is surrounded by a reactive growth of bone the process is probably a chronic infection (e.g., tuberculosis), however, there is also osteoblastic response around metastases from prostate and breast cancer. Aggressive lysis with no reaction favors malignant disease (such as those of renal, lung and thyroid metastases), while a smooth regular margin is probably both benign and quiescent (Coley 1950). Resorption of bone can also be caused by contact pressure erosion from tumors or along the path of dilated blood vessels. The latter occurs when an alternate pathway must be used after normal circulation is obstructed by congenital or acquired disease.

Deposition of bone is most frequently due to inflammation of the periosteum. This is most readily seen in the exuberant callus after a fracture, but it is also seen in more subtle injuries where muscular injuries next to the bone lift the periosteum away from the cortical bone and blood collects underneath the tissue. The blood clot is organized into scar tissue and may be converted into bone. Chronic infections (osteomyelitis) of bone rarely heal spontaneously, producing years of inflammation resulting in very large deposits of bone, often with a central sinus (cloaca), which continually drained pus. Tumors, especially primary bone cancers, produce wildly aberrant patterns of new bone admixed with areas of aggressive lysis. Hematologic disorders (thalassemia, sickle cell) are associated with exuberent bone proliferation causing radical enlargement of the diploë in response to the formation of additional marrow space in compensation to the anemia. The trabeculae become vertically oriented and produce a "hair on end" appearance in radiographs.

The above examples merely touch on the range and diversity of resorption, deposition, or both, each requiring interpretation of its pattern, speed and interaction for accurate diagnosis.

OSTEOARTHRITIS

A chapter on bone remodeling is not complete without a discussion of osteoarthritis (OA), also known as degenerative joint disease (DJD). Not only is OA chiefly recognized by remodeling changes in skeletal series, it is the most common manifestation of disease after dental caries. It is present in most persons older than 50 years, and in 90 percent of octogenarians, although pathological and biomechanical studies have shown that age alone is not a sufficient cause of OA, as there are enormous differences between the old and the arthritic joint (Bayliss 1991; Lasater and Groer 2000; Watt and Dieppe 1990). See also Alexander (1990), Brown et al. (2008), Felson et al. (1987), Resnick and Niwayama (1988).

As widespread as it may be, the causes of this disease are diverse and obscure. Osteoarthritis most frequently results in destruction of weight-bearing joints (vertebrae, hips, knees) although any joint may be affected. Two clinical patterns emerge, primary and secondary OA. Primary OA is idiopathic, of no known cause, but may be one of the many changes of aging. The spine is most often affected but the patient largely remains unaware of the process. Secondary OA is related to "wear and tear," while this concept continues to fuel debate and we all endure wear and tear over the years, not all of us "finish the race," so to speak, in the same condition or with the same degree of degenerative joint disease. Clearly we must consider other genetic, environmental, physiological behavioral, and biomechanical factors that contribute to our aging process. Traumatized joints (including repetitive micro trauma), congenital abnormalities, and other etiologies (remember KITTENS) are subject to the inflammation and repair process of any living, vascular tissues. Thus, the remodeling is limited to a specific degenerative process and hence the alternate term "degenerative joint disease."

A theory has been advanced to explain each clinical pattern. The biochemical theory seeks to explain primary OA as an aging phenomenon where the body gradually loses its ability to maintain joint cartilage. Stresses develop, damage the joint and the attendant inflammation in the synovium (synovitis) releases enzymes and inflammatory chemicals that attack the cartilage (Mankin 1967, 1968). Once the cartilage is damaged, the changes common to both clinical patterns follow and are irreversible.

A biomechanical theory emphasizes loading stresses directly injuring the cartilage cells, affecting their ability to maintain the matrix. The cartilage becomes soft, with fissures cracking the surface and flakes breaking off. The inflammation induced by this damage then chemically attacks the remaining cartilage and a cycle of accelerating damage develops. Changes in the underlying bone may accompany the process leading to microcysts beneath the cartilage, which then collapses from loss of support (Mankin 1967, 1974a, 1982).

Regardless of the mechanism, the final appearance is common to both. With the loss of chondrocytes and fragmentation of the softened cartilage, the subchondral bone becomes exposed. Exposed bone forms small callus rich blood vessels followed by an extensive remodeling process. Without cartilage or a synovial sack to protect the two bony surfaces from abrading, "bone on bone" wear produces a thick polished (eburnated) bone resembling ivory. Other areas show osteoclastic activity with microcysts and even small fractures. The cartilage spreading out at the joint margins and the synovial lining turn into ossified outgrowths known as spurs (osteophytes, osteocartilaginous growths), indicative of proliferative osteoarthritis, as opposed to erosive changes (for example, periarticular cysts and surface pitting). Chips of cartilage or broken bone spurs may float freely in the joint, called "joint mice" (see Figure 206).

The remodeled subchondral bone and osteophytes are obvious changes typical of OA. But hidden in the marrow cavity is another change. The trabecular pattern, which identifies the lines of stress, differs from the normal pattern and reflects changes in the stress loading of the bone. There is some debate as to whether this reflects cause or effects of the OA. Whichever it is, radiographs soon reveal its existence.

Radiological and clinical features of OA in the living are: (1) narrowing of the joint space, (2) osteophytes, (3) altered bone contour, (4) subchondral bony sclerosis and cysts, (5) periarticular calcification, and (6) soft tissue swelling (Dieppe et al. 1986). Most of these changes are helpful to the anthropologist. Questions anthropologists raise are primarily concerning the predictive value of the presence of OA in skeletal remains. Specifically, does OA reflect changes due to continuous stress, or trauma to a joint induced by characteristic activities ("overuse" hypothesis) (Alexander 1990)? What are normal ranges and the limits of the normal "wear and tear" for old age? Can differences be found between the dominant and nondominant hands? These questions are particularly pertinent as physical anthropologists examine remains, devoid of soft tissue and lacking medical histories, trying to interpret what activities resulted in joint lesions (Kennedy 1998; Capasso, Kennedy and Wilczack 1998). Archeological interpretations from historic references have been published; a few examples are: Lai and Lovell (1992), Stirland (1998), Weiss (2003a), and differential results have been presented (Hunt and Bullen 2007; Weiss et al. 2010). As the reader will see, no statement will be made that these questions can be adequately answered. We can only supply the reader with findings of several clinical studies focusing on the etiology and frequency of OA of contemporary groups.

A number of groups have reported comparing the incidence of OA in various occupations. For example, increased frequencies of OA were found in bus drivers and cotton pickers (Lawrence 1961, 1969), foundry workers (Mintz and Fraga 1973), persons (porters) carrying heavy loads on their heads (Jager et al. 1997), print setters (from flicking letters across the tray with their thumbs, Dieppe et al. 1986), and coal miners (Kellgren and Lawrence 1958). However, pneumatic hammer operators (Burke and Fear 1977), long-distance runners (Puranen et al. 1975), soccer players (Adams 1979; Klunder et al. 1980), and parachutists (Murray-Leslie et al. 1977) that should expect to experience increased frequencies of OA, don't. That physical activity is clearly one path to recovery and increased health and mobility is evident in that most rheumatologists endorse physical exercise for managing both degenerative and inflammatory joint disease (Burry 1987). Despite the voluminous research and medical attention given it, OA remains a topic of controversy and uncertainty. See Hunter and Eckstein (2009) for a good discussion on OA and physical activity.

A study of 134 individuals aged 53–75 from northern California, for example, addresses the hypotheses that increased frequencies of OA are associated with the dominant hand (93% were right-handed) or due to "wear and tear" (Lane et al. 1989). The subjects were studied using radiographs, rheumatologic evaluation, and questionnaire. Ninety-five percent of the individuals were classified as having occupations that were nonphysical. The subjects were separated into subgroups based on lifetime heavy hand use. The researchers found "no significant differences or meaningful trends in any subgroup between dominant and nondominant hands" (Lane et al. 1989). However, the authors concluded that since the subjects were in nonphysical occupations they might not have had enough chronic stress or trauma to accelerate joint degeneration in the dominant hand. This study as well as others showed that heavy use remains an open question. A broad range of wear may result in no

adverse effects to the joints or the development of OA (Lane et al. 1989; Panush et al. 1986; Wright 1980). Thus, interpretations of skeletal remains deserve no less caution.

This "overuse" hypothesis, widely held by anthropologists and clinicians alike, appears to be rooted in the belief that too much of anything can be bad, therefore too much physical activity causes osteoarthritis. Following this logic one would conclude that oarsmen, toolmakers, blacksmiths, sharpshooters, long-distance runners, and archers would develop osteoarthritis (not just soft tissue inflammation) from overusing their joints in these activities. This is an especially attractive way of thinking when reconstructing past behaviors in ancient groups through skeletal analysis. Interestingly, studies of professional keyboard and string musicians have failed to reveal increased incidences of OA in these activities practiced over long periods involving substantial forces on vulnerable joints (van Saase et al. 1989; Radin et al. 1971). The present authors have long believed that the "overuse" hypothesis is too often and too liberally used as an explanation for frequencies/patterns of osteoarthritis and activities leading to its development in ancient groups. It's likely that OA is a result of genetic and environmental variables rather than activity or "overuse" or "wear and tear" alone.

Readers wishing more detailed information on bone disease and remodeling should refer to Crelin (1981), Manchester (1983), Ortner and Putschar (1985), Ortner (2003), Steinbock (1976), or one of the many texts or journal articles on embryology and pathophysiology.

Chapter V

DISEASE OF INDIVIDUAL BONES

SKULL AND MANDIBLE

Osteoma (Button Osteoma, Hamartoma)

Small to large (ca. 1 cm), polished and roughly circular raised area(s) of dense bone that resembles a small mound, dome or mushroom – these growths are classified as benign (harmless) tumors but, in fact, are not true tumors, but benign condensations of bone that, according to Eshed et al. (2002), better fits a hamartoma (button lesion) than an osteoma or exostosis suggesting an evolutionary background (**Figure 1 and Plates 21 and 34**). A common occurrence in most populations (37.6% in modern groups and 41.1% in archaeological specimens) (Eshed et al. 2002; Brothwell 1967; Bullough 1965; Bushan et al. 1987; Grainger et al. 2001; Ortner and Putschar 1985; Steinbock 1976).

Porotic Hyperostosis

Small (0.5 mm) to large (2.0 mm) sieve-like holes involving the outer table and diploë accompanied by increased vault thickness – there is considerable debate over the etiology and proper classification of this condition (**Figure 2 and Plate 13**). Some populations will show an extremely high frequency of porotic hyperostosis (e.g., coastal Peruvian), while other groups show little or none (e.g., contemporary American whites and blacks). Hrdlička (1914), in examining 4800 crania, found porotic hyperostosis to be common among the prehistoric coastal peoples of Peru but absent in groups from the mountainous areas. Other names for porotic hyperostosis include symmetrical osteoporosis, osteoporosis symmetrica, cribra cranii, and less frequently, Tara Grufferty syndrome. See Moseley (1965) for an interesting example of "symmetrical osteoporosis."

Regardless of the proposed etiology (e.g., iron deficiency or hereditary anemia) (Miles 1975; Salvadei et al. 2001), nutrient losses associated with diarrheal disease (Walker 1986), vitamin or nutritional deficiencies (McKern and Stewart 1957), toxic causes (Hrdlička 1914), or infection, it is imperative that an accurate descriptive analysis be conducted. See discussion by Ortner (2003:55–56, 363–375). As the name implies, porotic hyperostosis should only be applied to those cranial bones that exhibit both porosis (pits and holes) and thickened (increased) bone. Ortner (2003:383–393) identifies a pattern of porosity on the parietals, frontal and especially porosity on the sphenoid, ascending ramus of the mandible and on the superior regions of the scapula as an indicator of scurvy. Porotic hyperostosis should not, however, be confused with hyperplastic conditions (**Figure 52**) or cancers that do not fit the geographical distribution on the skeleton or appearance of porotic hyperostosis.

Ectocranial porosis

Many crania will show tiny pits (porosity) of the parietals (most commonly), occipital, and frontal bone near bregma; however, no thickened bone will be present. Since porotic hyperostosis must by definition include the presence of thickened bone, the present authors have chosen

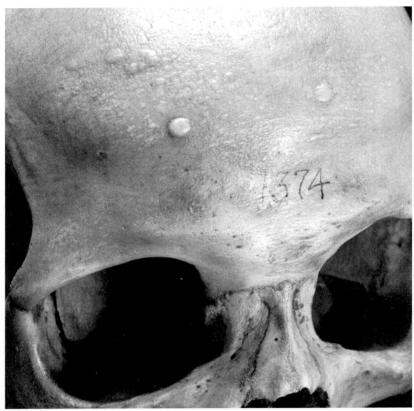

FIGURE 1. Button osteomas (ivory osteoma; button lesion; osteomata) (NMNH-T 1374). See Plates 21 and 34 for more examples.

to use the purely descriptive term ectocranial porosis (**Figures 3, 4, 6**) for pitting of the outer vault giving it an "orange-peel" texture, not accompanied by thickened bone. Some researchers (Carlson et al. 1974; Lallo et al. 1977) believe that cribra orbitalia (**Figure 13**) is an early form of porotic hyperostosis. Cf. Angel (1964), Carlson et al. (1974), Cybulski (1977), Dallman et al. (1980), El-Najjar and Robertson (1976), El-Najjar et al. 1975), El-Najjar et al. (1976), Hill (1985), Keenleyside (1998), Lallo et al. (1977), Lanzkowsky (1968), Mensforth et al. (1978), Ortner (2003: 363–382), Palkovich (1987), Papadopoulos (1977), Ponec and Resnick (1984), Rothschild (2001), Steinbock (1976:213–252), Stuart-Macadam (1985, 1987a–b, 1989, 1992), Vercillotti et al. (2010), and Walker (1986). See also Hengen (1971), Hershkovitz et al. (1997), and Walker et al. (2009).

The tiny pits in the outer surface of the vault are not accompanied with thickened bone. Although it is difficult to distinguish what the first author refers to as "ectocranial porosis" from conditions reflecting anemia (porotic hyperostosis [PH]), great care must be exercised when identifying outer vault porosity as PH. One way to avoid having to choose an interpretive "diagnosis" is to carefully describe the location, appearance and distribution of pitting in the outer vault. Many "normal" crania exhibit pinpoint porosities in the frontal, parietals and occipital bone. It is a common occurrence, mostly seen in middle-aged adults.

Ectocranial Accessory Vessel Sulci (Grooves)

One or more grooves are visible above the orbits for transmission of branches of the supraorbital vessels and nerves (Grant 1948) (**Figure 5 and Plate 14b**). These shallow grooves sometimes trail into the supraorbital notch or foramen (see also **Figure 4**). This is considered a normal

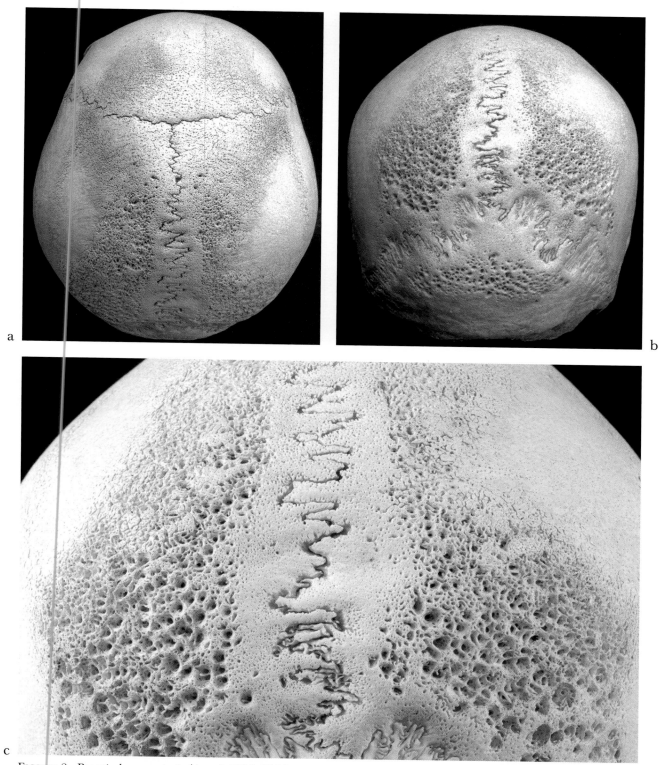

FIGURE 2. Porotic hyperostosis (Angel 1966; El-Najjar et al. 1976; Lallo et al. 1977; Palkovich 1987; Schultz 2001), symmetrical osteoporosis (Zaino, 1967), symmetric osteoporosis (Hrdlička 1914) (NMNH 264543). See Plates 13, 36 and 37 for color examples.

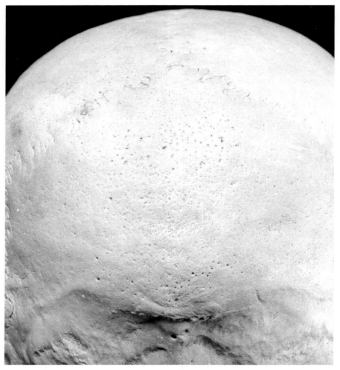

FIGURE 3. Ectocranial porosis on the occipital (NMNH).

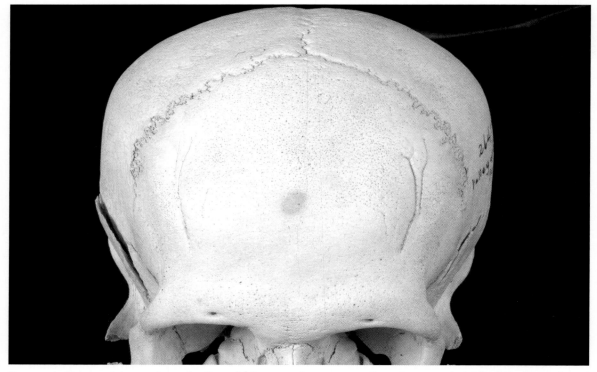

FIGURE 4. Ectocranial porosis on the frontal (also note the accessory frontal grooves/sulci for blood vessels and nerves) (NMNH 264776). See Plate 14b for an example of accessory frontal grooves.

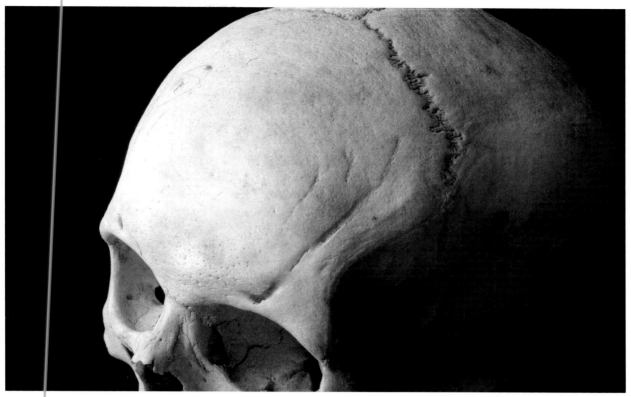

FIGURE 5. Normal accessory vessel grooves (sulci, frontal grooves) in the frontal bone sometimes mistaken for trauma, but actually a nonmetric trait and normal anatomical variant (NMNH 264827). See Plate 14b.

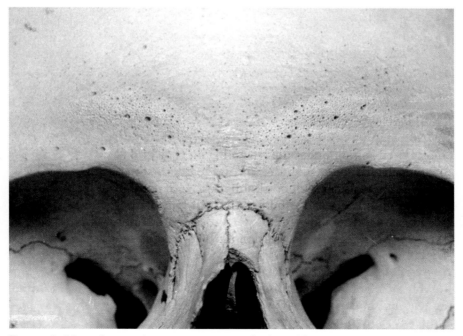

FIGURE 6. Normal pitting/pinpoint porosity in the frontal bone, often more pronounced in males (sometimes referred to as the male vermiculate pattern) (NMNH 264417).

anatomical variant (non-metric trait). See variants presented in Hauser and DeStefano (1989:48–50, Plate VII) for different degrees of expression. Rhine (1990) attributes this feature to be more frequently observed in American blacks.

Children's palates are usually more porous and pitted than adults. Children also have multiple canals in the anterior portion of the maxilla and mandible due to normal growth and development of the teeth. Some of these canals are remnants of dental crypts, while others, known as gubernacular canals that carry a cord connecting the deciduous and permanent tooth, will guide the permanent teeth through the palate through what some call the openings of the dental sacs (Dixon 1937) (**Figure 7**).

Fractures of the Cranial Vault ("Pond" Fractures and Depression Fractures)

Such fractures may be found in any area of the skull but are most frequently seen in the frontal and posterior or lateral portions of the parietals (**Figures 8 and 9**). The typical appearance of a depressed fracture is a concave defect in the outer vault, with or without radiating fractures. The size of the wound is usually about the size of a dime or nickel and circular or ellipsoidal, although any size and shape may be encountered. It is often difficult to distinguish a depressed wound from a healed lesion originating in the scalp. Depressed cranial fractures, such as in this example, often result in a "pond" fracture because of its shape. Frequent finding in many groups with high interpersonal violence and warfare. Cf. Walker (1989) discussion of Native American populations; Walker (2001) for an archaeological populations overview; and Boylston (2011:357–381) as example of British warfare trauma study.

Depressed fractures result from a variety of weapons and conditions. Although a large non-healed (perimortem) fracture of the skull does indicate trauma to the head, it usually will not leave any evidence of the specific type of weapon

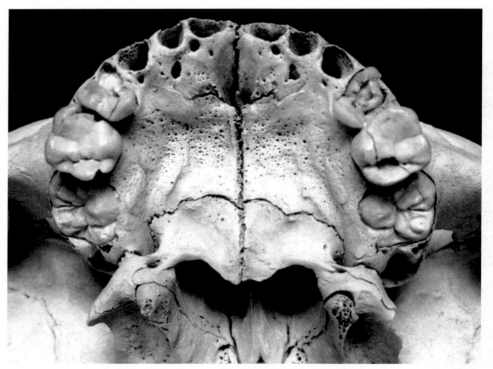

FIGURE 7. Normal "guiding" pitting and canals (known as gubernacular canals at the openings of the dental sacs) in the maxilla/hard palate of a 4–5-year-old child – these pits are also present in mandibles (NMNH). Cf. Dixon (1937).

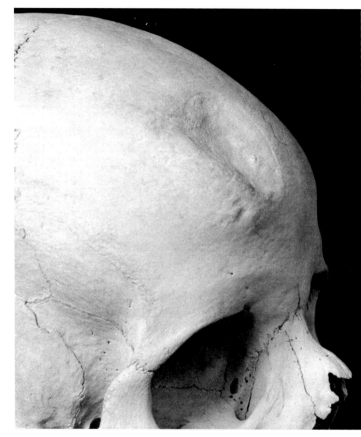

FIGURE 8 (*right*). Healed depressed ("pond") fracture to the right frontal (injury). Cf. Lovell (2008). See Plate 19 for a color example.

FIGURE 9 (*below*). Healed depressed fracture to the right lateral parietal (NMNH-H 321498). Also note the large Inion spike on the occipital bone.

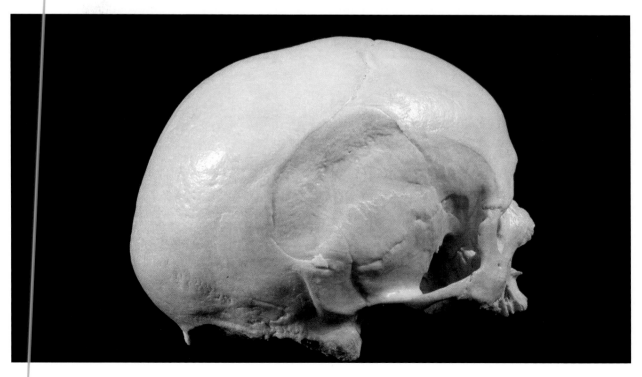

used or the circumstances of the traumatic event(s). Further, healed depressed fractures may cause the analyst/investigator to suspect that an assailant struck the individual, although this may not be an accurate interpretation. For example, the person might have fallen from a rock ledge, struck his or her head, only to die a week later. The fracture pattern could be identical if the person had been struck over the head with a blunt instrument (e.g., war club) resulting in immediate death. Obviously, a projectile point embedded in bone leaves no doubt that some form of conflict occurred. What is important is that bony trauma, whether the result of blunt force or sharp force (bladed or edged implement), must be carefully examined, described, photographed, and measured. The key phrase is "described and documented" with drawings and, preferably, photographs with scale. Many times, speculative interpretation is best not attempted. Be sure to examine the endocranial surface of the skull, as a blow to the outer vault often results in radiating fractures and/or displacement of a portion of bone endocranially. It might also be helpful to insert a flashlight into the foramen magnum or hold the skull up against a bright (100 watts or greater) desk light so that the light shines through the bone, illuminating any unusually thin areas, fractures, or displaced bone (**Figure 10a–b**).

Another important aspect of trauma is the pattern of the wound(s) in the population under study. For example, are all of the depressed fractures the same size and shape? Where are the wounds located? How many sites of trauma show healing suggesting that the victim lived for some time after the event? Again, the overall population picture is important in reconstructing the events surrounding the trauma (Jurmain 2001). Cf. Browner et al. (2003), Galloway (1999), Kimmerle and Baraybar (2008), and Lovell 2008.

Depressed fracture with no healing (perimortem) – look closely for any indication of healing has occurred. There will be small patches of periosteal/woven/fibrous new bone growth and possibly porosity in the area of the defect or along the fracture lines (**Figure 11 and Plate 40**). If death occurred immediately or soon after sustaining trauma (up to about two weeks), the bone may show no new growth (postmortem breakage, as opposed to perimortem fractures, differs in that the former has a "torn paper" or jagged appearance – perimortem fracture lines are smoother and sharper both visually and to the touch). As can be seen in **Figures 9 and 11**, any number of objects or implements could have been responsible for causing (the shape) the defect. The shape of the wound would depend, for example, on whether the person was hit with the pointed end of a club

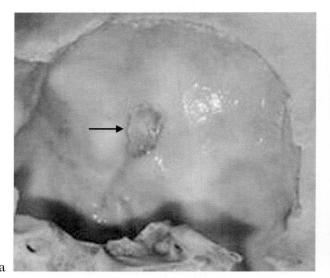 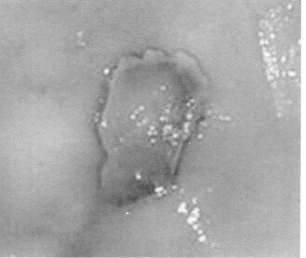

a   b

FIGURE 10. Blunt force injuries to the skull resulting in concentric fractures to the right temporal and parietal (a) and an "island" of bone (arrow) displaced but adhering to the endocranium opposite the area of impact in the left parietal.

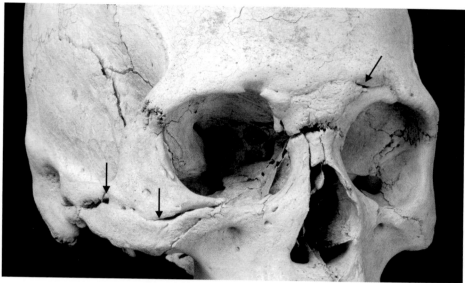

FIGURE 11. Healed radiating fractures of the frontal, malar and zygomatic arch. Healing is evidenced by the presence of several bridges of bone that span the fracture lines (NMNH 379259).

(small, circular depressed wound) or the blunt end (larger, oval or oblong depressed wound). Again, the pattern of the wounds in the population under study may help to clarify the type of weapon used. Research by one of the authors on documented cases of skeletal trauma suggests that the term perimortem, when based solely on the amount of healing visible in bone, must be used with caution, as osseous remodeling may not be grossly discernible until about ten days after sustaining trauma (see **Plate 57** for healed mandibular pseudoarthrosis). The rate of healing, however, is heavily dependent on the element affected and the individual's age as rapid healing occurs in children and slows with increased age. Cf. Barbian and Sledzik (2008).

One of the authors (RWM) has found it very useful to shade in the shape of a skull wound using a piece of paper and pencil or artist's charcoal. If the depressed wound is in the frontal bone it would be "shaded" using the following method: make a drawing (approximate anatomical size) of the frontal view of the skull, note the approximate position of the wound, and correspond this with the drawing. Hold the drawing against the frontal bone and use the side of the lead to rub across the defect. The result is an unshaded (white) area above the depressed area (the lead won't come in contact with the concave areas). This method renders a silhouette of the size and, more importantly, the shape of the wound. You then can compare the depressed wounds in all of the skulls to see if a size and shape pattern is present which may reflect the weapon(s) used.

Metopic Suture (Unfused Frontal or Interfrontal Suture)

The metopic suture (**Figure 12**) usually begins to obliterate at the end of the first year after birth, with fusion being complete not later than the fourth to sixth year (Limson 1924). In some individuals (1% to 12% of adults (Krogman and İşcan 1986), the two frontal bones fail to unite resulting in "metopism" or a persistent or patent (open) metopic suture extending from Nasion to Bregma. Use care not to confuse the supransal suture (SNS), a horizontally oriented zigzag suture situated 1 to 2 centimeters superior to the Nasion, with the less complex metopic suture. The SNS forms (secondary) a few years after the metopic suture obliterates/disappears. The complexity of the SNS may reflect the transmission of compressive forces due to mastication up through the maxilla and frontal bone. The SNS may remain visible throughout one's lifetime, long after the metopic suture has closed (Barnes 1994:

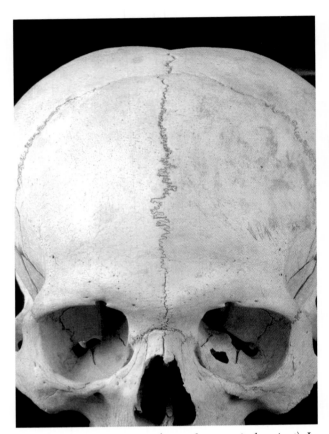

FIGURE 12. Metopic suture (normal anatomical variant). In children before the age of two, the metopic suture normally separates the two frontal bones down the midline (NMNH 264597).

148–152; Chopra 1957; Harbert and Desai 1985; Hauser and DeStefano 1989:41–44; Hess 1945; Latham and Burston 1966; Manzanares et al. 1988; Schultz 1918; Torgersen 1950, 1951).

Cribra Orbitalia

Cribra orbitalia usualy occurs bilaterally and appears as small to large holes with a sieve-like appearance in the upper surfaces of the orbits (**Figure 13**; see **Plate 7**). In children, the bone may actually be thickened and spongy-like while in adults only remnants of the holes (frequently only pits) remain. The frequency of this condition varies greatly by population and depends on a number of factors, many of which are under debate (e.g., iron-deficiency anemia, perhaps related to malnutrition, scurvy, chronic gastrointestinal bleeding, ancylostomiasis, and epidemic disease) (Hirata 1988; Ortner 2003:55–56, 363–375). See Walker et al. (2009) for a reappraisal of the iron-deficiency anemia theory that PH and CO may have different etiologies. Cribra orbitalia may or may not be accompanied by porotic hyperostosis (pitting and/or thickening) of the outer table of the skull. Stuart-Macadam (1989) suggested, "The similarity between porotic hyperostosis of orbit and vault with respect to macroscopic, microscopic, radiographic, and demographic features supports the idea of their relationship" (Carlson et al. 1974; Cybulski 1977; Fairgrieve and Molto 2000; Glen-Haduch et al. 1997; Guidotti 1984; Hengen 1971; Lanzkowsky 1968; Mittler and Van Gerven 1994; Robb et al. 2001; Schultz 2001; Steinbock 1976:239–248; Walker et al. 2009; Zaino and Zaino 1975). See also the section "Porotic Hyperostosis."

Nasal Septum Deviation

The nasal septum deviates to either side of the midline of the nasal aperture and usually presents no clinical problem except when the deviation is so severe that it blocks the nasal passage, causes chronic sinusitis, pain and possibly snoring – while this condition often goes unnoticed or unrecognized in anthropological analyses, a pneumatized turbinate is the most common normal variant (14–53.6%, Zinreich et al. 2003) in the nasal sinonasal anatomy (**Figure 14** shows a left deviated septum and a pneumatized right middle turbinate). When one or both of the middle turbinates (nasal conchae) enlarges, it may push against the nasal septum, markedly deviating it to one side (**Plates 5 and 9**) (Hatipoğlu et al. 2005). With normal asymmetry of the human face, some level of deviation in the nasal septum will be observed in all individuals. See also Collett et al. (2001), Guyuron et al. (1999), Kayalioglu et al. (2000), and Lidov and Som (1990).

Nasal Aperture Erosion

Typically, erosion of the nasal spine is one of the early bony changes of the nasal area, followed by erosion of the inferolateral border of

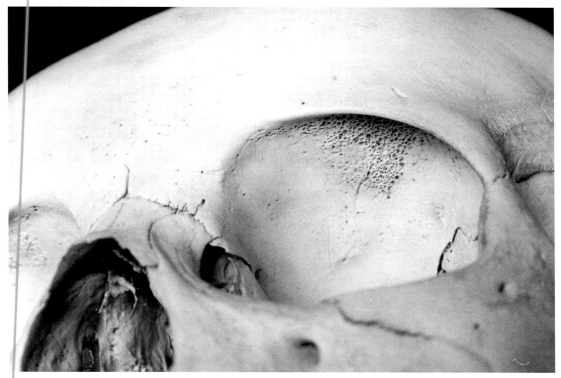

FIGURE 13. Cribra orbitalia of the porous and slightly raised type (Ursa orbitale. Møller-Christensen and Sandison 1963) (NMNH). See Plate 7 for more examples.

FIGURE 14. Deviated nasal septum with enlargement/pneumatization of the right middle turbinate (NMNH 264502). See Plates 5 and 9a for color examples.

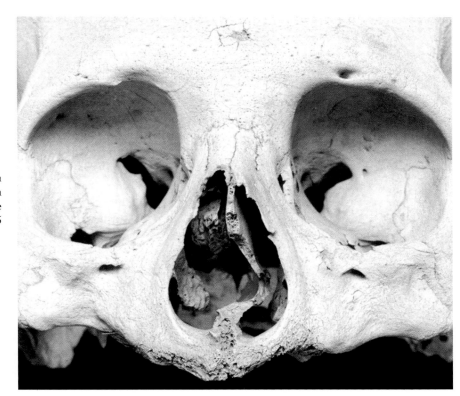

the aperture (rhinomaxillary change; Manchester 1989) (**Figure 15**). In archeological settings, the nasal spine often does not survive normal taphonomic erosion, and thus erosion in this area and the absence of the nasal spine should be carefully studied to determine whether it is pathological or pseudopathological. This is a rare finding in most populations and is uncommon to common in Europe (depending on the time period).

Erosion of the nasal area is one symptom of syphilis, *facies leprosa*, tuberculosis, and leishmaniasis among other conditions (Andersen 1969; Ell 1988; Manchester 1989; Møller-Christensen 1961, 1974, 1978; Reichart 1976; Queneau et al. 1982.). Extensive erosion is a rare finding.

Periapical Abscess and Erosion (Periapical Lesion)

The margins of a periapical lesion or abscess (hole) will exhibit some periosteal reaction (pitting), and a pocket may be visible at the root apex (**Figures 16, 17** and **Plates 6, 10b, 11a,**

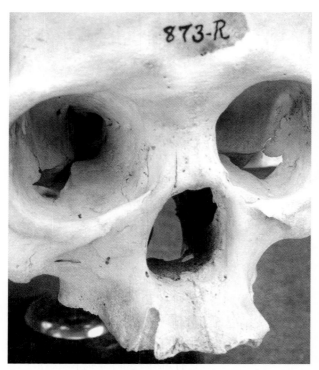

FIGURE 15. Eroded nasal spine and/or aperture (NMNH-T 873R).

23c–d, 26a, 29, 30 and 31). The margins of a healed abscess will be smooth, rounded and of similar texture as the surrounding bone. Although a periapical abscess is a common finding in most skeletal collections, care must be exercised in distinguishing a periapical abscess from a periapical granuloma and periapical cyst. The normal sequelae or developmental sequence, if chronic and untreated, is that an abscess becomes a granuloma, then a cyst – each can only be definitively diagnosed histologically. Cf. Sapp et al. (1997). Pulpal involvement and necrosis, however, is often a strong indicator of a periapical abscess.

Enamel hypoplasia and Amelogenesis Imperfecta

Hypoplastic lines (enamel hypoplasia) appear as shallow or deep grooves encircling the tooth crowns at the same level that are widely reported to be an indicator of nonspecific stress (**Figures 18 and 19**) (Crawford et al. 2007; Ogden et al. 2007). Figure 18 exhibits a severe form of amelogenesis imperfecta that is passed down as an inherited disorder on defective amelogenin genes (cf. Sciubba et al. 2002) (see **Plates 53 and 54**). Enamel hypoplasia ("Cuspal Enamel Hypoplasia") (Ogden et al. 2007) can be caused by many factors including periapical inflammation or trauma to a deciduous tooth, fever, disease, nutritional deficiencies (especially A and D), endocrine dysfunction, and generalized infection during odontogenesis (Robinson and Miller 1983). Note that hypoplastic events, grooves, or lines appear at the same horizontal level across several teeth, indicating the timing of the event and the individual's age of enamel disruption. A second form of hypoplasia, not shown, is represented by pits of various sizes in the enamel. Ogden and colleagues (2007), in an attempt to more accurately identify and classify the complexity of enamel hypoplasia, proposed to increase the classic two forms of hypoplasia from pits and furrow, to pits, furrow, plane and cuspal. This is a common finding in many populations. See also Blakey et al. (1994), Corruccini et al. (1985), Cucina (2002), Cucina and İşcan (1997), Goodman

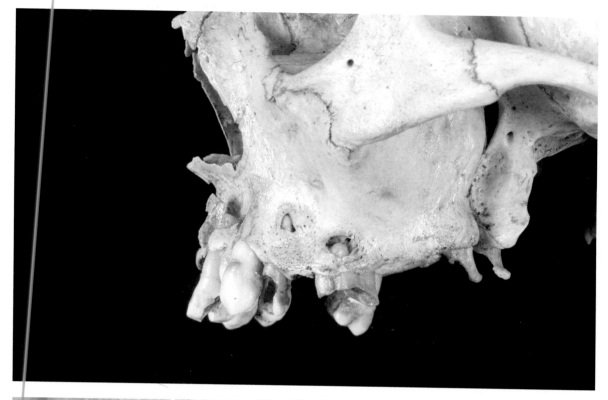

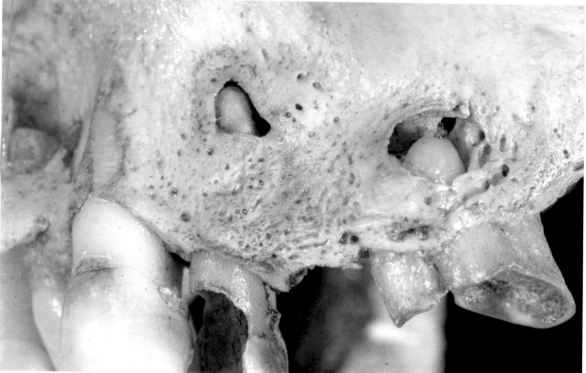

FIGURE 16. Two periapical abscesses – inflammation at the apex of a tooth root with sinus formation (large to small pocket) and penetration of the maxilla or mandible (NMNH-T 176R).

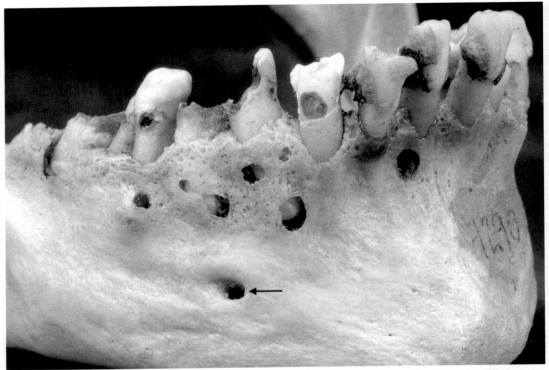

FIGURE 17. Multiple (at least 7) periapical abscesses (periapical lesions) in the mandible (arrow points to the normal mental foramen) (NMNH–T 1290). See Plates 6, 10b, 11a, 23c–d, 26a, 29, 30 and 31 for color examples.

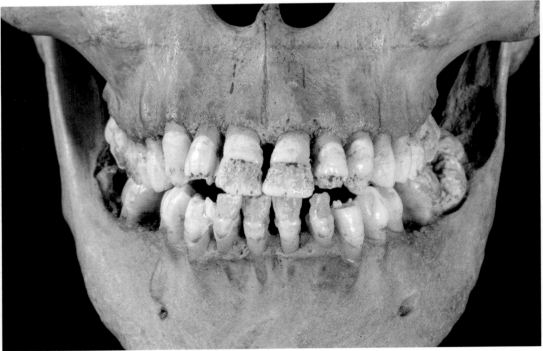

FIGURE 18. Amelogenesis imperfecta affecting large areas of the teeth. Cf. Crawford et al. (2007). See Plates 53 and 54 for color examples.

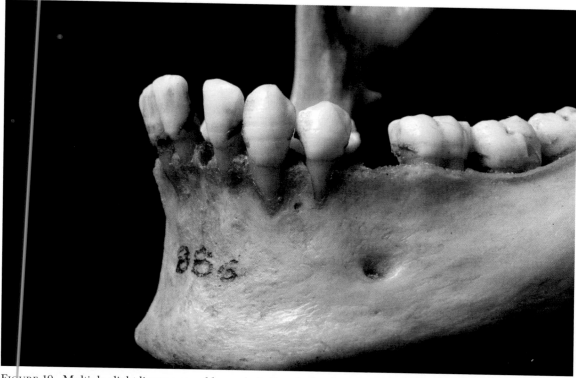

FIGURE 19. Multiple slight linear enamel hypoplasia lines in the dental enamel (NMNH-T 886). Cf. Littleton and Townsend (2005); Sarnat and Schour (1941).

and Armelagos (1985, 1988), Goodman and Rose (1990, 1991), Goodman and Song (1999), Goodman et al. (1980, 1987), Hillson (2005), Hillson and Bond (1997), Hutchinson and Larsen (1988), Lawson (1967), Littleton and Townsend (2005), Pindborg (1970), Reid and Dean (2000), Robb et al. (2001), and Sarnat and Schour (1941).

Shovel-shaped Incisors

Shovel-shaped incisors are so called because the lingual surface is shaped like a shovel (**Figure 20**). This is a dental trait found commonly in Asian groups (ranging from above 60% to 100%), while whites and blacks may exhibit the trait but to a lesser degree (generally below 5%). See also Dahlberg (1951), Devoto (1971), Devoto and Arias (1967), Hinkes (1990), Hrdlička, (1920), Hsu et al. (1997), Littleton and Townsend (2005), Lorena et al. (2003), Saini et al. (1990), Saunders and Mayhall (1982), Scott and Turner (1997), Sharma (1983), Tsai and King (1998), and Ubelaker (1999).

Nasal Fractures

Look for asymmetry of the nasal bones including depressions, small adhering bone fragments, and fractures with evidence of healing (**Figure 21**). Fractured nasal bones and adjacent maxilla are common findings in most populations. See Walker (2011) for an overview study.

Dental Ablation

Some ethnic groups intentionally remove (extract or knock out; ablation) one or more teeth for aesthetic and cultural reasons (Chimenos-Kustner et al. 2003; Fitton 1993; Gould et al. 1984; Handler et al. 1982; Nelsen et al. 2001; Stewart and Groome 1968; Sweet et al. 1963) (**Figure 22**). In questionable cases, histological (Schultz 2001) and radiological analysis can help distinguish resorption due to cellular activity verses erosion due to acidic soil or an abrasive object.

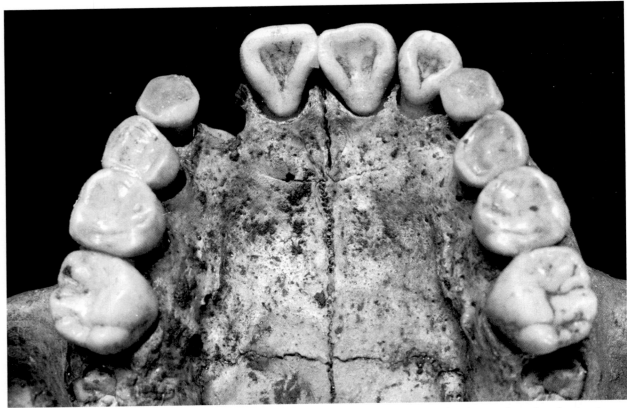

FIGURE 20. Shovel-shaped incisors in a child (NMNH). See Plate 48b for an example of enamel pearl and extension.

Cleft Palate

Cleft palate (**Figures 23, 24 and Plate 47a**), one of the most frequently encountered congenital malformations, is a common structural defect that results from faulty development (failure to fuse, interact or merge) of the soft (lip and soft palate) and hard tissues (maxilla/palate) of the oral cavity between weeks eight and twelve of pregnancy. Cleft lip with cleft palate occurs in about 1:1000 Caucasian births (O'Rahilly and Muller 1996), 1:1500 in Japanese and about 1:5000 in American blacks (Bergsma 1978). Females seem to be more prevalent to cleft palate than males (Fishbien 1963). The expression may be unilateral or bilateral, and appears as a separation, however small, between the tooth sockets, or a separation in the posterior portion of the hard palate (O'Rahilly and Muller 1996). Cleft lip without cleft palate occurs in about 1:2500 births (60–80% affected infants are male (O'Rahilly and Muller 1996) and can result from a variety of genetic and environmental factors that alter palatal growth and formation in the developing embryo (Bender 2000; Coleman and Sykes 2001; Johnston and Millicovsky 1985). While facial clefts and craniosynostoses often occur together, they differ in their distribution regarding sex, familial occurrence, race, and geography (Larsen 2001). See Barnes (1994; especially Figure 4.27 on page 186) for a drawing of the range of forms of clefting. For other references for this condition, see Goodman and Gorlin (1977), Ortner (2003; 456–459) and **Plate 47c** for a color example, and **Plate 46** for other developmental dental dislocations.

Endocranial Vessel Depressions, Pacchionian and Arachnoid Granulations

Although these depressions (**Figure 25**) vary in size, most are large, smooth bordered, and serve to house large clusters of arachnoid granulations (Pacchionian granulations, Pacchionian

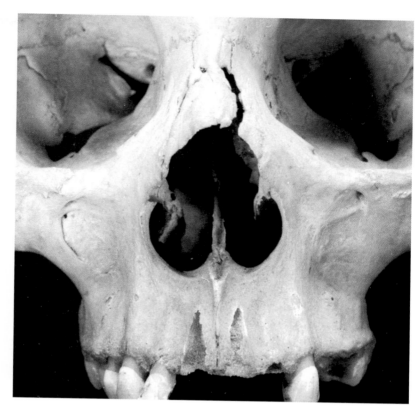

FIGURE 21 (*right*). Fractured nasal bone(s) and right maxilla (NMNH-T).

FIGURE 22 (*below*). Dental ablation of the anterior maxillary dentition in a historic American black (NMNH 387866).

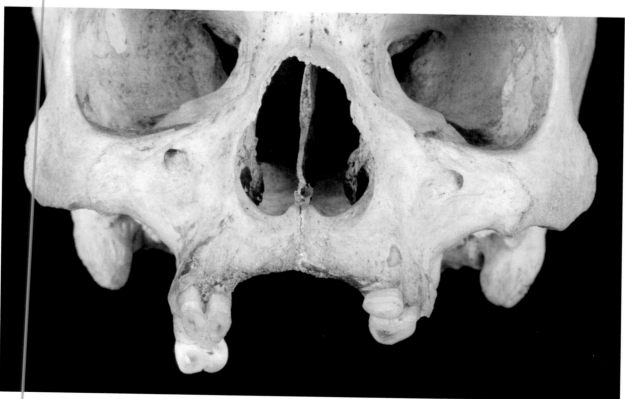

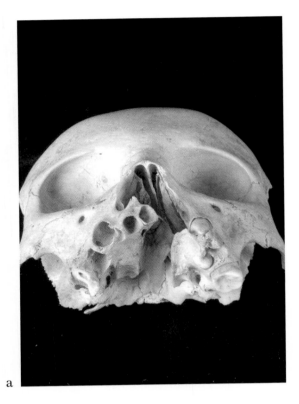

a

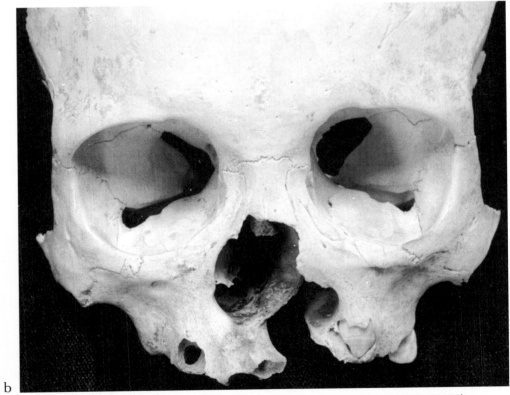

b

FIGURE 23. Cleft palate (uranoschisis) (unilateral expression) (NMNH 293252).

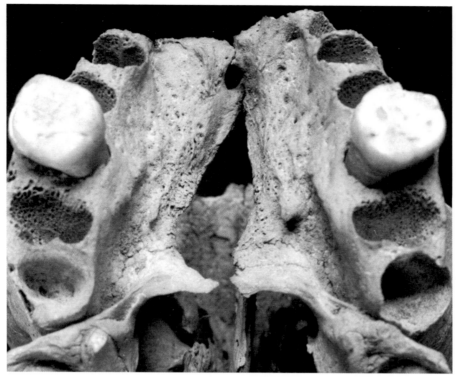

FIGURE 24. Cleft palate in the mid and posterior hard palate (NMNH 316482). See Plate 47a.

bodies; see **Plates 43 and 44**). Pacchionian pits, on the other hand, are small, sometimes clustered endocranial pits with sharply defined ("sharp-edged"; Le Gros Clark 1920) margins caused by erosion by arachnoid granulations. Lacunae laterales are always located post-bregmatic (in the anterior parietals only) while Pacchionian pits may be found in the parietals (possible within the lacunae laterals) and frontal. See also Fox et al. (1996). Be careful not to confuse effects of infectious disease (see Plates 41 and 42).

There are a number of theories as to the etiology for the erosion of the inner vault of the skull due to arachnoid granulations. It is known, however, that these granulations primarily serve to filter and return cerebrospinal fluid. In some cases the lacunae laterales may become eroded and result in localized protrusion and, rarely, perforation of the outer vault by the cauliflower-shaped and ossified arachnoid granulations. Both depressions (lacunae) and pits are common in all populations, increase in number and depth with

age, and are usually considered normal variants (Lang et al. 2002). Perforations of the outer vault that are not due to postmortem erosion is rare (see Pacchionian pits). Pits resulting from erosion of the inner table of the vault are due to enlargement and ossification of arachnoid granulations. In young individuals arachnoid granulations are villous and small. During old age the granulations enlarge, become cauliflower-shaped, and erode the cranial vault resulting in varying sized pits. Pacchionian pits appear as relatively small (2 mm) to large (5 mm) pits with sharply-defined margins that are mostly confined to the parietals. The depth and frequency of these lesions increase with age and, possibly, disease, although Le Gros Clark (1920) commented that he never saw an adult skull without them; their presence, therefore, is a normal finding, but their frequency, size and depth likely increase with age. This is a common finding in all populations. See Le Gros Clark (1920) for a good discussion of their development and pathogenesis and Leach et al. (2008) for findings on MR.

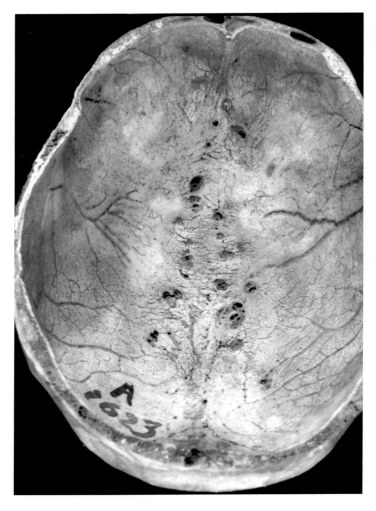

FIGURE 25. Lacunae laterales (Grant 1972) – shallow endocranial depressions (possibly a raised "mound" ectocranially) located on either side of the sagittal suture. The pits house arachnoid granulations that circulate and filter cerebral spinal fluid (AFIP A1623). See Plates 43 and 44 for color examples of Pacchionian pits.

Vascular/arterial grooves – shallow endocranial grooves, usually symmetrically situated in both parietals, lateral and posterior to the parietal foramina.

Venous lakes – normal vascular structures situated at the end (commonly) or along the path of one or more arterial grooves. Grossly, these irregularly shaped depressions in the inner table often resemble cauliflower and as radiolucencies radiographically. They often penetrate through the inner table into the diploë and rarely, if ever, perforate the outer table.

Nuchal Crest and Inion Hook

This area may either be flat (typically in females) or developed and projecting inferiorly in some males (**Figure 26**). Heavily developed nuchal crests (muscle attachments for the neck) may have a shelf-like ridge and an inferiorly oriented "spike/spine" or Inion hook of bone in response to muscular activity and bone stimulation and is found more commonly in males (63%) and 4.2 percent in females (Gulekon and Turgut 2003). This is a common finding.

Suprainion Depression

While the precise etiology of the suprainion depression remains unknown, it is often observed on crania which have had frontooccipital or occipital cranial deformation (**Figure 27**) (K. Murray 1989, personal communication). See also Stewart (1976). Several factors, including biomechanics

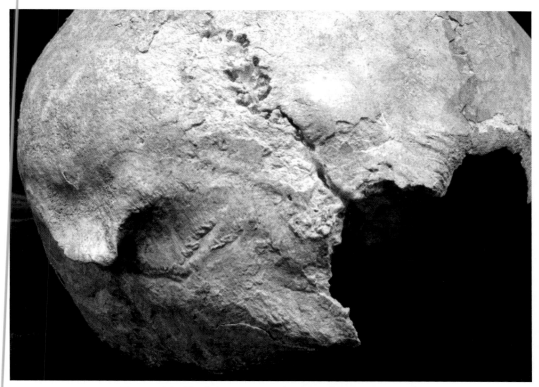

FIGURE 26. Nuchal or Inion spike/hook (NMNH 387865).

FIGURE 27. Large suprainion depression (NMNH 264444).

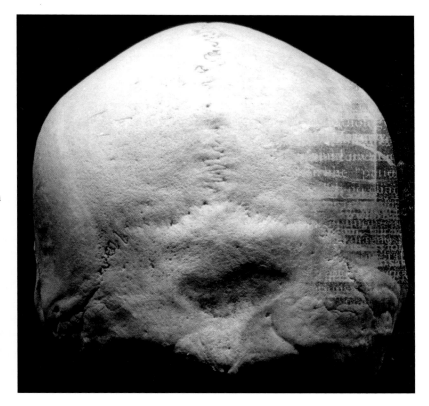

in response to artificial cranial deformation (although these depressions have been found in undeformed skulls), parasitic infection, ulcers, and necrosis have been postulated as causative agents (Curtin 2007).

Craniosynostosis (Premature Suture Closure)

Craniosynostosis is a term commonly used to refer to any suture that fuses at too early an age (**Figure 28 and Plates 14 and 37b**). While most authors use the terms craniosynostosis and craniostenosis interchangeably, others (Prokopec et al. 1984) use the latter term only when the cranial capacity is diminished. One point of clarification is that, by definition, "premature" craniosynostosis refers to a suture that closes too early, while craniosynostosis merely refers to "closure" of a suture – one refers to early closure and the other (to some researchers), both normal and abnormal closure. When the sagittal suture fuses prematurely, the skull continues to grow and results in a long-headed individual (scaphocephaly or hyperdolicocephaly). Premature closure of the coronal suture results in a high (pointed) skull known as steeple skull, tower skull, or oxycephaly (Silverman and Kuhn 1993). Craniosynostosis does not refer to artificial cranial deformation from such practices as cradle boarding or head binding (Dingwall 1931; Ozbek 2001; Stewart 1941). Craniosynostosis is an uncommon finding in most skeletal populations and may be congenital, hereditary, or the result of metabolic disturbances. Craniostenosis is more common in males (Cecil and Loeb 1951) and has been associated with Crouzon syndrome (Gorlin et al. 1976). See Barnes (1994:152–175) for an in-depth discussion of developmental problems causing agenesis and stenosis with reference to paleopathological cases and photographic illustrations. See also Aufderheide and Rodriguez-Martin (1998:52–55), Chopra (1957), Cohen (1986), and David et al. (1982). See Khanna et al. (2011) for a good radiographic pictorial of various types of craniosynostosis, as well as Moore (1982), Prokopec et al. (1984), Stewart (1975, 1982), and Turvey et al. (1996). See Webb (1995:251) for severe deformation from scaphocephaly in an Australian aboriginal.

Scaphocephaly

Scaphocephaly is one of the most common forms of craniosynostosis. (The authors have come across this form in numerous crania from around the world). Some of the traits associated with this condition consist of a bulbous, projecting frontal bone, a raised keel of bone running the length of the sagittal suture, and low-set eye orbits in relation to the frontal bone. Bilateral fusion of only the coronal suture is sometimes referred to as brachycephaly in the old literature. Unilateral fusion of the coronal is referred to as plagiocephaly (**Figure 28a–b and Plate 14**). See also Barnes (1994:152–157), Cohen (1986), Cohen and MacLean (2000), David et al. (1982), Moore (1982), Prokopec et al. (1984), Tod and Yelland (1971), and Turvey et al. (1996).

Otitis Media and Mastoiditis

Inflammation and subsequent infection of the middle ear (otitis media) may result in perforation and resorption of the mastoid process or other portions of the temporal bone (**Figure 29 and Plates 25, 51b, and 52**). The middle ear is an air-filled space within the petrous portion of the temporal bone lying immediately behind the tympanic membrane (eardrum). Otitis media is very common in infants beyond the neonatal period (after 28 days of age) and shows a decline in incidence after the first year of life (Bluestone and Klein 1988). Skeletal involvement of the mastoid, however, is uncommon to rare in most skeletal samples. Care must be taken not to mistake the normal fissures and squamo-mastoid sutures (see Hauser and DeStefano 1989:196–207 for normal variants) in the outer surface of the mastoid for a pathological condition. Also be careful not to confuse the numerous bony air cells (compartments) of the normal mastoid for disease (see **Plate 55** for normal morphology of the mastoid air cells).

When mastoiditis is suspected, use radiographs to look for sclerosis and pocket formation and consult with a radiologist or otolaryngologist. See also Dugdale et al. (1982), Edwards (1988), Flohr and Schultz (2009), Gregg and Gregg (1987),

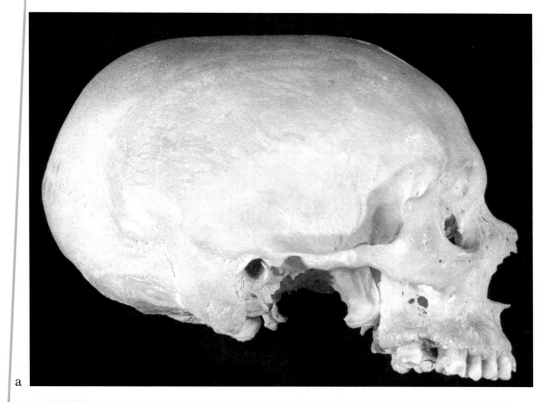

a

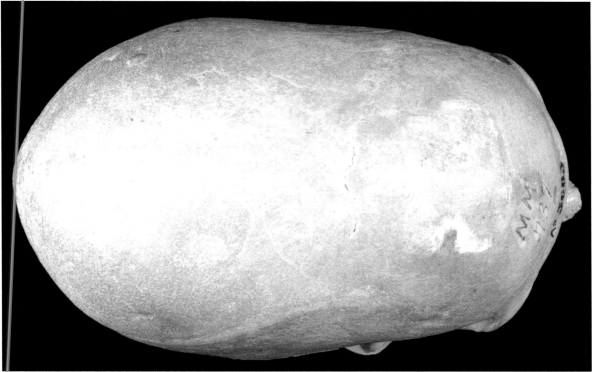

b

FIGURE 28. Scaphocephaly (craniosynostosis, craniostenosis, premature craniosynostosis, premature suture closure) (AFIP MM737/A3603). See also Figure 40a–c and Plate 14b.

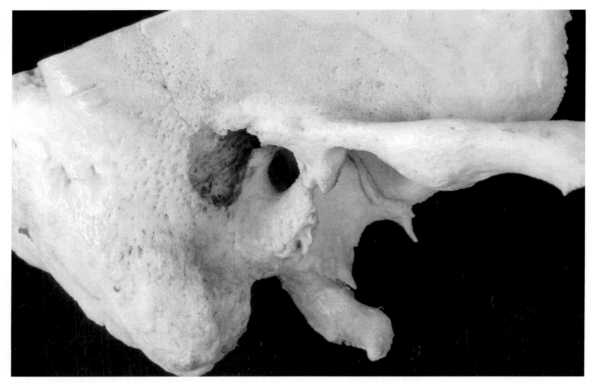

FIGURE 29. Mastoiditis resulting in otitis media and perforation of the temporal bone (NMNH-H). See Plates 25 and 51b for color examples.

Paparella et al. (1980), Schultz (1979), and Teele et al. (1984).

Auditory Exostosis

Auditory exostosis, a benign tumor (**Figure 30 and Plate 24**) first identified by Professor Selig-man in 1864 (Turner 1879), is easily visible as a rounded lump/exostosis ranging in size from small (barely visible) to large, virtually filling the opening. Rarely the external auditory meatus may be absent at birth and closed in its entirety; no external canal will be seen in the temporal bone (the authors have encountered two such cases). Hrdlička (1935) identifies exostoses in much higher frequencies in males than in females and that they appear to increase in size with age. Auditory exostoses, which likely form as an osteoblastic (periosteal) response to trauma due to a variety of factors including thermic shock and the "refrigeration action of the meatus" (Okumura et al. 2007), are uncommon to common findings found in higher frequencies in swimmers and divers, especially in cold water (Arnay-de-la-Rosa et al. 2001; Filipo et al. 1982; Kroon et al. 2002) and have been attributed to both genetic (Hanihara and Ishida 2001a) and behavioral traits (Kennedy 1986), including artificial cranial deformation. Otolaryngologist DiBartolomeo (1979) found 70 cases of auditory exostosis in 11,000 patients over a period of 10 years (6.4 per 1,000 patients) in a coastal region. He reported that "irritation nodules" in the external auditory canal are painless until the tenth year of aquatic exposure, at which time obstruction of hearing may occur. For other case reports see references by Pulec and Deguine (2000, 2001a–b) and Deguine and Pulac (2001a–b). Wong et al. (1999) found a relationship between the length of time spent surfing and the severity, location, and preva-lence of auditory exostoses. Godde (2010), in comparison, reported six of 744 archaeologically-derived Nubians with exostoses in a geographical region where there is no evidence of dependency

FIGURE 30. Auditory exostosis (torus) – benign bone tumor(s) of the ear canal (NMNH). See Plate 24 for another example.

on marine resources or prolonged cold water exposure in these agriculturist/pastoralist groups. The author cautions against attributing low frequencies of auditory exostoses as the result of cold water exposure and suggests that non-water factors may contribute to their development. See also Fenton et al. (1996), Godde (2010), Graham (1979), Gregg and Bass (1970), Gregg and Gregg (1987), Hanihara and Ishida (2001a), Hrdlička (1935), Hutchinson et al. (1997), Kennedy (1986), Kroon (2002), Longridge (2002), Pulec and Deguine (2001a–b), Sheehy (1982), Steinbock (1976:329–333), Turner (1879), and Van Gilse (1938).

Cranial Vault Lesions

To identify the cause of perforations in the cranial vault, look for the formation of new bone encircling the lesion (both endo- and ecto-cranially), increased vascularity (tiny, smooth-bordered pits), and remodeling (healing or filling in) along the margin of the lesion (**Figure 31**). Some of the more common diseases that produce

these lesions are metastatic carcinoma, tuberculosis (usually a single lesion), multiple myeloma (**Plate 4**), eosinophilic granuloma, treponematosis (**Plate 5**) and fungal infections. All of these lytic lesions cause destruction of bone that can be mistaken for postmortem damage (see **Figure 56b**). Seek the assistance of an experienced researcher before drawing any conclusions. See also Apley and Solomon (1988), Grupe (1988), Marks and Hamilton (2007), and Mays et al. (1996), Olmsted (1981), Ortner (2003), Sefcakova et al. (2001) (for an interesting case and presentation of metastatic carcinoma in ancient remains), Silverman and Kuhn (1993), Steinbock (1976), and Webb (1995).

Caution must be exercised when making a differential diagnosis based solely on examination of the skull (**Plates 1, 4, 15, 16 and 17**). Carefully describe the appearance and distribution of the lesions throughout the skeleton. As an example, in comparison to multiple myeloma, histiocytosis X of the skull usually appears as a single lytic lesion with a central island of bone/sequestration. There are a number of other diseases that produce

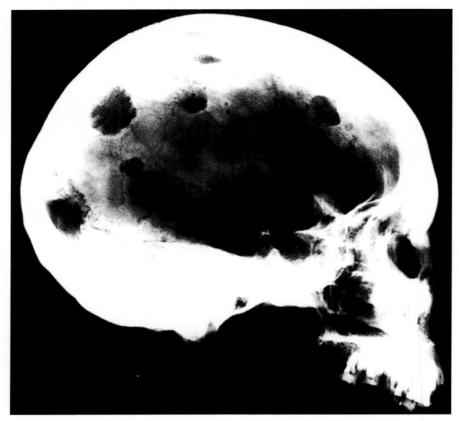

FIGURE 31. Radiograph illustrating perforation(s) due to disease of the cranium — diagnosing diseases based solely on examination of lytic lesions in the skull is extremely difficult. See Plates 1 and 4 for color examples.

similar lesions including histiocytosis X (more correctly called Langerhans cell histiocytosis (Ladisch and Jaffe 1989; Taylor and Resnick 2000), multiple myeloma, fungal infections, and tuberculosis. The distribution of the lesions in the skeleton must be considered, as well as their radiographic appearance. The sex of the individual may also affect the expression; in males, prostate cancer causes osteoblastic response at the metastisis, while in females the metastisis will have both osteoblastic and osteoclastic responses from breast cancer. Age may determine the cancer. For example, if a child of less than two or three years of age exhibits these lesions, the acute phase of histiocytosis X (Letterer-Siwe) might be suspected, while in adults, Hand-Schuller-Christian disease (the chronic stage of histiocytosis X) might be the proper diagnostic term (Lichenstein 1970; Taylor and Resnick 2000). Bone involvement in histiocytosis X is 80 percent (Ladisch and Jaffe 1989). See also Coley (1950), David et al. (1989), Jaffe (1975), Lichenstein (1953), Moseley (1963), Olmsted (1981), Ortner and Putschar (1985), Steinbock (1976), and Wroble and Weinstein (1988).

Biparietal Thinning (Symmetrical Osteoporosis)

In biparietal thinning, the parietal bones at the lateral bossing and superior posterior regions will be extremely thin, fragile, and translucent (**Figures 32 and 41 and Plates 22 and 35**). Cederlund and colleagues (1982) studied 3636 patients and found parietal thinning in 86 patients (2.37%) and that it was more common in women with a mean age of 72 years and males 63 years. The authors concluded that parietal thinning is not an anatomical variant or dysplasia of the diploë, but rather a slow, progressive disease of middle-aged and older people.

The process of atrophy begins in the outer table (Wilson 1944). This fairly rare condition, which was reported as present and unexplicable in adult orangutan and elderly human women in 1873 (Humphry 1873), has an unknown etiology but appears to be age related (Grainger et al. 2001). Alternative theories that this is the result of endocrine imbalance has also been postulated. The youngest affected individual known to the authors is 40 years of age. An interesting note is that the thinning usually avoids the parietal foramina leaving approximately 1 cm of unaffected bone encircling these foramina. The authors have not encountered any cases where the outer table of the vault has perforated due to this condition. See also Barnes (1994:146–148), Brothwell (1967), Hauser and DeStefano (1998: 83–84), Lodge (1967), Ortner (2003:415), and Shepherd (1893).

Elongated Styloid Process and Eagle Syndrome (Megastyloid)

An elongated styloid process can result from natural elongation of the styloid itself (primary) or ossification (calcification/mineralization) of the stylohyoid ligament (secondary elongation) (Sikanjic and Vlak 2010). **Figure 33** reveals a segmented form (see also **Plate 27**) – the styloid has attachments for three muscles and two ligaments attaching to the hyoid and mandible. While the normal length of the styloid process is 2.5–3.0 centimeters and can achieve a length of 6 centimeters or longer, most physicians classify styloids longer than 3 centimeters as enlarged, elongated, "Eagle" or megastyloid (Llguy et al. 2005; Rateitschak and Wolf 2002; Sikanjic and Vlak 2010). Note that in the present case, the calcification has been interrupted and a pseudarthrosis formed at the junction (Type II). Type I defects consist of uninterrupted styloids. Type III ("segmental") defects consist of elongated styloids with multiple interrupted segments of ossified stylohyoid ligaments. Even in severe cases of stylohyoid ligament ossification, more than 50 percent of individuals are asymptomatic (see **Plate 28**)

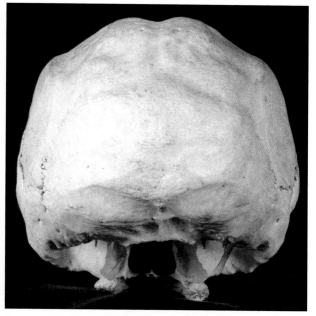

FIGURE 32. Biparietal thinning (Epstein 1953); senile atrophy (Wilson 1944) – symmetrical depressed areas in the posterior outer table of the parietals (NMNH-T 878). See Plates 22 and 35 for color examples.

(White and Pharoah 2000). Others, however, may be accompanied with pain, impingement of the carotid artery upon rotating the head, sore throat, the sensation of a "foreign body" lodged in the throat, painful swallowing, pain in the ear, vertigo, and other symptoms that fit Eagle's syndrome (Chi and Harkness 1999; Eagle 1948; Kay et al. 2001; Keur et al. 1986; Langlais et al. 1986; Monsour and Young 1986; Prasad et al. 2002; Restrepo et al. 2002; Sivers and Johnson 1985; Takada et al. 2003; Thot et al. 2000). Unexpectedly, the length of the elongated styloid and severity of pain does not appear to be correlated. Two points of consideration are: First, some individuals who exhibit no symptoms, discomfort or pain, have styloids measuring longer than six (6) centimeters. Second, many asymptomatic individuals are born with agenesis (absence) of one or both styloid processes (Kilgore and Van Gerven 2010). Symptoms of Eagle's syndrome show no sex predilection and appear after age 30. One study revealed calcification of the stylohyoid ligament in 40 percent of children, with a mean age of 11 years (Camarda et al. 1989). Review of 1771 panoramic

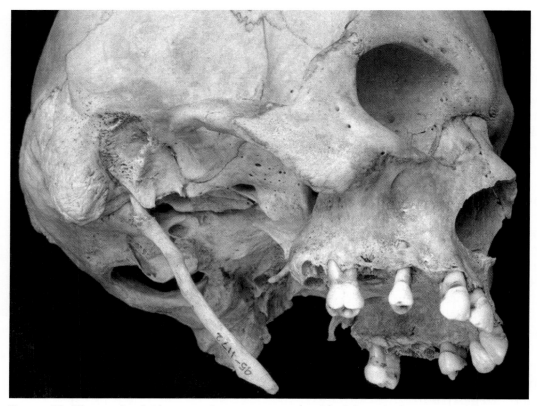

FIGURE 33. Elongated styloid process showing junction (circle) of the styloid process and the ossified ligament (AFIP 95-1172). Refer to Plate 27 for a color example and Plate 55b for normal internal structure of a temporal styloid.

radiographs revealed that while 18.2 percent of the individuals had mineralized styloid processes, only 1–5 percent of the patients were symptomatic (Correll et al. 1979). While it is not possible to reliably ascertain whether an individual suffered symptoms of Eagle's syndrome from skeletonized remains, the styloid should be described and measured (cf. Basekim et al. 2005; Kilgore and Van Gerven 2010). See also Douglas (1952), Feldman (2003), Ferrario et al. (1990), Omnell et al. (1998), and Patni et al. (1986). Styloid processes, especially elongated forms, are prone to fracture and healing. These fractures are difficult to detect due to the irregular shape and appearance of the process (Haidar and Kalamchi 1980).

Mandibular Torus (Torus Mandibularis)

Rounded, usually symmetrical bony growths (hyperostoses or hamartomas) may be found along the superior lingual border of the mandible below the premolars, but may extend further along the mandibular corpus (**Figure 34** and **Plates 64c and 65**). The tori arise from the cortical plate and are developmental anomalies that may extend in the form of two bulging ridges that nearly touch one another behind the incisors. If the tori are quite large, restriction of the tongue may occur. Chronic gum disease (gingivitis), however, may also stimulate similar, albeit less pronounced growths. Jainkittivong and Langlais (2000), in a study of 960 Thais, found that exostoses were more common in men than in women, suggesting an interplay of multifactorial genetic and environmental factors. The presence of tori increases in frequency with age. Tori are most commonly found in Eskimo populations but can range in frequency depending on the population from nine to 66 percent, depending on ethnicity (Seah 1995; Hauser and DeStefano 1989:184–5).

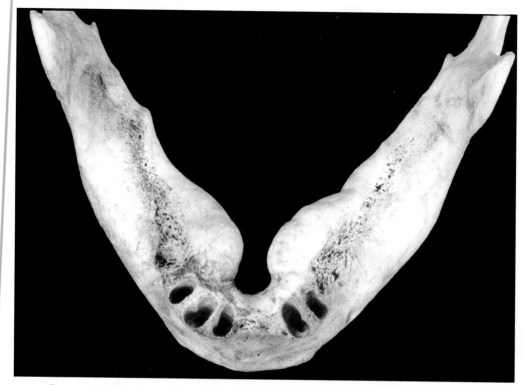

FIGURE 34. Torus mandibularis (NMNH). See Plates 64c and 65 for color examples.

Shah et al. (1992) reported that tori in a sample of 1000 patients in India were rarely present before 10 years of age. Eggen and Natvig (1986) found a positive correlation between torus mandibularis and the number of functioning teeth in a sample of 2010 dental patients over 10 years of age. See also Antoniades et al. (1998), Hauser and DeStefano (1989:182–185), Pynn et al. (1995), and Seah (1995).

Stafne's Defect (Static Bone Cavity)

Stafne's defects vary in size and shape as: (1) a circular or oval, smooth-walled concavity varying in size from 1–2 mm to more than 1 cm in diameter and located in the lingual surface of the mandible inferior to the mylohyoid line and molars; or (2) a shallow, circular, or oval and roughened defect less than 1 mm deep in the location noted above (see **Figures 35 and 36 and Plates 60b and 63** for more examples). While dentist Edward Stafne (1942) described lingual defects located inferior to the molars, the term

Stafne's defect has since become synonymous with resorptive pits and concavities in the anterior (Sisman et al. 2010; Turkoglu and Orhan 2010) and posterior mandible, as well as the ramus and sulcus (Mann and Tsaknis 1991). Radiographically, the defect often has a "folded" sclerotic border reflecting buccal remodeling of the normal lingual plate. The location and radiographic appearance of these lesions are highly suggestive, although not pathognomonic (diagnostic), of Stafne's defects. Although the etiology of Stafne's defect is unknown, nearly all have proven to be benign defects, not tumors or cysts (Thawley et al. 1987), containing normal submandibular salivary gland tissue, fatty, vascular, lymphoid, or muscular tissue. Although some defects have been found to be empty, the soft tissue occupying them was probably dislodged during exploratory surgery, rendering the cavity "empty." The osseous defects likely result from contact pressure erosion of the mandible by the submandibular salivary gland, or duct, or adjacent soft tissues. Interestingly, the mandible is

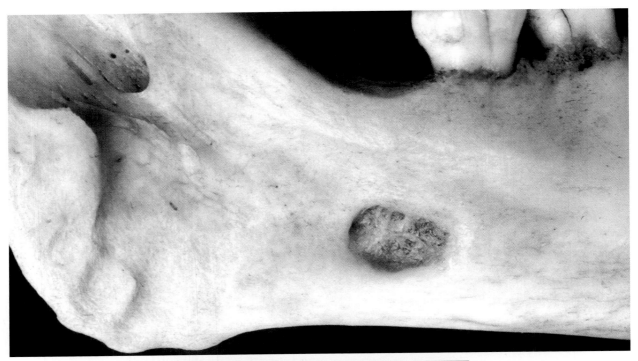

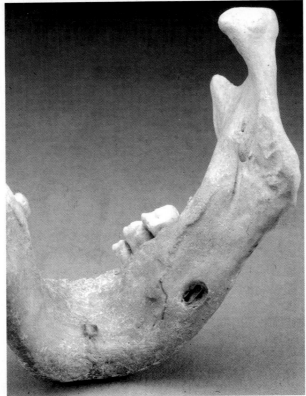

FIGURES 35 (*top*) and 36 (*bottom*). Stafne's defect (static bone defect, posterior lingual depressions, lingual cortical mandibular defects, salivary gland defect, latent bone cyst). Figure 36 photographed by Chip Clark. See Plates 60b and 63 for color examples.

among only a few areas (e.g., Pacchionian pits in the inner vault) in the skeleton that forms bony concavities in response to pressure by soft tissue.

Preliminary examination (RWM) of over 5000 dry-bone mandibles from historic and prehistoric sites revealed 91 individuals with defects, of which 81 were males (17 individuals from the same site in Alaska). Numerous researchers have confirmed the predominance of this trait in adult males; most individuals are in their forties and fifties when the defects are first detected. Stafne's defects are developmental rather than congenital or traumatic in origin and fit the expected frequency of an X-linked recessive trait (Mann 2001). The youngest known individual with a Stafne's defect was an 11-year-old boy from Sweden (Hansson 1980). Stafne's defects are usually unilateral. This is an uncommon finding in most skeletal samples. See also Aguiar et al. (2011), Correll et al. (1980), Finnegan and Marcsik (1980), Gorab et al. (1986), Harvey and Noble (1968), Lukacs and Rodriguez-Martin (2002), Mann (2001), Minowa et al. (2009), Pfeiffer (1985), Quesada-Gómez et al. (2006), Philipsen et al. (2002), Shields and Mann (1996), Shimizu et al. (2006), Stafne (1942), Tolman and Stafne (1967), Uemura et al. (1976), and Vodanovic et al. (2011).

Inca Bone and Lambdoidal Ossicles

Although a true Inca bone is one in which the mendosal suture has been retained (a hypostotic trait) and extends from asterion to asterion, Inca bones are identified by most researchers as any large accessory bone(s) in the lambdoidal region of the occipital bone; many researchers, however, distinguish an Inca bone from lambdoial ossicles (a hyperstotic trait) and an "ossicle at Lambda," if positioned at the Lambda (**Figures 37, 38, and 39**). The common presence of Inca bones in most world populations, including Sub-Saharan Africans, indicate that this trait is not uniquely Asian (Hanihara and Ishida 2001b). Inca bones are found to be more prevalent in males (Berry 1975, Hauser and DeStefano 1989:103). Some researchers believe there is a correlation of the occurrence and severity of lambdoidal ossicles to cranial deformation. Wilczak and Ousley (2009),

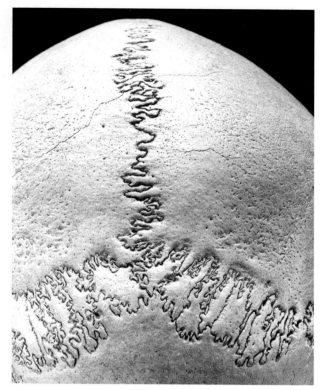

FIGURE 37. Wormian bones (lambdoid or lambdoidal ossicles, sutural bones) — small to large bones that may persist as separate ossicles or unite with the parietal and occipital bones — are accessory bones and, therefore, are hyperostotic traits (Hauser and DeStefano 1989:84–103). See Figure 15 for range and location of ossicles in the lambdoidal suture, and see Plate 40 for another example of lambdoidal ossicles.

however, did not find this correlation in a study of Southwestern Native Americans. See also Barnes (1994:140), El-Najjar and Dawson (1977), and Hauser and DeStefano (1989:99–103) and see Figure 12 for illustrations of different degrees of expression, as well as Matsumura et al. (1993) and O'Rahilly and Twohig (1952).

Enlarged Parietal Foramina

Enlarged parietal foramina are usually bilateral bone defects in the posterior portion of the parietal bones in the region known as Obelion (**Figures 40 and 41**). These defects, caused by incomplete or faulty ossification, range in size and shape from small to large circular or oval perforations measuring several centimeters in

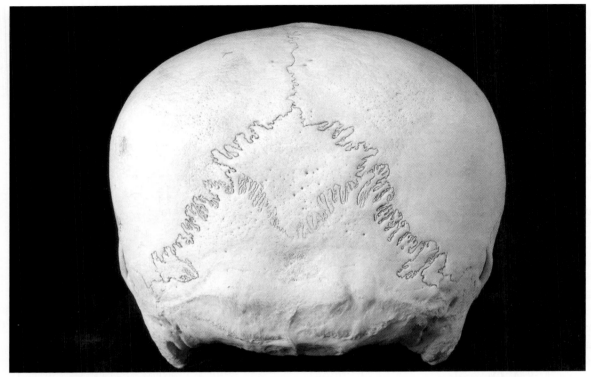

Figure 38. Ossicle at Lambda. The ossicle at lambda is a hyperostic trait (NMNH 264539).

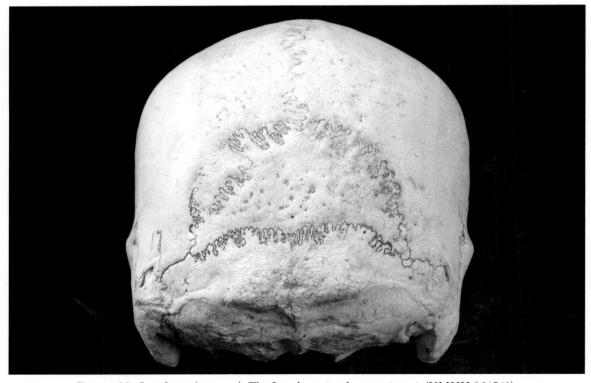

FIGURE 39. Inca bone (os incae). The Inca bone is a hypostatic trait (NMNH 264541).

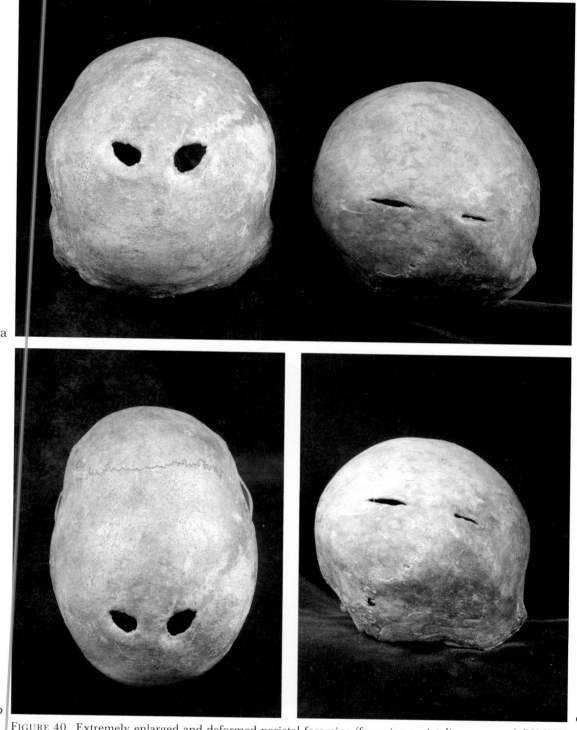

FIGURE 40. Extremely enlarged and deformed parietal foramina (foramina parietalia permagna) (NMNH 276981, left and 276982, right). Figure 40b is a superior view of NMNH 276981 and Figure 40c a posterior view of NMNH 276982. Notice sagittal stenosis in the adult and in the subadult additional stenosis of the right lambdoidal suture with skewing of the right temporal bone. See discussions on parietal foramina by Goldsmith (1922) and Hoffman (1976).

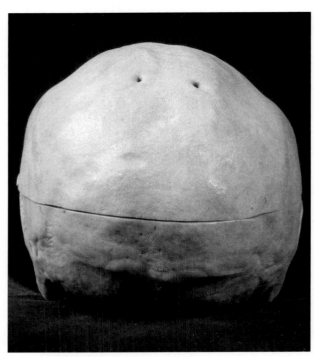

FIGURE 41. Enlarged paritetal foramina in an older adult with biparietal thinning/symmetrical osteoporosis (NMNH-T 1463).

diameter, to slits. Although usually benign and asymptomatic, enlarged parietal foramina have been found in association with other malformations. Greatly enlarged foramina (foramina parietalia permagna) reflect a hereditary condition, which may be transmitted in family members as an autosomal dominant trait (Dzurek et al. 1956; Fein and Brinker 1972; Goldsmith 1922; Hoffman 1976; Sjovold 1984). This variant is found more often in males. While small perforations are found in 60 percent to 70 percent of normal skulls, variants larger than 5 mm are less common with a prevalence of 1:15,000 to 1:25,000 (Kortesis et al. 2003). See also Barnes (1994:143–146), Dzurek et al. (1956), Fink and Maixner (2006), Hauser and DeStefano (1989:78–82), Hollender (1967), Kutilek et al. (1997), Lodge (1967), Mann (1990), Murphy and Gooding (1970), O'Rahilly and Twohig (1952), Pang (1982) Pepper and Pendergrass (1936), Rasore-Quartino et al. (1985), Sapp et al. (1997), Stallworthy (1932), Stibbe (1929) Symmers (1895), Wuyts et al. (2000), and Zabeck (1987).

Palatine Torus (Torus Palatinus)

A raised plateau (**Figure 42**) of bone varying in size along the midline of the palate (not to be confused with the bony buildup along the anterior median palatine suture and exostoses) (see **Plate 12**). The frequency and severity of this trait varies (see **Plates 1, 32b and 48a** for color examples). Chohayeb and Volpe (2001), for example, studied the presence of torus palatinus in five ethnic groups (448 women) residing in Washington, D.C. and found the highest frequencies in African Americans, followed by Caucasians, Hispanics, Asians and Native Americans. They found no statistically significant relationship between age and the presence of torus palatinus. A study by Belsky and colleagues (2003) found a significant correlation between postmenopausal Caucasian women with bone density and size of torus palatinus; interestingly, many of the subjects were unaware they had a torus palatinus, even those with large tori, until their fifties or older. Patients who were aware they had a torus, however, reported it was present their whole life and that the size, number and appearance of the torus remained the same. See also Antoniades et al. (1998), Belsky et al. (2003), Eggen and Natvig (1994), Gorsky et al. (1998), and Hauser and DeStefano (1989:174–179) (Plate XXVI illustrates different degrees of expression), Jainkittivong and Langlais (2000), MacInnis et al. (1998), Sapp et al. (1997), and Vidic (1966).

Tympanic Plate Dehiscence (Foramen of Huschke)

There can be perforation of the tympanic plate in the temporal bone (**Figures 43, 52 and 67b**). The dehiscence appears as an irregular circular hole covered by a membrane, posterior to the temporal fossa (TMJ). This is a developmental aberration with failure of closure of the foramen of Huschke, also known as foramen tympanicum, that is seen in all young children that only occasionally persists beyond five years of age (Anand et al. 2000; Berry and Berry 1967; Wang et al. 1991). Wang et al. (1991) found the foramen of Huschke in 7 percent of adult skulls in a sample of 377 dry skulls. There appears to be no sex specificity, seen

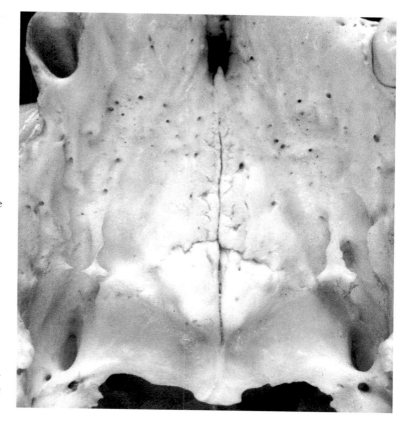

FIGURE 42 (*right*). Small torus palatinus. See Plates 1, 12c and 48a for color examples.

FIGURE 43 (*below*). Bilateral tympanic dehiscences (foramen of Huschke) (NMNH). See Plate 67b for a color example.

in higher frequency in males by some investigators (Krogman 1932; Laughlin and Jorgenson 1956) while higher frequencies were observed by others (Corruccini 1974; Dodo 1972; Molto 1983). This is an uncommon to common finding. See also Gerszten et al. (1998), Hauser and DeStefano (1989:143-147), Herzog and Fiese (1989), Lacout et al. (2005), and Sharma and Dawkins (1984).

Torticollis (Wry Neck Syndrome)

Premature closure of the cranial sutures (e.g., lambdoidal, temporal or coronal) affects the overall structure and morphology of the cranium and can also result in an asymmetrical cranial base (**Figure 44**). Also referred to as torticollis or Wry Neck syndrome (Canale et al. 1982; Klepinger and Heidingsfelder 1996) and "twisted neck," this asymmetric postion of the cranial base and

occipital condyles (the atlas and axis vertebrae may fuse to the cranial base, a condition known as occipitocervical synostosis – see **Plate 66** for an example of this condition) affects the postion of the body and thus changes in the spine may also be observed. It is an uncommon finding in most populations but in a higher frequency in Eskimos (experience of the authors). See also Barnes (1994:136-137) and Canale et al. (1982).

Temporomandibular Joint Erosion and Porosity

Normal temporal fossae typically have undulating surfaces but will not have the porosity or bony buildup (**Figures 38b, 45 and 46 and Plate 68b**). Since the TMJ is a paired joint that cannot function alone, OA, if present, is often symmetrical. Lesions are usually detectable earlier and are more

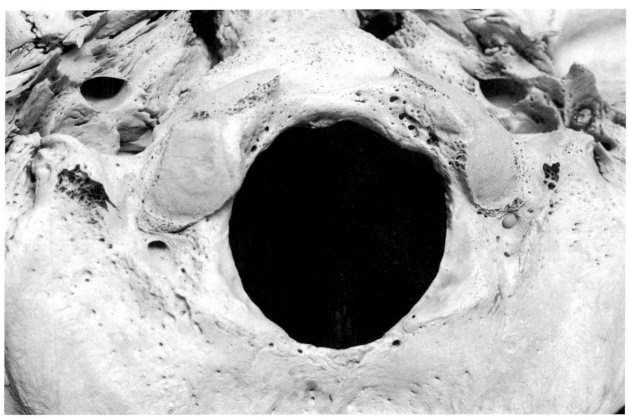

FIGURE 44. "Tilt" head-twisted and asymmetrical occipital condyles and foramen magnum referring to a condition initially identified in Hawaiian skulls by Snow (1974:61–67). See Plate 66 for a color example of atlanto-occipital fusion and occipitocervical synostosis, some of which resulted in "tilt" head deformity.

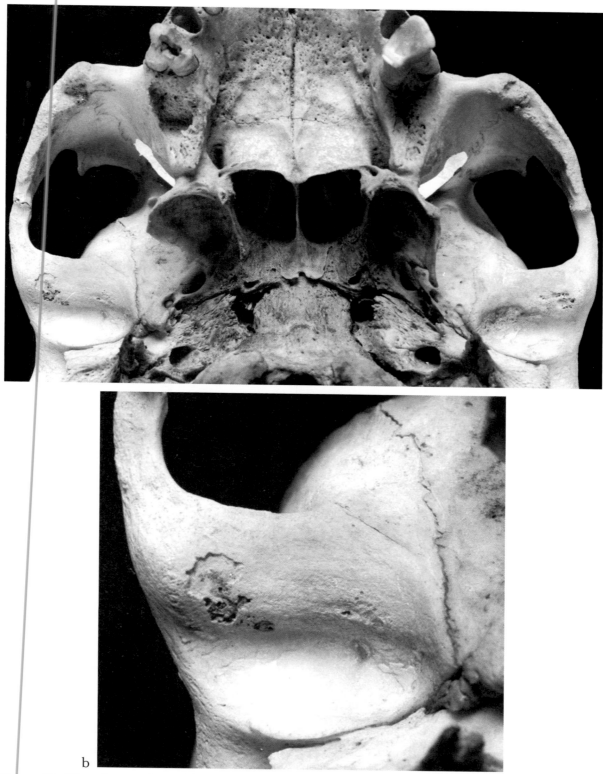

FIGURE 45. Bilateral osteoarthritis of the temporomandibular joint (TMJ) – tiny to large pits and/or osteophytes on the articular surface, margin and eminence of the temporal fossa. See Plates 38b and 68b for color examples.

severe in the temporal surface than in the mandibular condyle. If only one joint exhibits severe bone
loss or bony remodeling, look for signs of infection
or trauma. OA of the TMJ is a common finding
in most populations. See also Blackwood (1963),
Markowitz and Gerry (1950), and Ryan (1989).

The first indication of porosity (pits) can usually
be seen in the middle of the fossa or on the articular eminence (the raised area just anterior to the
fossa; when the head of the mandible anteriorly
dislocates, it rides forward and up on this eminence). Usually there is a well-defined rim with a

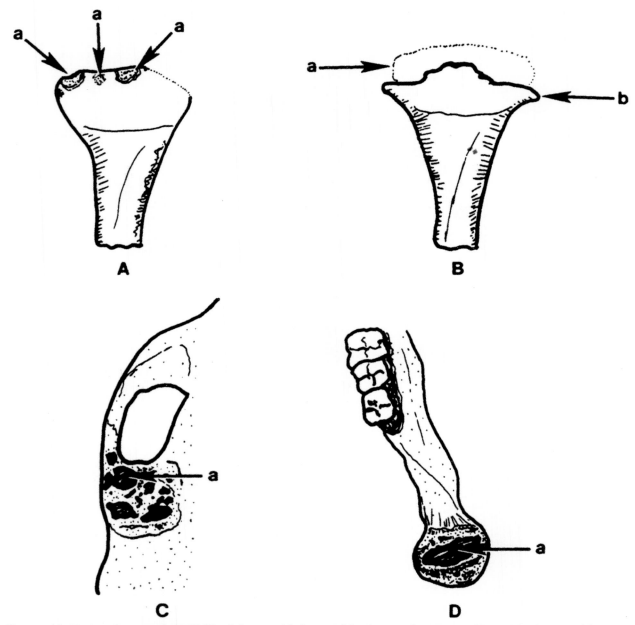

FIGURE 46. Forms of osteoarthritis/DJD of the mandibular condyle, fossa and eminence. Erosion and pitting (a) of the
bone surface from slight pits (featured in A), to larger sclerotic edged erosion as featured in C and D. Complete loss of the
joint surface from erosion and abrasion is found in more severe cases (featured in B and D). Osteophyes will form (b) along
regions where reactive bone growth occurs. See Plate 61a for an example of severe OA and partial fusion of the TMJ.

central depressed area representing early OA. Moderate OA are those cases where large areas of bone are missing due to erosion (flattened) or porosity. Severe OA of the temporal fossa presents when most of the fossa has been eroded, with or without the formation of osteophytes.

Precondylar Defect and Ossicle

A small concavity in the precondylar area is one variant produced in the precondylar developmental formation (**Figure 47 and Plate 69**). Caudal shifting in the development of the spine affects the morphology of this region, either producing a fossa or odontoid-like tubercle(s) (Barnes 1994:83–90; Hauser and DeStefano 1989:134–136). Incidence is low, generally not greater than 5 percent in a population (Barnes 1994:83–94).

This developmental trait should not be confused with an occasionally observed pseudarthrosis at this site produced by the dens when

collapse or distortion of Cl allows the dens to ride on the precondylar region (see example in Barnes 1994:97, Figure 3.29) or by arthritic enlargement by osteophytic action of the dens (see **Figure 68**) and the Cl articulation.

Look for irregularly shaped margins and extensions of bone, especially along the posterior margin of the condyles. **Figure 48** is a close-up of left condyle from cranium pictured in **Figure 47**.

Hyperostosis Frontalis Interna (HFI)

Irregular, undulating, thick and heavy bony growths located on the inner surface of the vault, heavily predilected to females (>95%) over 40 years of age (**Figure 49 and Plates 1 and 43**). While the etiology of hyperostosis frontalis interna (HFI) is unknown, Jaffe (1975) suggests the bone formation is a response to endocrine inbalance since it is seen in greater than 60 percent of postmenopausal women. Classic involvement results in bilateral and symmetrical bony

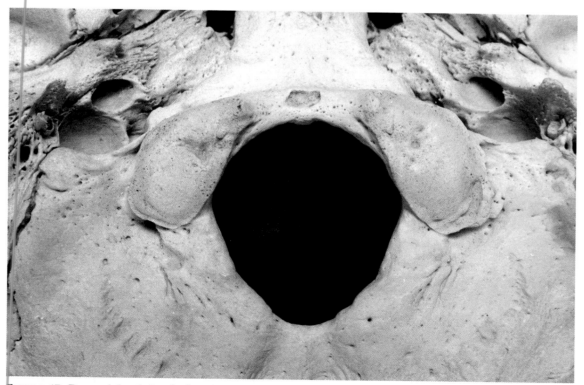

FIGURE 47. Precondylar defect (pit) at the anterior margin of the foramen magnum – note the postero-lateral extension along the margin of the occipital condyles, probably reflecting DJD.

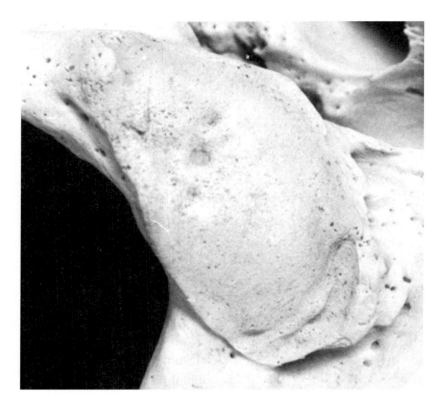

FIGURE 48 (*left*). Marginal osteophytes on the occipital condyle likely reflecting osteoarthritis/DJD.

FIGURE 49 (*below*). Hyperostosis frontalis interna (Hyperostosis cranii. Moore 1955) (NMNH-T 472). See Plates 1 and 43 for color examples.

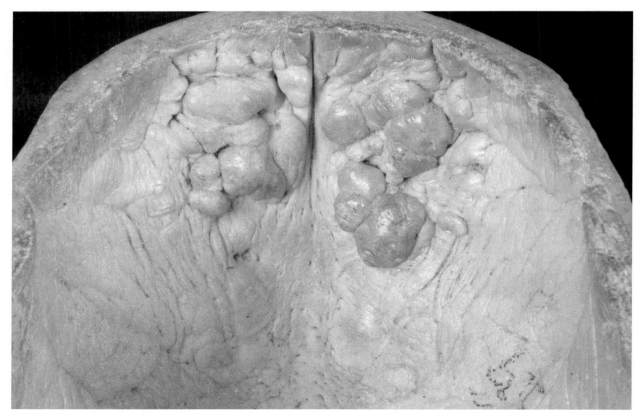

growths on endocranial surface of the frontal bone. Humphry (1873) described what appears to have been biparietal thinning and HFI in a 73-year-old woman, the latter of which was probably associated with, or the result of, shrinking of the brain. See also Hershkovitz et al. (1999), Mulhern et al. (2006), and She and Szakacs (2004). It is an uncommon finding, most prevalent on the frontal bone, followed by the parietals (Gershon-Cohen et al. 1955; Perou 1964; Resnick and Niwayama 1981, 1988).

Paget's Disease (Osteitis Deformans)

Paget's disease is a chronic inflammatory condition that results in proliferation, thickening of the spongy bone in which this less organized deposition produces softened bone that may affect any or all bones in the skeleton (**Figure 50**). Mirra (1987) hypothesized that a unique slow-virus infection of osteoclasts causes Paget's. In the early stages of this disease the lesions are typically lytic (resorptive), originate in one focus of bone, and slowly spread until the entire bone is affected (Mirra 1987). Bone involvement generally occurs unilaterally at first. Late phases result in grossly enlarged, dense bones, especially noticeable in the skull and extremities. The disease seldom appears before the age of 40 years. In European populations, frequencies are found generally around 2 percent but increase to approximately 10 percent as population age increases. Only about 5 percent of the individuals with the disease will express the manifestations of Paget's in the skeleton (Russell 1984). Both sexes are vulnerable to the disease, but males are more often affected (Jaffe 1975). (See Plates 2 and 3 for other hyperostotic effects). Differential diagnosis should include metastatic carcinoma, hyperparathyroidism, leontiasis ossea and saber shin from syphilis (Zimmerman and Kelley 1982:58). See also Auferdeide and Rodriguez-Martin (1998:413–417), Barry (1969), Kaufmann et al. (1991), Lawrence (1970b), Mirra et al. (1995a), Mirra et al. (1995b), Moore et al. (1990, 1994), Nugent et al. (1984), Ortner (2003:435–443), Ortner and Putschar (1985), Singer (1977), Wells and Woodhouse (1975), Whyte (2006), and Zimmerman and Kelly (1982:56–58).

Ectocranial Table Erosion

Extensive destruction (cavitation) of the outer table and diploë, stellate scarring (radiating grooves), and nodules (mounds) in the outer vault caused by a variety of treponematoses (not to be confused with lesions caused by hemangioma or meningioma as in **Figure 52 and Plate 5**) – the classic triad of cranial syphilis consists of nodes, cavitations and stellate scars; bone infection is sometimes referred to as syphilitic osteomyelitis or osteitis. Note the smooth, raised, and rounded nodules (healed) as well as the depressed lesions that have eroded both the inner (rare) and outer tables. The late stage of syphilis (tertiary) (**Figure 51**) can affect any number of bones but is most frequently seen in the tibiae (thickening periostosis and "sabre shin" tibia) (see **Figures 184, 191 and 192**) and skull. Erosive destruction of the skull may be evident in any area of the outer vault, maxilla, palate, malars, and nasal aperture. See also Aufderheide and Rodriguez-Martin (1998:157–164), Crane-Kramer (2000), Crosby (1969), Kaye (2008), Ortner (2003:280–283), Ortner and Putschar (1985), and Stirland 1991a.

Carefully consider the processes which have caused erosion of the bone. A hand lens or magnifying light should be used to look for indications of taphonomic processes or osteolytic activity. As Schultz (2003) suggests histological evaluation should be employed to more accurately determine the cause of the lesion, or pseudo-lesion (**Figure 53**) (Schultz 1997, 2003; Wells 1967).

Coronal Suture Stenosis

Premature closure of one-half of the coronal suture (**Figure 54**) resulting in an oblique (skewed) skull form (Aufderhede and Rodriguez-Martin 1998:53–4; Cohen and MacLean 2000; Perou 1964; Schendel et al. 1996; Turvey et al. 1996). Stenoses are more often seen in males (Cecil and Loeb 1951). This is an uncommon to rare finding (the authors have encountered only one case in an American Indian). See also Barnes (1994:152–157).

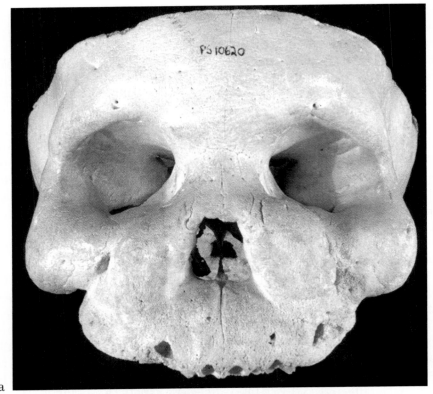

a

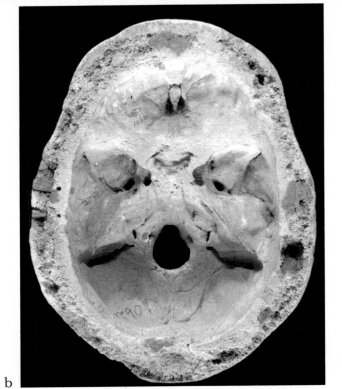

b

Figure 50. Paget's disease (osteitis deformans, Paget 1877) (AFIP PS10620).

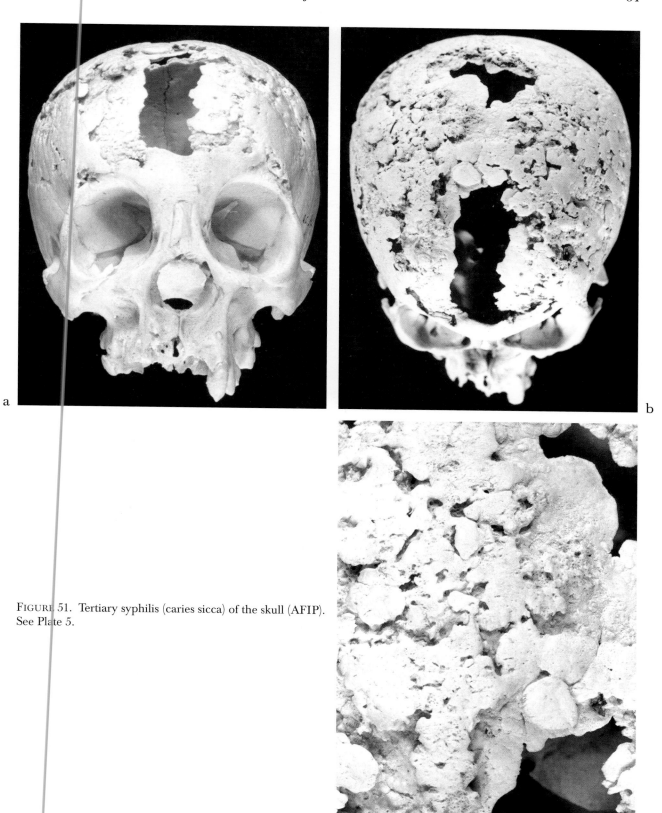

FIGURE 51. Tertiary syphilis (caries sicca) of the skull (AFIP). See Plate 5.

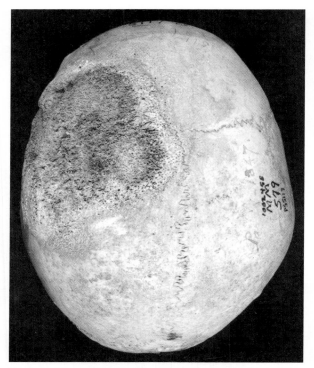

FIGURE 52. Exuberant hemangioma on the left frontal and parietal (AFIP MM579).

FIGURE 53. Postmortem erosion of the left parietal and frontal bone resulting in pseudopathology resembling resorptive lesions similar to treponematosis (NMNH 387867).

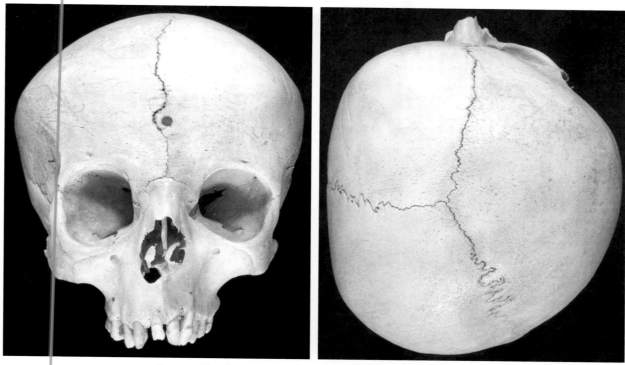

FIGURE 54. Plagiocephaly (frontal and superior view) (AFIP).

Hydrocephaly and Macrocephaly

This condition results in an unusually large and often wide cranium and is in many cases incorrectly referred to as "hydrocephaly" when in fact the cause is unknown (**Figures 55 and 56**). The cranial bones generally are thin and the regions of the fontanelles may have ossicles. Macrocephaly may be a response to a number of factors, possibly congenital since macrocephaly is more often identified in males (Zimmerman and Kelly 1982:26) but may be due to physical effects such as juvenile hydrocephaly ("water on the brain"). Macrocephaly is a rare finding that is often difficult to discern from proportionately large, yet normal skulls in children. Macrocephaly should be differentiated from acromegaly where the cranium is proportionally large as a result of hyperpituitary growth. See also Zimmerman and Kelley (1982:23–26).

Microcephaly

Microcephaly is a rare, neurological disorder where the circumference of the head/skull is smaller than the average for the sex and age in the population (**Figures 56 and 57**). The general range of microcephaly is considered to be from 4–700 cc (compared to 11–1400 cc for normal crania) (Zimmerman and Kelley 1982:22). Microcephaly may be congenital or it may develop within the first few years of life. Infants may be born with normally sized heads that later fail to develop. Microcephaly is due to failure of the brain to grow. Cranial sutures should not have premature closure to be considered true microcephaly versus malformation due to stenosis. The face develops normally, resulting in a proportionally enlarged face relative to the rest of the cranium, producing a receding forehead with a small vault. Compare this morphology to artificial cranial deformation due to cradle-boarding, wrapping and other devices for cultural purposes (see **Figures 58, 59 and 60 and Plates 37a, 38a and 39**). Life expectancy for microcephalics is low. See also Barnes (1994: 157–159), Krauss et al. (2003), McKusick (1969), Nyland and Krogness (1978), and Zimmerman and Kelly (1982:21–23).

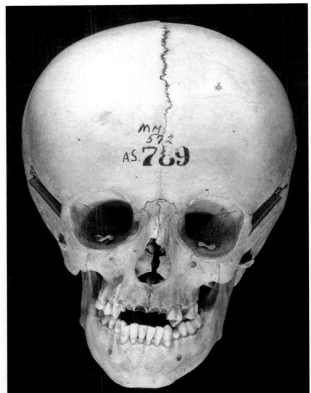

a

FIGURE 55. Macrocephaly (frontal and left lateral view)
(AFIP MM572/AS789)

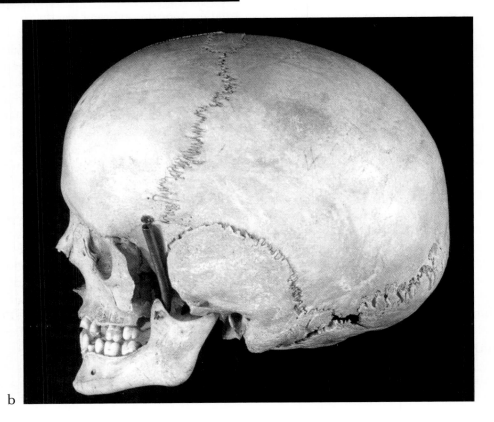

b

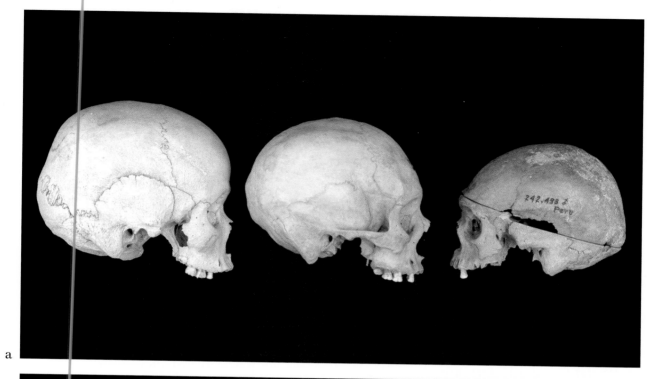

a

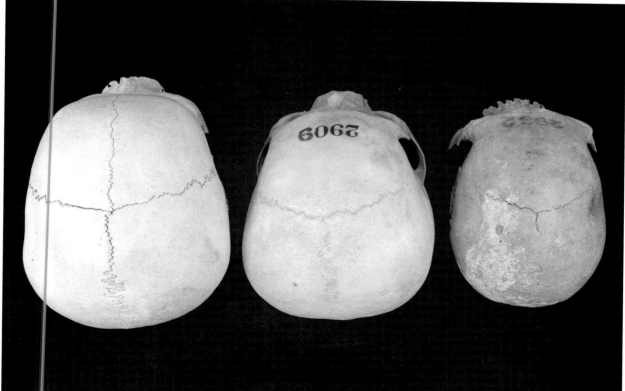

b

FIGURE 56. (a) Macrocephaly (1910cc), left; normal cranium, central; and (b) microcephaly (910cc) from Peru (NMNH 293381, left; 242549, center; 242498, right).

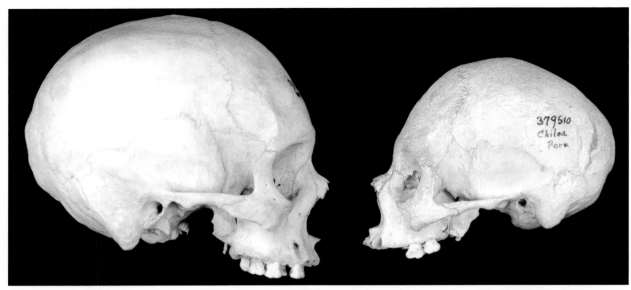

FIGURE 57. Normal cranium compared to an extreme microcephaly (485cc) from Peru (NMNH 242549, left and 379510, right).

Cranial Vault Cultural Modification (Artificial Cranial Deformation)

A significant number of references to cranial deformation in cultures throughout the world are in the literature, discussing the morphological design as well as cultural and physiological significance (**Figures 58, 59 and 60 and Plates 37a, 38a and 39**). The crania illustrated here are only to show the amazing plasticity of the human cranium when intentionally restricted or bound not to encompass the complete range of this practice and its results. Illustration of eleven different types of head wrappings found in mummies from South America is presented in Allison and Gerszten (1982). See also Allison et al. (1981), Aufderheide and Rodriguez-Martin (1998:34–36), Dean (1995), Gerszten (1993), Hrdlička 1912), Moss (1958), Stewart (1941, 1973), and Ubelaker (1999:96–99), cf. Wilczak and Ousley (2009).

Sharp Force Trauma and Gunshot Trauma to the Skull

Sharp instrument defects on the bone are generally identifiable by a well-defined margin where the blade or sharp edge has cut into the bone (**Figure 61**). Pronounced chopping (**Figure 61**) may be seen in butchering and defleshing procedures, deeper as the blade removes large portions or wedges of the bone, as represented in this illustration. Cf. Fiorato et al. (2001), Ortner (2003:137–142), and Symes et al. (2002).

Gunshot wounds are sometimes misidentified as lesions and vice versa (**Figures 62 and 63**). Gunshot wounds, in the present case, an exit wound in the left side of the frontal bone with radiating and concentric fractures, will almost always exhibit at least one beveled margin if the projectile passes through bone. The beveled margin, which can be seen here, reflects the direction the bullet was traveling through the body and is characterized by an irregular/porous/jagged defect encircling the gunshot wound. In the present case the bullet entered the person's head from the right mandible and traveled upward and to the victim's left, exiting the frontal bone. Radiographs may reveal evidence of small metallic bullet or casing fragments in or around the gunshot wound, or scattered metallic fragments in other parts of the skull as a result of the bullet breaking up as it traveled through the head. Remote fractures may also be revealed in the radiographs where normal visual observation is not possible (e.g., the orbits ["blowout" fractures], sphenoid, internal regions of the cranial base, TMJ and endocranial regions). The number, frequency

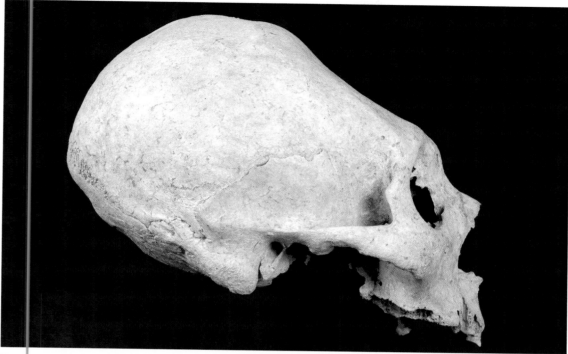

FIGURE 58. Type of artificial cranial deformation (NMNH 378586). See Plates 37a, 38a, and 39 for color examples of artificial cranial deformation.

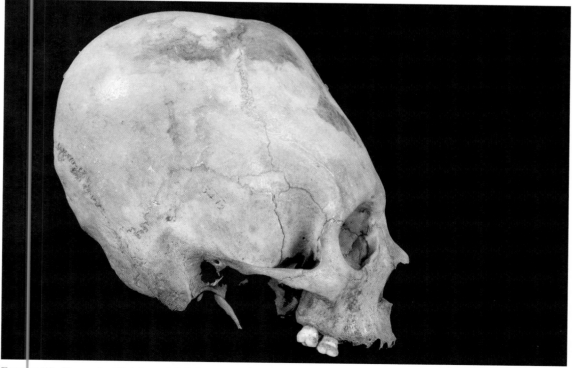

FIGURE 59. Type of artificial cranial deformation (NMNH 276106). See Plates 37a, 38a and 39 for color examples of artificial cranial deformation.

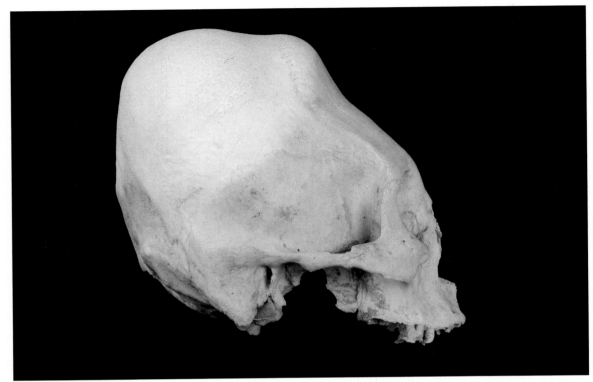

FIGURE 60. Type of artificial cranial deformation (NMNH 293689). See Plates 37a, 38a and 39 for color examples of artificial cranial deformation.

(sequence of injuries) and directionality of gunshot wounds in bone, especially those in the skull, can be interpreted based on the presence of internal and external beveling, as well as the pattern and termination of radiating and concentric fractures. Check for separation or widening of one or more sutures indicating diastatic fracture (transmission of energy along the sutures), usually in young

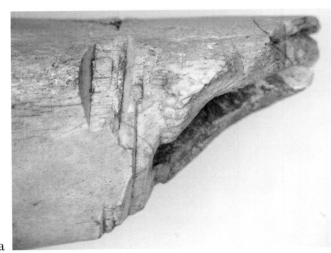

a

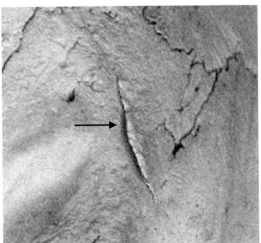

b

FIGURE 61. Perimortem chopping injuries to a long bone (a) and slicing injury to skull above right mastoid process. (b Photo courtesy Greg Berg, Killing Fields, Cambodia.)

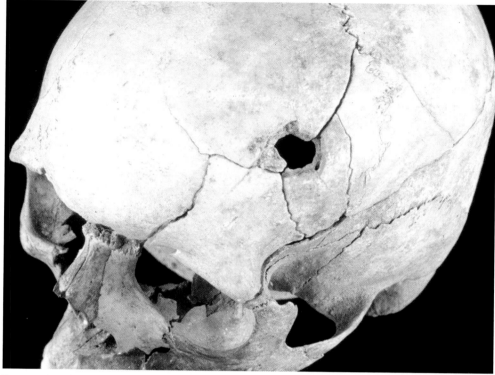

FIGURE 62. Gunshot exit wound to the frontal bone showing radiating and concentric fractures as well as diastatic fracture through the coronal suture (Historic North Korea).

and middle-aged individuals with open or partially fused sutures (Leestma 2008). For a description and biomechanics of bone associated with gunshot wounds in bone, refer to Berryman and Gunther (2000), Berryman and Symes (1998), Di Maio (1999), Dixon (1982), and Smith et al. (1987, 1993). See also Browner et al. (2003), Galloway (1999), Grandmaison et al. (2000), Kimmerle and Baraybar (2008), Mann and Owsley (1992), Quartrehomme and İşcan (1999), Ross (1996), and Spitz (1980).

Cranial Trephanation (Trepanation, Trephination)

There is a broad spectrum of literature concerning trephanation to the cranium in prehistoric, historic and modern populations, and therefore review of these discussions is not necessary for this volume. To present this change to normal skeletal morphology, the figures presented here illustrate the scraping method (**Figure 64a and Plate 33**) and the cutting method (**Figure 65**) of

trephanation. They also illustrate both perimortem activities as well as the healed forms of this surgery, and in particular, **Figure 65** shows the

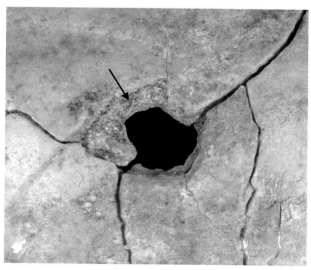

FIGURES 63. Close-up of exit gunshot wound in left frontal bone; note beveling/cratering (arrow) encircling the defect.

typical undulating pattern of healing on the ecto-cranial table from the laying back of the scalp and damage to the periosteum during surgery. For an extensive overview and illustrations on the different types of trephination, see Lastres and Cabiese (1960), Aufderhide and Rodriguez-Martin (1998:31–34), Ortner (2003:169–174) and a series of 42 slides of trephined crania from the San Diego Museum of Man (Rogers and Merbs 1980). See also Margetts (1967), Rifkinson-Mann (1988), Stewart (1957), and Tyson and Dyer (1980).

Trephination is an ancient technique of re-moving a portion(s) of the cranium by scraping (**Figure 64**), drilling, or sawing (see **Plate 33** for color examples from Peru). Although rare in North America, this practice is quite common in prehistoric skulls of South American (in particular, ancient Andean populations) as well as many other groups throughout the world (John Verano, personal communication 1989). Webb (1988) re-ported on two skulls from Australia that may represent the first such cases from that continent. Piek et al. (2008) reported on a Neolithic skull from Germany that exhibited a healed scraping-type of trephination in the frontal bone and a healed fracture of the left temporo-occipital region. The authors concluded that the trephination was done for medical reasons. Care must be exercised when trying to differentiate lytic lesions (e.g., in-fectious diseases, tumors), axe and sword wounds, and blunt trauma from trephinations in the skull. Look closely for cut marks and abrasions around the defect that would indicate trephination. Healed trephinations will exhibit rounded margins, scar-ring, tiny spicules or roughened (possibly porous) areas surrounding the defect where infection or healing has occurred. In this example, one of the defects has not fully remodeled, as evidenced by visible diploë and relatively smooth (unscarred) margin. For additional background and references and illustrations of healed, healing and peri-mortem trephination, refer to Ortner (2003:169–174) and Aufderheide and Rodriguez-Martin (1998:31–34). See also Lisowski (1967), Sanan and Haines (1997), Shaaban (1984), Steinbock (1976:29–35) Stewart (1957), and Velasco-Suarex et al. (1992). See **Plates 8, 18, 19a and 20** for modern trephination surgeries.

Scalped skulls, in comparison, will exhibit patterned cuts that usually circumscribe the entire cranium. Typically the cuts extend across the middle of the frontal bone from temporal to temporal. Also look for deep or superficial cuts and scratches on any raised area of the skull where the scalp was removed by cutting through the muscles along the temporal lines, zygomatics, supramastoid crests, and mastoids. See also Steinbock (1976:24–30).

Animal Taphonomy (Rodent Gnawing)

Rodent gnawing (**Figure 66**) may also be present in the skull as well as any of the long bones. Although it is sometimes difficult to dis-tinguish rodent gnawing from defleshing marks (e.g., cultural practices before burial), the former are usually located above the eye orbits, nasal area, zygomatic arches, mandible and other raised ridges and margins. Rodent gnawing appears as short but deep *parallel* cuts localized to raised areas of the bone. Rodents can destroy entire bones, as can carnivores that may also scatter the remains. See also Wells (1967:10–11), Steinbock (1976), and Ubelaker (1999:105).

Mandibular Condyle Erosion

Erosion and pitting (a) of the mandibular condyle-pitting and erosion of the condyle are common findings in all populations although the severity varies greatly from one group to another generally due to differences in the speed and severity of attrition. Porosity appears as a localized area(s) of bone loss, usually with a sharply defined margin. The common sites for such destruction are noted in **Figures 46 and 67**. The dotted line denotes bone loss due to erosion or, simply, the wearing away of the joint surface. Alteration of the mandibular condyle may range from a very small area of pitting (porosis) to complete de-struction of the articular surface. Diet, mechanical factors related to chewing, dental wear, caries (cavities), disc abnormalities, facial morphology, and abscesses all contribute to destruction of joint. See also Richards (1987, 1988), Richards and Brown (1981), and Ryan (1989).

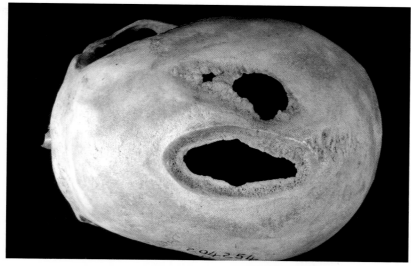

a

FIGURE 64. Trephinations (trepanation) of the skull. Superior view of cranium and close-up of perimortem surgery by scraping (b) and close-up of a significantly healed trephination (c) (NMNH 204254). See Plate 33 for a color example.

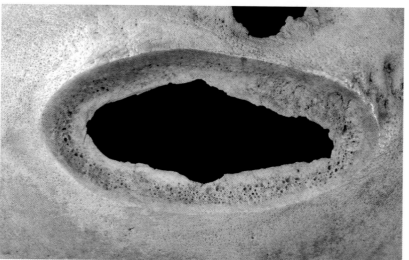

b

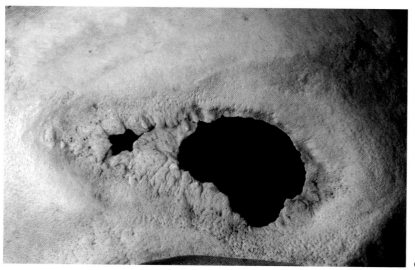

c

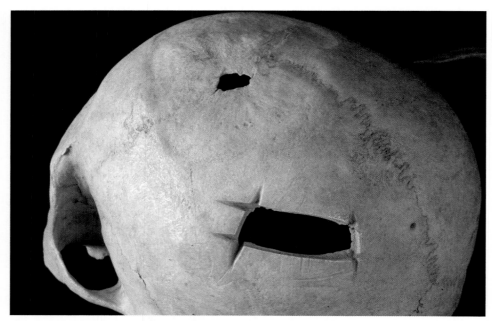

FIGURE 65. Trephinations of the skull. Perimortem cutting/sawing trephination on the left lateral parietal with cutmarks around the periphery of the surgery. The healed trephination located near Bregma has an undulating area around the healed opening. This is remodeled bone from the lifting of the scalp and periosteum during the surgery (NMNH 293778).

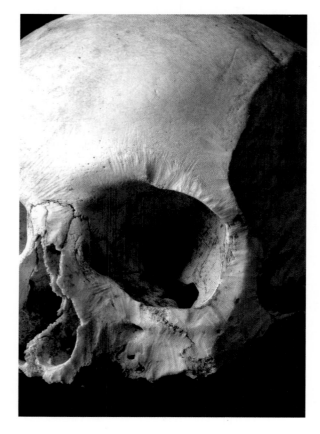

FIGURE 66. Rodent gnawing on bone. Note the parallel striations on the bone surface from the rodent's incisors (NMNH).

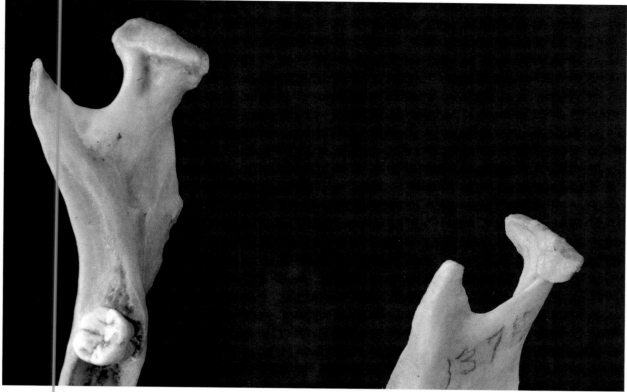

FIGURE 67. Pathological changes (OA) of the temporomandibular joint (TMJ) (Markowitz and Gerry 1950) of the mandibular condyles (NMNH-T 1374). See Plate 47b for an unusual example.

Erosion (a) and osteophytes (b) of the mandibular condyle—when looking for erosion of the condyles compare the shape and size of both heads for symmetry. There may be slight asymmetry in the condyles that is developmental or genetic and not representative of a pathological condition. Slight erosion is a common finding in most populations while severe forms appear to be age-related (elderly) or the result of trauma or chewing stress.

VERTEBRA

Osteoarthritis and Erosion

The atlas may exhibit marginal osteophytes and surface porosity on the articular facets or dens (odontoid process) (**Figure 68**; see **Plates 83c and 85** for an example of os odontoideum). In some instances an elderly individual will exhibit eburnation (polishing) of the anterior surface of the dens and ossification of the apical ligament (most superior portion of the dens). All of these changes are consistent with OA/DJD. Erosion of the dens, however, is one of the criteria associated with rheumatoid arthritis (RA). (In advanced cases of RA the entire dens may erode and fracture resulting in death of the individual.) The first cervical vertebra may also exhibit corresponding changes. OA is a common finding although unequivocal cases of RA in archaeological specimens are rare. See also Bland (1994) and Sager (1969). Cf. Klepinger (1979). See also os odontoideum (**Plate 85**).

The most common finding of OA in the cervical vertebrae is marginal osteophytes (osteophytosis) of the anterior and posterior inferior margins of the body (**Figure 69 and Plate 84**). Although the

FIGURE 68. Osteoarthritis of the second cervical vertebra (axis) – note the resemblance of this vertebra with a seated person (NMNH-H). See Plates 83c and 86a–b.

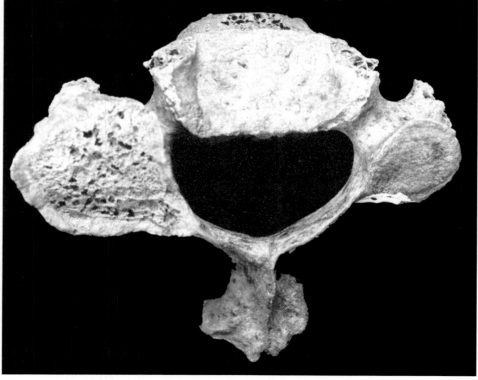

FIGURE 69. Osteoarthritis of a cervical vertebra (NMNH-H).

anterior border may be slightly irregular in normal vertebrae, osteophytes appear as bony extensions and spicules projecting inferiorly. Marginal osteophytes and surface osteophytes on the articular facets may also be present (**Figure 70**). In more severe cases, the vertebral body may exhibit macroporosity and distortion of the endplates. See also Bland (1994), Lestini and Wiesel (1989), Ortner (2003:555–558), Rogers et al. (1997), and Sager (1969).

Note the irregular margins encircling both of the superior and inferior facets, as well as moderate–severe pitting/porosity (what some refer to as "macroporosity") of the superior facets. These features are characteristic of any joint in the body affected by degenerative joint disease/osteoarthritis.

"Hangman's Fracture"

"Hangman's fracture" traditionally refers to traumatic spondylolisthesis of the axis with separation of the body and pedicles in judicial hangings due to hyperextension of the upper cervical spine and distraction resulting in lethal damage to the spinal cord. Fracture of the axis, most commonly transversely across the entire vertebra posterior to the dens, can result from a variety of situations including hanging, motor vehicle accidents, a fall on the head, or being struck with an object (for example, a wooden beam) (Junge et al. 2002; Smith et al. 1993; Williams 1975). James and Nasmyth-Jones (1992) examined the cervical vertebrae of 34 victims of judicial hangings and found fractures in only seven cases – six of the axis/C2 and one to C3. Hangman's fracture, therefore, may be less common than previously thought (**Figure 71**). See also Foreman (2001) and McCort and Mindelzun (1990).

Degenerative Joint Disease

This is the form of bone growth that develops in degenerative joint disease. Note its resemblance to a bird's beak (**Figure 72**) and degree of curvature as it bridges the two vertebral bodies

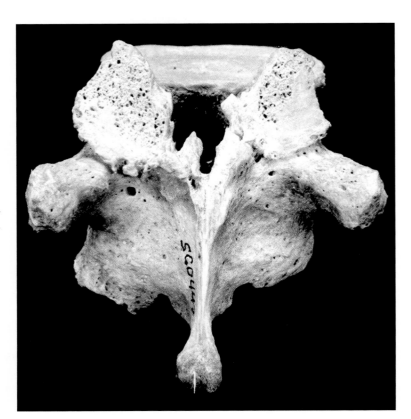

FIGURE 70. Osteoarthritis of the superior articular facets of a thoracic vertebra (NMNH-H).

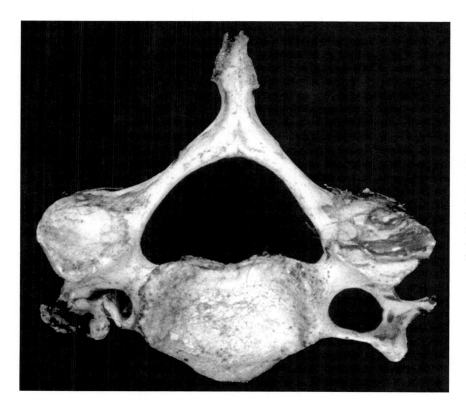

FIGURE 71. Compression and collapse of the vertebral formaen (foramen transversarium) in the axis (C2) vertebra of a man who was hanged. Collapse of the foramen was the result of traumatic constriction of the noose; note that the left foramen is normal.

FIGURE 72. Osteophytes – "beak-shaped" new bone that buttresses weakened vertebral bodies in the elderly (osteoporosis) and/or trauma to the spine, in this case a pair of osteophytes along the anterior centra (NMNH-H). See Kasai et al. (2009) for an interesting discussion of the development and location of vertebral osteophytes.

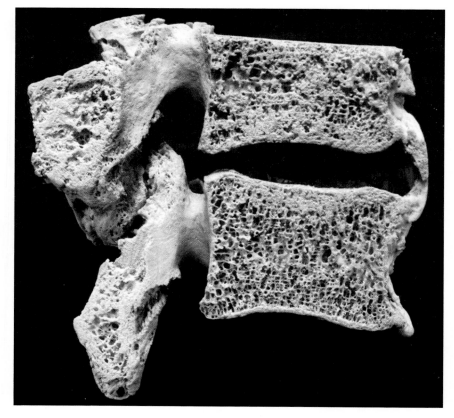

(see also **Figure 78**). Common in all populations. See also Hilel (1962).

Diffuse Idiopathic Skeletal Hyperostosis (DISH)

Although the etiology is unknown, diffuse idiopathic skeletal hyperostosis (DISH) (**Figure 73**) is a common disease and form of arthritis in middle-aged and elderly individuals, with males affected 2:1 over females (Utsinger 1984). Although DISH frequently affects the spine, other peripheral skeletal sites may be involved and exhibit "whiskers" or "whiskering" periostosis (Burgener and Kormano 1991; Utsinger 1985),

or irregular ossifications known as enthesophytes (bony spikes or projections) reflecting inflammation at the insertions of tendons and ligaments on the iliac crest, ischium, greater and lesser trochanters, trochanteric fossa, patella, calcaneus, ulna, and linea aspera, to name a few. See also Aufderheide and Rodriguez-Martin (1998:97–99), Ortner (2003:559–560), and Scutellari et al. (1995). (See Chapter X for a thorough discussion of enthesophytes.)

Criteria (Resnick and Niwayama 1976) for distinguishing DISH from ankylosing spondylitis include: (1) flowing calcification (resembling melted candle wax) and ossification along the anterolateral surfaces of at least four contiguous

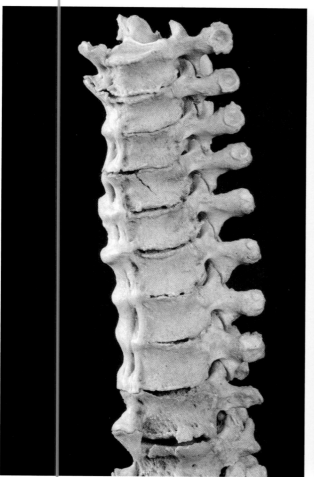
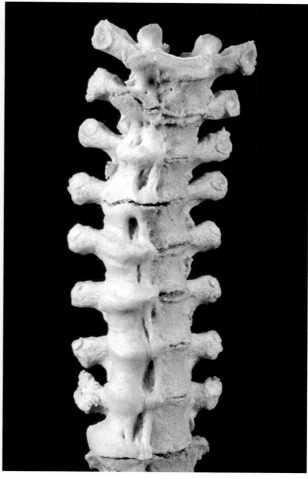

a
b

FIGURE 73. Calcification and ossification of the anterior longitudinal ligament associated with diffuse idiopathic skeletal hyperostosis (DISH) (NMNH–T 236R). Note the melted "candle wax" and flowing appearance of the bone along the right side of the spine, opposite to the aorta. See Plates 87 and 88a for color examples.

vertebrae, (2) relative preservation of intervertebral disc space in the involved segments, and (3) the absence of apophyseal joint (facets) bony ankylosis and sacroiliac erosion, sclerosis or osseous fusion (compare to **Figure 87** of an ankylosed spine). Utsinger (1985) revised the diagnostic criteria of Resnick and Niwayama as follows: continuous ossification along the anterolateral surfaces of at least two contiguous vertebral bodies, primarily in the thoracolumbar spine (broad, bumpy, buttress-like band of bone) and symmetrical enthesopathy of the posterior heel, superior patella or olecranon. See also Garber and Silver (1982).

Rarely ankylosing spondylitis and DISH co-exist in the same individual. An extensive radiographic survey of 8993 persons over the age of 40 by Julkunen et al. (1973) revealed a standardized prevalence of 3.8 percent in men and 2.6 percent in women. Interestingly, the Pima Indians of Arizona showed an incidence of 34 percent in males and 6.6 percent in females (Henrard and Bennett 1973). Weinfeld et al. (1997), in examining 2364 patients 50 years or older, found the highest incidence in whites, followed by blacks, Native Americans and Asians, indicating some population genetic factors. Ankylosing spondylitis is inherited as a Mendelian dominant trait with 70 percent penetrance in males and 10 percent in females (Sharon et al. 1985). There appears to be a metabolic component as well, evidenced from the findings of Julkenen et al. (1971) that 25 percent of the survey patients in Finland with DISH also suffered from adult-onset diabetes. This is an uncommon finding.

Note that a narrow flowing osteophytosis–thickened, flowing ossification is generally found along the right anterolateral aspect of the thoracic vertebrae (see **Plates 87 and 88a**). The left side of the spine is usually spared due to the presence of the aorta. Note that the "ribbon" (Scutellari et al. 1995) of ossification is clearly visible, extends across at least four contiguous vertebrae, and appears as if a thick, undulating, and smooth surfaced band or belt of bone has been applied to the right side of the spine. Some of the thickened ribbon of bone is the ossified anterior longitudinal ligament. See also Aufterheide and Rodriguez-Martin

(1998:97–99), Julkenen et al. (1971), Ortner (2003: 559–560, 571–577), Rogers and Waldron (1995), Sreedharan and Li (2005), Steinbock (1976:296–297), Utsinger (1985), and West (1949).

McKern and Stewart (1957) reported that laminal spurs (**Figure 74 and Plate 88b–c**), also referred to as ossified ligamentum flavum or flava, were most often found in the upper and lower thoracic vertebrae, then the lumbar and, with rare exception, the cervicals. Ossification of the ligamentum flavum, sometimes reported as normal variants of the spine, are often seen associated with increasing age, are believed to accompany strenuous activity, have been reported as occurring in 3.8 percent of 1736 Chinese volunteers (Guo et al. 2010) and can lead to spinal complications, including paraplegia. Regardless of etiology, the authors have encountered large spurs in children. This condition can be scored non-metrically either as present or absent, or as mild, moderate or severe. Some researchers have measured the lengths, but this practice is difficult to reproduce accurately. It is a common finding in all populations. See also Allbrook (1955), Davis (1955), Naffsiger et al. (1938), and Shore (1931).

Vertebral Body Collapse

Compression fractures and collapse of the vertebral centra produce forward bending of the spine (kyphosis) manifested as a hump back, most often observed in the elderly (**Figure 77 and Plates 92 and 93**). The varying loss of centrum height most frequently results from collapse of the superior endplate (which possibly can be also associated with Schmorl's nodes – see **Figures 83 and 84**), not the inferior plate. As the intervertebral disc protrudes into the endplates, the body of the vertebra gives way and collapses over time. Arthritic response to the exuding of the intervertebral disc anteriorly and laterally will also produce osseous spicule formation or osteophytes along the rim of the centrum (see **Figure 84**). This is a common finding in the elderly. See also Albright et al. (1941), Cummings et al. (1985), Dolan et al. (1994), Frymoyer et al. (1997), Horal et al. (1972), Kraemer (2009), Raisz (1982), Tehranzadeh et al. (2000), Twomey & Taylor

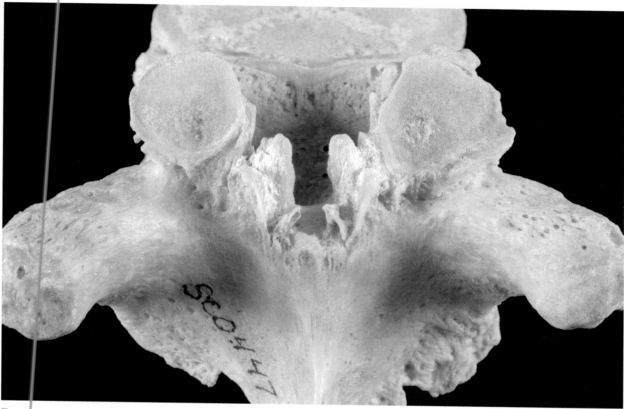

FIGURE 74. Laminal spurs (ossified ligamentum flavum) – spikes along the superior border of the "V"-shaped neural arch (lamina) that serve for attachment of the inferior ligamentum flavum (Hilel 1959) (NMNH-H). Cf. Guo et al. (2010). See Plates 79a and 88 for color examples.

(1987), Ubelaker (1999:84–87), and Wagner et al. (2000). See **Plates 97 and 98** for color examples of Schmorl's depressions.

Abnormal porosity and foramina involving the body of the vertebrae should not be confused with normal nutrient pits and foraminae necessary for vessels feeding the interior of the developing corpus. These foraminae and pits are prevelant in all sides of the vertebra, especially on the anterior and in the neural canal (see **Figures 75 and 76** for normal vertebral development in children).

Osteophytes and Enthesophytic Lesions on the Vertebral Rim

Rounded or radiating finger-like projecting bony growth(s) commonly found on the rim and vertebral bodies of older individuals. However, these formations are not age specific but do tend to increase in size and number with age, being found in association with younger individuals who have either sustained trauma to the spine or due to any number of spinal diseases such as tuberculosis, mycotic infections, etc. Osteophytes will result in small to large rounded protrusions at two or more adjacent vertebral bodies, often appearing in "opposing pairs" as small, raised irregularities along the margins or midsections of the centra and that gradually become larger, having a "parrot-beak" or "claw" appearance until bridged fusion of the bodies occurs. When two vertebrae become fused (ankylosed), the large osteophytes may be referred to as kissing or bridging osteophytes (**Figures 72 and 78**). This is a common finding. Kasai and colleagues (2009) studied the formation and direction of 14,250 pairs of anterior lumbar osteophytes and in 2850 patients and found that: (1) pairs of osteophytes

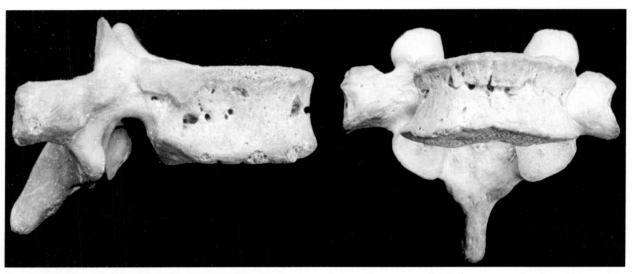

FIGURE 75. Normal nutrient pit for passage of the nutrient vessels into the centrum/body (more prominent and noticeable in subadults) (NMNH-H).

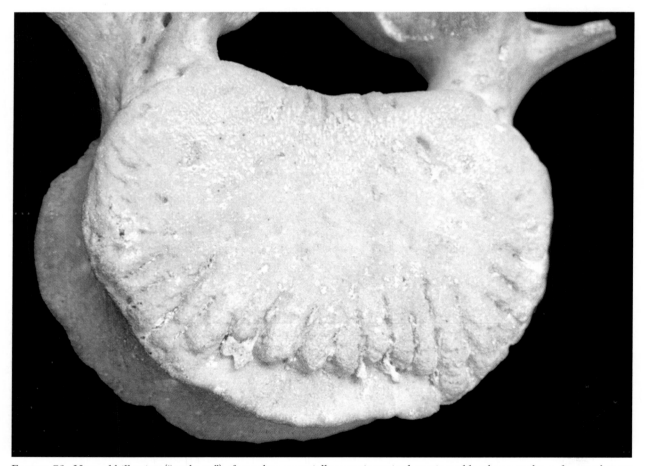

FIGURE 76. Normal billowing ("sunburst") of vertebra, especially prominent in thoracic and lumbar vertebrae, for attachment of the endplates and epiphyseal rings (NMNH-H).

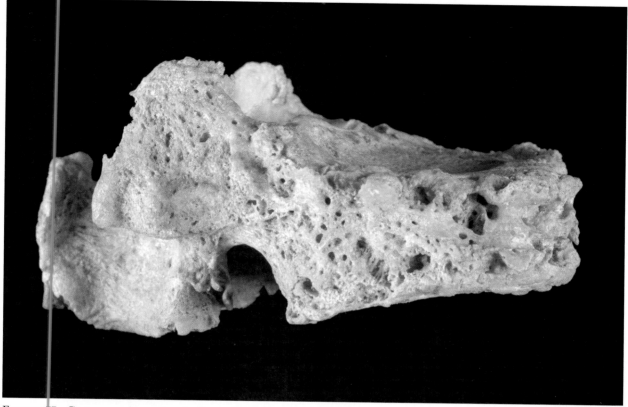

FIGURE 77. Compressed vertebra (collapsed vertebra, wedged vertebra) results from bone loss accompanying old age (osteoporosis), infection, or trauma such as a compression injury (NMNH-H). See Plates 92, 93 and 97 for color examples.

often formed in the direction of the opposing disc in the upper lumbar vertebrae and away from the disc margins in middle and lower lumbar vertebrae; and (2) biomechanical and anatomical differences in vertebrae affect the direction of formation of intervertebral spurs. See also Hough and Sokoloff (1989) and Stewart (1958).

A radiographic study by Papadakis et al. (2010) of 112 adult females (aged 40–72 years) in Crete revealed no association between women with the degree of sacral lordosis and the presence or absence of lumbar osteoarthritis. The authors concluded that lumbar lordosis is neither an outcome nor a contributing factor to spinal osteoarthritis.

Spondylolysis (Separate Neural Arch)

Pronounced *spon-di-loli-sis* (Thomas 1985) or spon-dee-low-lie-sis is separation of the vertebral body from the posterior vertebral arch, usually at a junction known as the pars interarticularis or isthmus (**Figures 79 and 80 and Plate 89**). Some researchers believe this condition to be congenital in origin while others state that stress plays a major role in causing the neural arch to separate, or basically a stress/fatigue fracture and nonunion (Merbs 1983; Newman and Stone 1963; Stewart 1953, 1956). Spondylolysis is seen in between 3 percent and 10 percent of the adult population (Collier et al. 1985), but occurs primarily in children with an increasing incidence until 20 years (Ohmori et al. 1995; Roche 1949; Roche and Rowe 1951, 1953) and an arrest of occurrence after age forty years (Stewart 1953; Wiltse 1962).

Activity, especially shearing stress, involving the lower spine does seem to contribute to the presence of this trait. Wynne-Davies and Scott (1979) reported that the frequency of spondylolysis is highest in athletes, and the highest risk

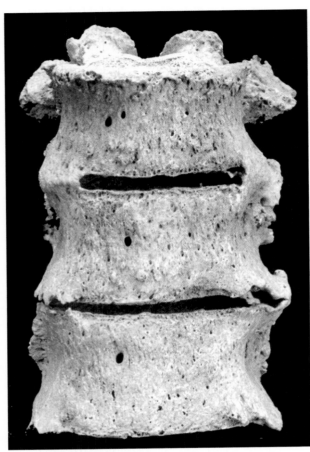

FIGURE 78. Osteophytes (osteophytosis, spondylitis deformans, spondylosis deformans) accompanying degenerative joint disease. Cf. Dieppe et al. (1986) (NMNH–H). See Plate 87 for a color example.

young people because they frequently engage in strenuous activity at a time when their intervertebral discs are more elastic and their neural arches may not be completely ossified. See Fehlandt and Micheli (1993) for report on lumbar stress fractures in ballet dancers and Brukner et al. (1996) study of 180 sports athletes. Thieme (1950) hypothesized that spondylolysis results from trauma during the developmental stage to final fusion. The lack of normal repair callus at the lesion, even in unilateral instances where movement of the "separating" arch is held at a minimum, suggests against postfusion breaks. Garth and Van Patten (1989) reported a case of bilateral fracture of the lumbar laminae, not all, but near, the pars interarticularis that showed no bony union after two months. Friedman and Micheli (1984) reported spondylolisthesis, a forward sliding or slipping of the centrum following bilateral spondylolysis, due to stress concentration on the neural arch from scoliosis surgery. A radiographic study by Wang et al. (2008) of 131 individuals aged 6–20 years revealed that the anatomy of the sacrum, including sacral kyphosis, in children and adolescents with spondylolysis contributed to developmental spondylolisthesis of L5–S1.

Most commonly, the fifth lumbar is affected, but the fourth and third may also show this trait. Spina bifida is a common finding associated with separate neural arch defects (Rowe and Roche 1953). Eisenstein (1978) found the defect in 17 skeletons (3.5%) of 485 examined; 1 example in L3, 4 in L4, and 13 in L5. Libson et al. (1982) reported that of 1598 patients with lumbar spondylolysis, only two individuals had lesions in the upper three lumbar vertebrae. However, spondylolysis in the lower two lumbar vertebrae is more common. Rowe et al. (1987) found the defect in the lower lumbar in 2 percent to 10 percent of active young individuals in the United States. The male to female ratio is 2:1 from age six to adulthood (Fredrickson et al. 1984), and the condition has never been reported as present in newborns. The youngest individual on record with spondylolysis is a 3.5-month-old infant (Seitsalo et al. 1988).

Although the etiology of spondylolysis remains unresolved, including genetic and congenital

groups are Alaskan Native American and a first-degree relative with spondylolysis at a risk level of 50 percent in development of spondylolysis. See Gerbino and d'Hemecourt (2002) for discussion of lumbar involvement due to football sports. The frequency of spondylolysis is found as high as 40 percent in Eskimos (Stewart 1953; Boyer et al. 1994). As it relates to activity, the incidence of spondylolysis in divers was found to be 63 percent and 32 percent in gymnasts (Gerbino and Micheli, 1995). Gerbino and Micheli (1995) contend that spondylolysis is the result of mechanical forces – primarily repetitive hyperextension of the lumbar spine – on the pars interarticularis (Omey et al. 2000). Cyron and Hutton (1978:238) hypothesized that spondylolysis is more likely to occur in

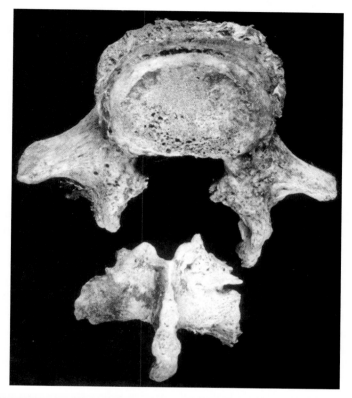

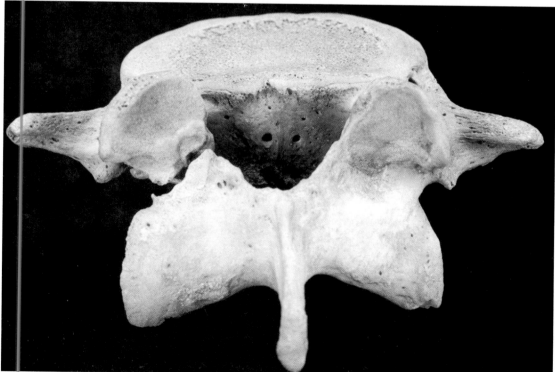

FIGURES 79 and 80. Bilateral spondylolyis (*top photo*) and unilateral spondylolysis (*bottom photo*) in two lumbar vertebrae (NMNH). See Plate 89 for color examples.

factors (Blackburne and Velikas 1977; McKee et al. 1971), repeated stresses and trauma to the lower back appear to play a major role in its development (Kono et al. 1975). Research by Rankine and Dickson (2010) suggests that asymmetry of the joint facets contributes to unilateral fracture by increasing the point loading of the pars interarticularis on the side with the greatest coronally-oriented facet. Spondylolysis is a common finding and is strongly associated with spina bifida and repeated hyperextension of the lower spine (Barnes 1994:46-53). Separation of the neural arch may be incomplete, complete, unilateral, or bilateral. Care should be exercised not to confuse spondylolysis with postmortem breakage, which is common at this site. See also Aufderheide and Rodriguez-Martin (1998:63), Barnes (1994:46–50, 259–265), Bradford (1978), Bridges (1989), Eisenstein (1978), Floman et al. (1987), Fredrickson et al. (1984), Griffiths (1981), Gunzburg and Szpalski (2006), Harris and Wiley (1963), Hitchcock (1940), Hoppenfeld (1977), Lamy et al. (1975), Laurent and Einola (1961), Letts et al. (1986), Libson et al. (1982), Lowe et al. (1987), Miles (1975), Nathan (1959), Newman and Stone (1963), Ortner (2003:147–149), Pecina and Bojanic (2004), Ravichandran (1981), Rosenberg et al. (1981), Stewart (1931,1953 and 1956), Thieme (1950), Troup (1976), Wiltse (1962), and Wiltse et al. (1975).

Spondylolisthesis

If the vertebral body slides forward on the sacrum (which may be detected by the presence of bridging osteophytes between L5 and S1), the condition is known as spondylolisthesis and presents a more complicated condition (Congdon 1931; Merbs and Euler 1985; Pedersen and Hagen 1988; Ruge and Wiltse 1977; Thieme 1950; cf. Wang et al. 2008; Wiltse 1962; Wiltse et al. 1975). Look for (1) fissures below the superior articular facets with or without evidence of healing (callus) that may or may not impinge on the neural canal and (2) unusually large/hypertrophied or asymmetrically-shaped inferior articular facets on the side of the spondylotic defect. It is an uncommon to common finding depending on the

population and sample studied. See also Bradford (1978), Farfan et al. (1976), Lester and Shapiro (1968), Merbs (1995, 1996, 2001), 2002), Miki et al. (1991), Nathan (1959), Neugebauer (1976), O'Beirne and Horgan (1988), Porter and Park (1982), Ruge and Wiltse (1977), Saifuddin et al. (1998), Waldron (1992, 1993), Willis (1924, 1931), and Wiltse et al. (1976).

Spina Bifida

Spina bifida is a congenital malformation of unknown etiology (**Figures 81 and 82 and Plate 94**) resulting in incomplete closure of vertebral neural arches. See the discussion in Barnes (1994:117–125) on the embryonic developmental problems which produces these anomalous conditions. When examining dry bone (no soft tissue or a patient history), it is extremely difficult, if not impossible, to determine if the condition was occulta (asymptomatic), cystica, or aperta. Cystica implies that the skin was involved and the lesion became cystic, while the term aperta refers to an open lesion (Strassberg 1982). Dickel and Doran (1989) present an interesting case of a 15-year-old child dating to 7500 BP diagnosed as having aperta. The most common site of neural arch deformity is S1 with a reported incidence of 9 percent in females and 13 percent in males (Cowell and Cowell 1976), but the entire sacrum can be involved (see **Figures 107 and 108**). Care must be exercised when examining and interpreting archaeological or ancient dry-bone specimens, as herniation of the spine can occur in even mild cases of spina bifida (James Vailas, personal communication 1989). See also Barnes (1994) Ferembach (1963), and Saluja (1988).

Schmorl's Depressions (Nodes/Pits)

Schmorl's or Schmorl depressions can be a circular, linear or combination of the two, depressed lesion(s), usually with a sclerotic floor in either of the centra endplates (**Figures 83 and 84 and Plates 97 and 98**); they often resemble letters of the alphabet such as V, H, L, and U. In some cases only a small circular depression or

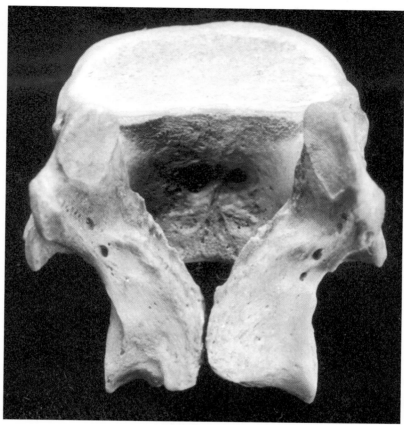

FIGURE 81. Bifid neural arch (spina bifida, split or cleft arch) (NMNH-H).

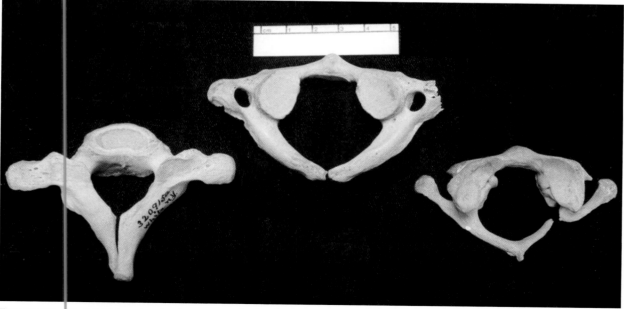

FIGURE 82. Cleft/bifid arches in thoracic and atlas (C1) vertebrae. Note that the cleft is along the midline in one C1 and unilateral in the other (Barnes 1994:120–124) (NMNH-H).

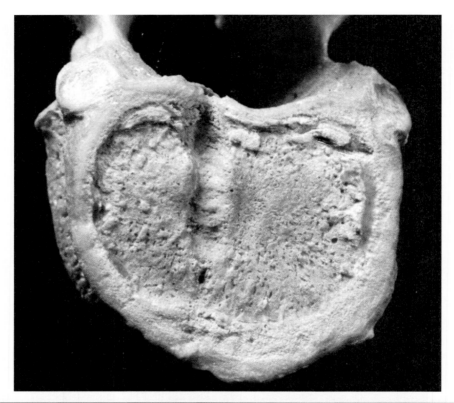

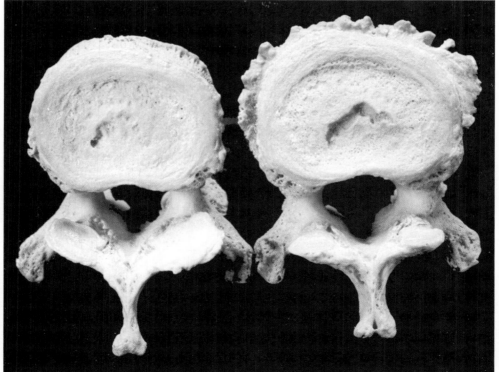

FIGURES 83 and 84. Schmorl's nodes/depressions in a thoracic (*top photo*) and lumbar vertebrae (*bottom photo*). See Plates 97 and 98 for color examples.

shallow pit will be present in the center of the centrum – such lesions should also be scored as present. Schmorl's depressions result from herniation of the nucleus pulposus, the partially liquid central portion of the intervertebral disc. This herniation of the nucleus can take place in any direction but will bulge where the bone (frequently the endplate) or annulus fibrosus is weakest.

Schmorl's depressions are common findings in the elderly and result from degenerative disc disease. However, the presence of such nodules/depressions can be seen in adolescents but is uncommon (only 2% of all herniated discs occur in children and adolescents) (Bunnell 1982). Schmorl's depressions in subadults result from trauma from such activities as a fall from height, heavy lifting, trauma during physical exercises, and similar activities (Bolm-Audroff 1992). Approximately 2 percent of all individuals, however, will have a clinically significant herniated nucleus pulposus at some time in their lives (Frymoyer 1988).

Schmorl's depressions are common (2% to 76%) in all populations (Yochum and Rowe 1996; cf. Schmorl and Junghanns 1932, 1971a–b) with an approximately equal frequency in elderly and younger individuals (Pfirrmann and Resnick 2001). Dar and colleagues (2009), in a study of 240 adult spines (T4–L5) found Schmorl's nodes in 48.3 percent (N = 116) of the individuals. The authors further found Schmorl's nodes to be age independent, but sex and ethnicity dependent, supporting a possible genetic component. When scoring its presence, the count can be based on either the number of vertebrae with the trait, how many total lesions are present in both the superior and inferior endplates, or the exact position of the lesions (e.g., between L3 and L4). For a good comparison of the frequency of herniated discs among populations, see Thieme (1950). In a study of sectioned (slab) and radiographed human cadaver vertebrae, Pfirrmann and Resnick (2001) noted 225 vertebral endplates with Schmorl nodes of a total of 3300 T1 to L5 endplates examined. Perhaps surprisingly, the authors found: (1) that Schmorl nodes were associated with moderate but not advanced degenerative changes to the vertebrae and (2) Schmorl nodes are probably not a significant factor in the

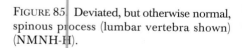

FIGURE 85. Deviated, but otherwise normal, spinous process (lumbar vertebra shown) (NMNH-H).

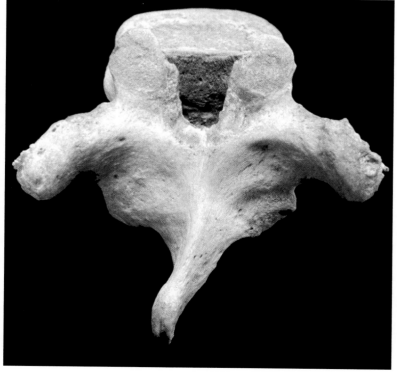

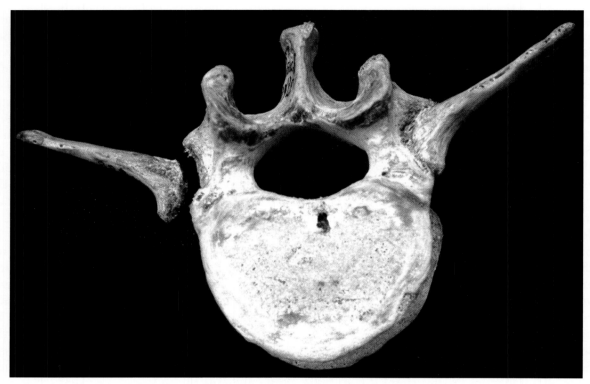

FIGURE 86. Separate transverse process (NMNH-H 317688).

development of spinal disease. Cf. Schmorl and Junghanns (1932, 1971).

Peng and colleagues (2003), using histology and a variety of imaging modalities, reported that Schmorl's nodes are the result of ischemic necrosis beneath the cartilaginous endplates and that herniation of the disc (nuclear pulposus) is secondary – they found similarities between the development of avascular necrosis of the femoral head and Schmorl's nodes. To be precise, these authors found "Loosely textured fibrous tissue with multiple small blood vessels completely replaced the marrow of the vertebral body underlying the cartilaginous endplate. There was an increase in reactive woven bone with thickened trabeculae and prominent osteoclasts and osteoblasts" (pp. 881–882). See also Bulos (1973), Campillo (1989), Gibson et al. (1987), Giroux and Leclerq (1982), Jayson and Dixon (1992), Lindblom (1951), Pfirrmann and Resnick (2001), Postacchini and Rauschning (1999), Resnick and Niwayama (1978), Saluja et al. (1986), Schmorl and Junghanns (1932, 1971), Tehranzadeh et al. (2000), and Weinstein et al. (1977).

Separated/Separate Transverse Process

A condition resulting in a separate/divided transverse process (**Figure 86**). A smooth-surfaced articular facet is associated just inferior to the superior articular facet at the base of the mammillary tubercle. This is different from a lumbar rib which attaches to the centrum and has a separate transverse process to the neural arch. This developmental variant is the result of caudal shift in the developmental pattern of the spine and affects the formation of the lumbar vertebra. The authors have encountered only one unilateral case in a first lumbar vertebra. Rarely a transverse process, usually a lumbar vertebra, may fracture and separate (Hoppenfeld 1977). This is a rare finding. See also Barnes (1994:104–108).

Ankylosing Spondylitis (AS) ("Bamboo" Spine)

Ankylosing spondylitis (AS) (**Figure 87**) is a chronic inflammatory disorder of unknown etiology (Borenstein et al. 1995) but has been linked (95%) to the histocompatability complex antigen HLA-B27 (Aufderheide and Rodriguez-Martin 1998:102; see Gerber et al. 1977 for alternative view) and is identified as a Mendelian dominant trait (Sharon et al. 1985). The disease is predilected to males (70%, 10% in females) as reported by Sharon et al. (1985) to 90 percent in males as reported by Aufderheide and Rodriguez-Martin (1998). The prevalence of AS varies by population from absent in Africans, blacks and Australian aborigines to 4.2 percent in adult male Haida Indians (Masi and Medsger 1989), but Caucasian males are the most predilected for the disease (Aufderheide and Rodriguez-Martin, 1998:102). Contemporary populations show an incidence of about 1 per 1000 individuals (Resnick and Niwayama 1981, 1988) to 1 per 2000 (West 1949) with a male to female ratio varying from 4:1 to 10:1 (Resnick and Niwayama 1981; West 1949). The typical age of onset is 15 to 35 years but children may also be affected with juvenile-onset AS (Jaffe 1975). Bony changes typically begin in the sacroiliac region (sacroilitis and subsequent ankylosis) and extend up the spine. As the spine becomes more involved it takes on an undulating contour ("bamboo spine") due to the development of extensive syndesmophytes between the vertebral bodies (Resnick 2002; Resnick and Niwayama 1981). In extreme cases, the ribs will also be involved in the osseous fusion. For a classic case of AS with rib involvement, see Ortner (2003:576, Figure 22.10). AS often results in ossification at tendinous and ligamentous attachment sites (enthesophytes) to bone and in some cases may be difficult to distinguish from DISH (see **Figure 73a–b**). While not focusing specifically on DISH, it is noteworthy that an MR and macroscopic study of the ligaments of the sacroiliac joint revealed them to be gender-dependent (Steinke et al. 2010). See also Calin (1985), Dieppe (1986), Long and Rafert (1995), Ortner (2003), Resnick and Niwayama (1988), Epstein (1976), Gibson et al. (1987), Giroux and Leclerq (1982), Schmorl and Junghanns (1932, 1971a–b), Steinbock (1976:294–298, 304–309), and West (1949).

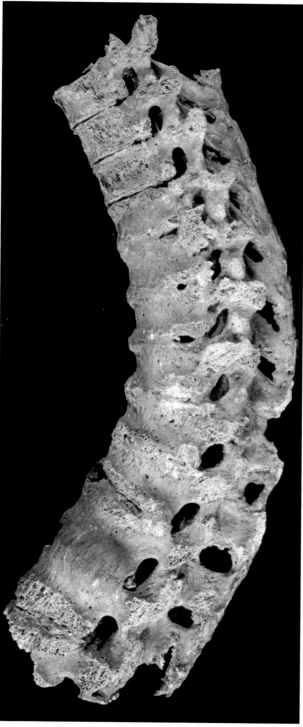

FIGURE 87. Bony changes associated with ankylosing spondylitis (AS), and degenerative joint disease (spondylosis) of the spine (Hough and Sokoloff 1989; Kahn 1984; Meisel and Bullough 1984; Ortner 2003:571-577; Wei Yu et al. 1998).

Spondylosis Deformans

Ossified tri-radiate ligaments. Resulting in fusion of the ribs to the vertebrae are typical of ankylosing spondylitis. This is an uncommon to rare finding.

Syndesmophyte (endesmophytes). Vertically-oriented bone growth typical of AS that forms from within the *margins* (annulus fibrosus) of the vertebral bodies (Francois 1965). This is a common finding, depending on the population under study. See also Baraliokos et al. (2007) and Resnick and Niwayama (1988).

Osteophyte. Horizontally-oriented, rounded bone growth associated with degenerative spinal disease (osteophytosis), increasing age, and trauma (see **Plate 87**). This is a common finding in all populations.

Spondylosis deformans (SD) is the most common degenerative disease of the spine that affects males much more frequently than females, is occupation related, and is found in nearly all individuals over the age of sixty (**Figure 88**). SD is characterized by small to large bridging osteophytes (arrows) that bulge at the level of the intervertebral disc and serve to reinforce the centra. The proper use of the term osteoarthritis refers only to degeneration and involvement of the apophyseal joints (articular facets) whereas osteophytosis refers to osteophyte formation along the vertebral bodies and degenerative disc disease (Hough and Sokoloff 1989). The following description given by Norman (1984) best illustrates the process of osteophyte formation commonly associated with old age:

> The early changes of spondylosis deformans affect the anterolateral margin of the vertebrae where the annulus fibrosus inserts. Tearing of the fibers weakens the annulus. The restraints on the nucleus pulposus are lost, and the disk protrudes forward. Further stress will lift the anterior longitudinal ligament from the vertebral body, and a buttress of periosteal new bone fills in the area of separation. The osteophytes form in a horizontal direction and curve to bridge the intervertebral disk space....

Many researchers attribute osteophytosis to many years of wear and tear that necessarily accompanies old age (Keim 1973). Trauma, heavy physical stresses to the spine, and obesity, however, may also result in osteophyte formation. See also Dieppe et al. (1988), Hough and Sokoloff (1989), Jaffe (1975), Moskowitz et al. (1984), Ortner and Putschar (1985), Steinbock (1976: 287–289), Stewart (1958), and Trueta (1968).

Aortic Aneurysm

An aortic aneurysm may result in large, scooped-out lesions of the vertebral bodies, sometimes accompanying syphilis (**Figure 89**). These lesions are caused by positive charges from the exterior surface of an expanded aorta, the eroded portions of the vertebral bodies resulting from the expansion of the aorta. Typically, males over the age of 50 are affected (Keim 1973). The lesions are restricted to the left anterolateral surface of the spine. Distinguishing aneurysmal from tubercular erosion in an isolated vertebra is difficult and may require the assistance of a radiologist. Rare finding in most populations (infrequent finding in cadaver and necropsy collections). A good review of aneurysms from various diseases and their bony effects as well as illustration of involvement of vertebrae and sternae is found in Aufterheide and Rodriguez-Martin (1998:78–81). See also Aydogan et al. (2008), Chao et al. (2003), Choi and Harris (2001), Diekerhof et al. (2002), El Maghraoui et al. (2001), Leung et al. (1977), Mii et al. (1999), Ortner (2003:356, Figures 13-11 and 13-12), Semrad et al. (2000), and Zimmerman and Kelley (1982:73–74).

Spinal Curvature (Kyphosis, Lordosis, Scoliosis)

Note that the drawing depicting a normally curved spine (figure left to right), kyphosis (humpback), and lordosis (swayback) are viewed from the side, while that of scoliosis is seen from the front (**Figure 90**). These conditions can only be detected by articulating the vertebrae in the approximate anatomical position (by aligning the articular facets). Remember that each intervertebral disc accounts for approximately 0.5 to 1.0 cm space (in young to middle-aged

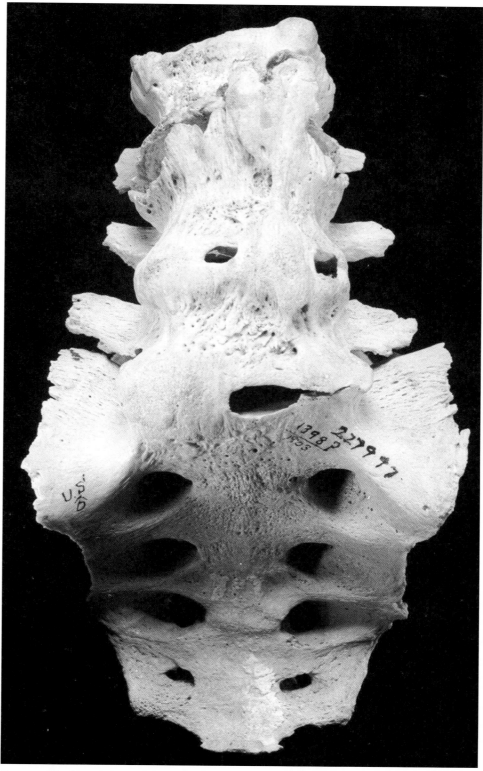

FIGURE 88. Spondylosis deformans (spondylitis deformans, degenerative hypertrophic spondylitis, osteoarthritis, degenerative spondylosis) of the spine (NMNH-H 227977).

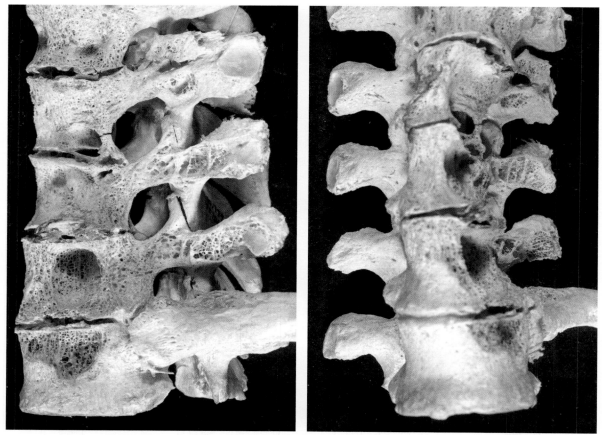

a   b

FIGURE 89. Aortic aneurysm of the vertebrae.

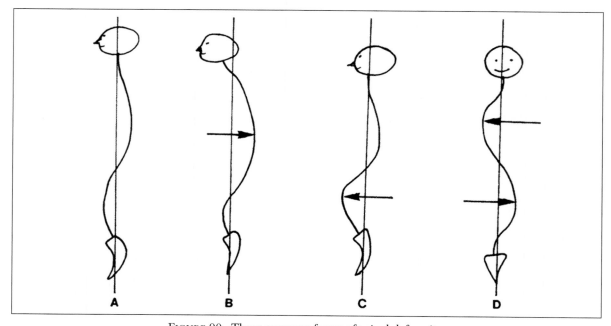

A   B   C   D

FIGURE 90. Three common forms of spinal deformity.

individuals) between adjacent vertebrae. Kypho-scoliosis, forward and to the side bending, is a common finding in the elderly. The dowager's hump of the elderly, usually most marked in females, is a common form of kyphosis. With increasing age, the vertebrae lose bone mass (osteoporosis), collapse, remodel, and become anteriorly wedge-shaped (see **Figure 77**). Physi-ological scoliosis in which the vertebral bodies are greatly distorted and asymmetrical is a rare finding (**Figures 91 and 92**). A number of etiologies may result in scoliosis due to loss of bone and support by lytic bone destruction in-cluding tuberculosis, non-specific osteomyelitis, and mechanical fatigue from acute trauma, os-teoporosis, or osteoid osteoma (Haibach et al. 1986; Savini et al. 1988; Swank and Barnes 1987). See also Barnes (1994: 38–40, 59–71), Benzel (2001), Davis (1955), Kilgore and Van Gerven (2010), and Zink et al. (2001). See **Plates 90 and 91** for color examples.

Scoliosis is defined as "one or more lateral-rotatory curvatures of the spine" (Keim 1972). Idiopathic or genetic scoliosis accounts for ap-proximately 70 percent of all scoliosis and is divided into infantile, juvenile, and adolescent. Scoliosis is a common finding in the elderly, especially women (postmenopausal or senile os-teoporosis) above 65 years of age. This form, however, is due to osteoporosis and collapse of the vertebral bodies and is not congenital in origin. Moderate to severe scoliosis in young individuals is a rare finding in most skeletal samples. See also Apley and Solomon (1988), Jackson (1988), Raisz (1982), and Trueta (1968).

In examining a spine for physiological scoliosis look for twisted (asymmetrical) vertebral bodies (**Figure 92**), irregularly shaped (asymmetrical) and positioned articular facets, missing transverse processes and distorted and misshapen ribs. The upper half of the centra may appear to be shifted in one direction while the lower half is shifted in the opposite direction. When the vertebrae are correctly articulated the spine will spiral, sometimes to an extreme degree. Severe sclolio-sis is associated with a number of pathological conditions including cerebral palsy, Marfan's syndrome, neurofibromatosis, and Klippel-Feil

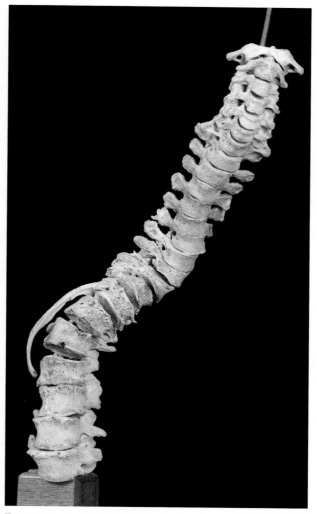

FIGURE 91. Scoliosis of the spine (lumbar used as example) (NMNH-T 1636).

syndrome (Klippel and Feil 1912; Prusick et al. 1985). See also Barnes (1994), Dickson (1985), Farkas (1941), Masharawi et al. (2008), Moe et al. (1978) Riseborough (1977), and Simmons and Jackson (1979).

Hemivertebrae

This vertebral form (**Figure 93**) is caused by the failure of fusion of the lateral halves of the vertebral body (Edelson et al. 1987; Epstein 1976; Moe et al. 1978; Pfeiffer et al. 1985). This rare condition is classified as a congenital defect of formation (developmental anomaly in the early

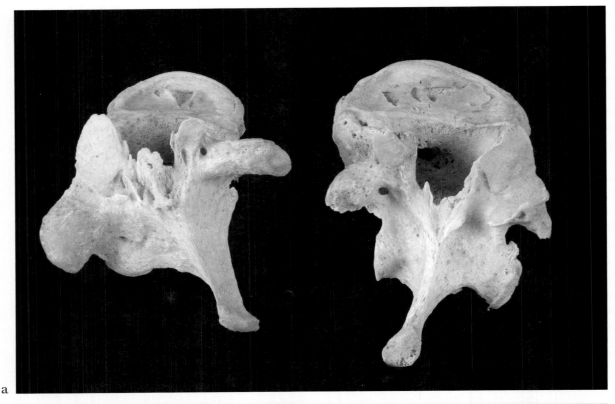

a

b

FIGURE 92. Another example of lumbar scoliosis of the spine (NMNH-H).

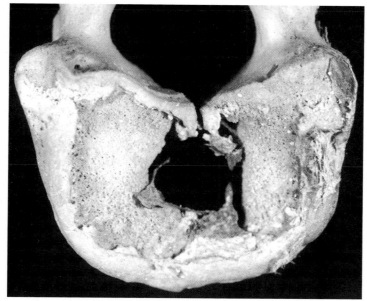

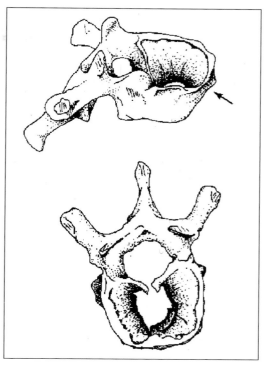

a

FIGURE 93. "Butterfly" vertebra (sagittal cleft vertebra, drawings of oblique and superior view). (Vertebra from Peru, courtesy of John Verano.) The term "butterfly" refers to the antero-posterior shape of the vertebra on radiograph.

b

embryonic period) (Muller et al. 1986) resulting in a variety of malformed centra (Barnes 1994:36–40) and is reported that it is more frequent in males than in females (Aufderheide and Rodriguez-Martin 1998:62). The body of the vertebra may be cleft down the midline resulting in two hemivertebrae or thin stands of bone may transverse the cleft (Epstein 1976). Other bony changes may include widely spaced pedicles, malformed neural arches, and marked disarray of the posterior elements. The intervertebral disc may bulge into the defect. Butterfly vertebrae, so called because of their radiographic appearance when taken on the coronal plane (when viewing an anteroposterior radiograph, the lateral halves of the vertebra looks like the wings of the butterfly and the dark central defect in the centrum looks like the butterfly's body), result in kyphosis (forward bending) due to the diminished height of the anteriorly wedged vertebral body. Note in the hemivertebrae in **Figure 94 and Plate 86c** that two of the lumbar vertebrae failed to develop properly and are anteriorly wedged. Also note the large bridging osteophyte joining two of the vertebrae along their anterior surfaces. A police

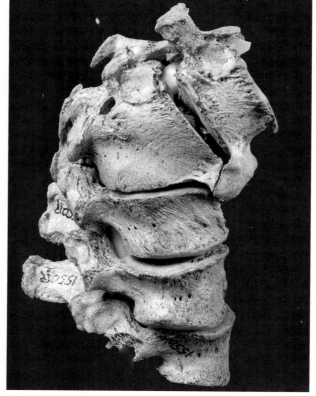

FIGURE 94. Hemivertebrae and scoliosis (NMNH–T 1550R).

case involving a coronal cleft vertebra noted radiographically was initially suspected to reflect an abusive fracture in an infant (Aronica-Pollak et al. 2003). See also Anderson (2003), Aufderheide and Rodriguez-Martin (1998, p. 62, Figure 4.9), Barnes (1994: 60–67) (cites several paleopathological cases from New Mexico), Colquhoun (1968), De Graaf et al. (1982), Mann and Verano (1990), Ozek et al. (2008), Taylor and Resnick (2000), Yochum and Rowe (1996), and Zimmerman and Kelley (1982:30). See **Plate 86c** for a color example.

Spinal Tuberculosis (Tuberculosis in the Spine, Pott's Disease)

Tuberculosis in the spine (Pott's disease, vertebral tuberculous spondylitis) as is represented in this illustration shows the typical cavitation of the vertebral body by erosive degeneration from infection (**Figures 95 and 96 and Plates 90 and 91**). Destruction of the vertebrae usually first begins in the anterior, inferior portion of the vertebral body causing loss of vertebral height, loss of definition of the endplates, collapse and subsequent

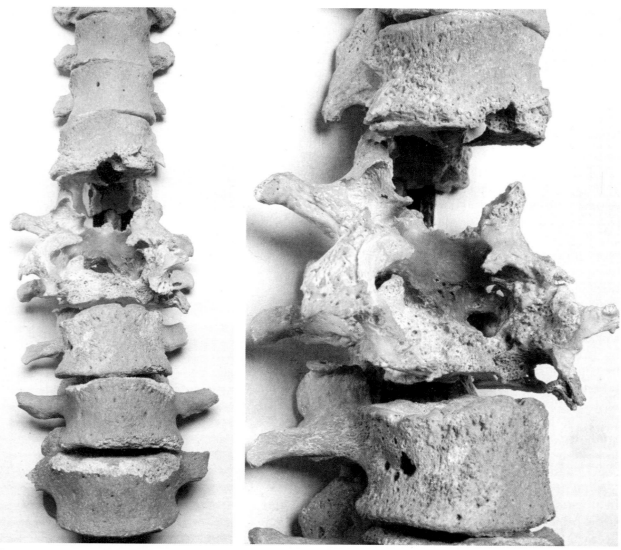

a

b

FIGURES 95. Tuberculosis of the spine (thoracic and lumbar vertebrae), also known as "Pott's" disease. (NMNH–T 1124R). See Plates 90 and 91 for color examples and also Figure 96.

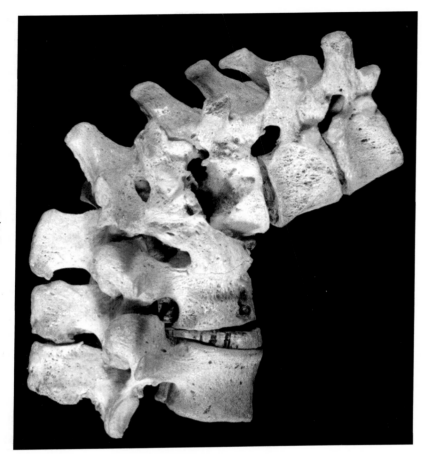

FIGURE 96. Pott's disease with collapse and forwarding bending (kyphosis) of the lower spine (AFIP). See Plates 90 and 91 for color examples.

severe angulation of the spine. Localized destruction in the form of cavitation and abscessing of the vertebral bodies is an inflammatory response to invading tubercle bacilli resulting in the formation of tubercles that stimulate erosion of the trabeculae and cortical bone. Note the minimal new bone formation in comparison to the amount of bone that has been resorbed. Adjacent vertebrae may show destruction where the inflammatory stimulus spreads beneath the anterior longitudinal ligament (sometimes the disease will "skip" vertebrae and continue above or below the initial vertebral abscess). Other frequent sites of tubercular involvement are the inner surfaces of the ilium (psoas abscess through the spread of the disease along the psoas muscle) (see **Figure 118**), skull, joints (usually only an isolated joint is affected), and ribs. Tuberculous spondylitis occurs in approximately 1 percent of patients with tuberculosis and is the most common form (25% to 60%) of skeletal tuberculosis (Lindahl et al. 1996; Patankar et al. 2000). Pott's disease (spinal infection) is the most common form of musculoskeletal tuberculosis, associated with nearly 50 percent of all cases (Kramer et al. 2004). It is an uncommon finding depending on the population under study. See also Apley and Solomon (1988), Arriaza et al. (1995), Aufderheide and Rodriguez-Martin (1998:121–124), Buikstra and Williams (1991), Fletcher et al. (2003) Ganguili (1963), Garcia-Lechuz et al. (2002), Hallock and Jones (1954), Hodgson et al (1969), Kelley and El-Najjar (1980), Madkour (2004) (good coverage), Mays et al. (2001), Moore and Rafii (2001), Morse (1961, 1969), Ortner and Putschar (1985), Ortner (2003: 230–235), Pálfi et al. (1999), Park et al. (2007), Rom et al. (2004), Steinbock (1976:176–182), Thijn and Steensma (1990) (radiographic study), Tuli (1975), Zimmerman and Kelley (1982:102–108), and Zink et al. (2001).

RIB

Osteoarthritis of the Transverse Articulation

Raised, irregular bony growths where the rib attaches to the transverse process or facet on the centrum (**Figure 97**). This is a common finding accompanying middle and old age. See Plates 72 and 73 for other possible bony growth on ribs.

Pseudarthrosis of the Rib

The formation of an irregular "false" joint is produced due to muscle tension and/or when there is continued motion between the two ends of a complete fracture. In some cases when the broken ends fail to properly unite, cartilage develops instead of bone (callus) and a more permanent pseudarthrosis results (**Figure 98**). Movement of this false joint may properly function for many years after its formation. See also **Figure 148** in this volume for pseudarthrosis of the ulna and **Plate 57** for mandibular pseudarthrosis, and Plate 74 for clavicle. For a fracture site to properly heal, the segments of the bone must be properly realigned and restricted from movement (immobilized in a cast) and muscular tension and motion at or near the site must be minimized (traction). It is an uncommon to rare finding (Ortner 2003: 151–153; Steinbock 1976:21–22).

Periosteal Thickening/Enlargement

Enlarged and thickened periosteal bone (**Figure 99**) confined to the cortical surfaces of the ribs gives them a clubbed appearance, especially at the vertebral ends. Small, roughly ovoid and well-defined areas of reactive bone along the dorsal (pleural/lung side) surface of the ribs may be indicative of tuberculosis (Marc Kelley, personal communication 1989; Pfeiffer 1991; Roberts et al. 1994, 1996). Mays et al. (2002), however, found no DNA association between rib lesions and *M. tuberculosis*. The etiology of ribs that are greatly thickened down the majority of the shaft is unknown, although tuberculosis and other chronic pulmonary diseases are suspected. Carefully examine the entire skeleton (especially the spine) for other indications of infection or malformation. This is an uncommon to rare finding in archaeological samples but common in early twentieth-century specimens (Charlotte Roberts, personal communication 1989). See also Kelley and Micozzi (1984), Lambert (2002), Roberts et al. (1994, 1996), and Santos and Roberts (2001). See Wilbur et al. (2009) and Donoghue et al. (2009) for a discussion of interpreting tuberculosis in ancient remains with DNA.

Effects of Osteomalacia

These ribs (**Figure 100**) are noticeably straight as a result of chronic kyphosis from severe osteo-

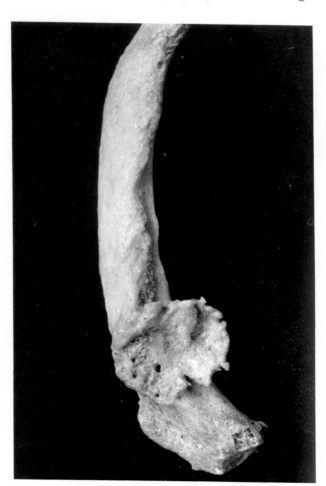

FIGURE 97. Osteophytes and porosity of the articular facets of a rib (OA).

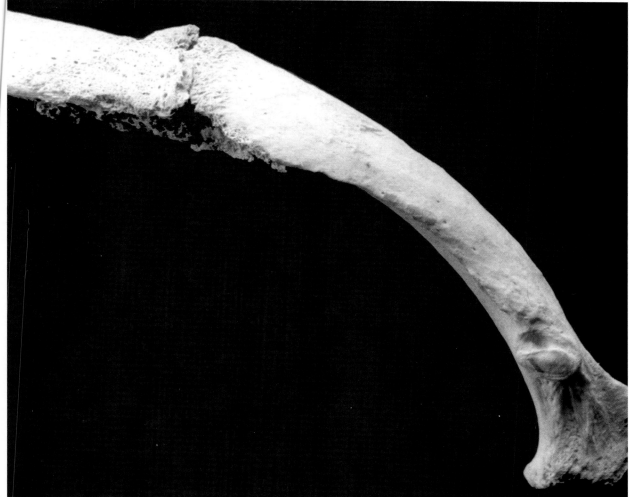

a

b

FIGURE 98. Pseudarthrosis (malunion, false joint, incomplete union, nonunion) (NMNH Egypt).

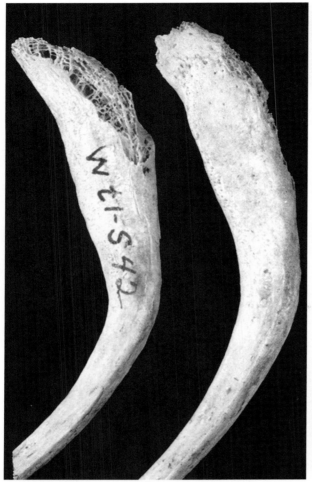

FIGURE 99. Active or healed periostosis (possibly associated with tuberculosis or other intrathoracic disease) (Wtl-S42).

porosis (osteomalacia) or otherwise diseased individuals. The ribs lose their natural C shape when the torso is bent forward and causes the internal organs to push against the rib cage, thus straightening the ribs. See also Nugent et al. (1984), and Pitt (1981, 1991).

Bicipital and Bifid Ribs

Bicipital ribs and bifid (sometimes called "two-headed ribs") consist of what should have been two ribs that fuse to form one (Resnick 2002; Turner 1883). Bicipital and bifurcated ribs, although sometimes used interchangeably, differ in that the former (**Figure 101**) are true fusions of what should have been two independent ribs, usually those associated with the first and second thoracic vertebrae (Aufderheide and Rodriguez-Martin 1998:68–69), while a bifurcated rib has a single head and body that divides or forks at the sternal end. In this example, the two vertebral ends of the ribs are separate and their bodies are fused along the middle segment. Uncommon to common finding. Not to be confused with congenital fusion of many ribs associated with certain syndromes (e.g., Klippel-Feil 1912). Barnes (1994: 71–77) covers the various forms of segmental developmental disturbances which produce bifurcation and bicipital ribs from caudal shift of the cervicothoracic border (Popowsky 1918).

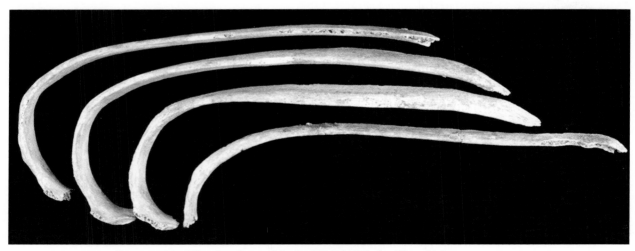

FIGURE 100. "Shepherd's crook" ribs (NMNH-T). See Plate 91 for comparison.

a

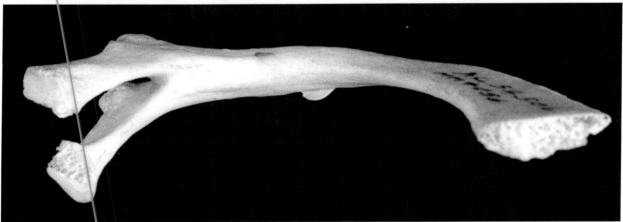

b

FIGURE 101. Variations of bicipital and bifid/bifurcated ribs (irregular segmentation of ribs; cf. Barnes 1994) (NMNH-H).

Cervical Ribs

Cervical ribs (**Figure 102**) are enlarged costal elements (the bony bridges that span from the centrum to the rudimentary transverse process of cervical vertebrae) of the seventh, sixth or fifth cervical vertebra (Grant 1972; Turner 1879, 1883; Yochum and Rowe 1996). Such ribs may be rudimentary or fully developed and roughly similar in shape to normal first ribs but smaller and less

U shaped. The developmental changes are due to shift in the cervicothoracic border in early developmental growth (Barnes 1994:99–104). Radiologically, cervical ribs are found twice as often in females, 50 to 66 percent occur bilaterally, and in 5.6 per 1000 patients (Grant 1972; Grainger et al. 2001; Yochum and Rowe 1996). While Moore and Persaud (2003) report these ribs as occurring in approximately 0.5 to 1 percent of people; they are uncommon findings in archaeological samples.

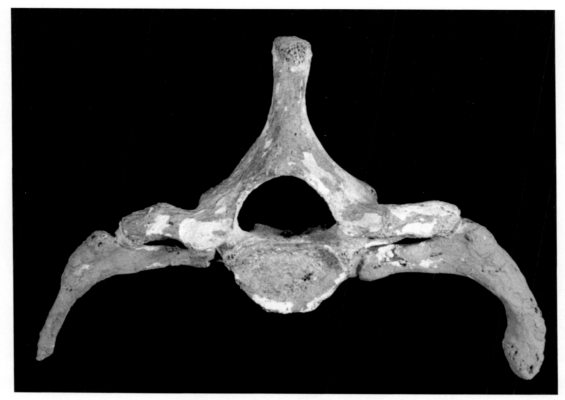

FIGURE 102. Cervical rib (accessory rib).

Erken et al. (2002) found a strong relationship between the presence of cervical ribs and sacralization. See also Aufderheide and Rodreguez-Martin (1998:68–69, Figure 4.13), Barnes (1994: 99–104), Brannon (1963), Cave (1941), Dwight (1887), Gladstone and Wakeley (1931–32), Hertslet and Keith (1896), Henderson (1914) (reports on 31 cases of cervical ribs), Lucas (1915), Magee (2002), Purves and Wedin (1950), Resnick (2002), Schaeffer (1942), and Todd (1912).

SACRUM

Sacralization

Sacralization (**Figures 103, 104, 105, and 106**) is the result of a congenital condition that causes partial or complete fusion of the most inferior lumbar vertebra (fifth or even sixth lumbar) to the sacrum (or when the lowest lumbar vertebra resembles or has features of a first sacral segment) (Barnes 1994:108–113; Yochum and Rowe 1996). This condition has been found in approximately six percent of individuals in clinical studies (Grainger et al. 2001). Occasionally, a vertebra will exhibit characteristics of a lumbar vertebra on one side and sacral morphology on the other resulting in the so-called hemilumbarization or hemisacralization (Bergman et al. 1988); this condition is often referred to as lumbosacral transitional vertebrae (LSTV) (cf. Konin and Walz 2010). Although a common clinical finding, these conditions are uncommon in most archaeological samples. See **Plate 109b** for a color example.

Lumbarization

Lumbarization occurs when the first sacral segment has shifted cranially in the developmental

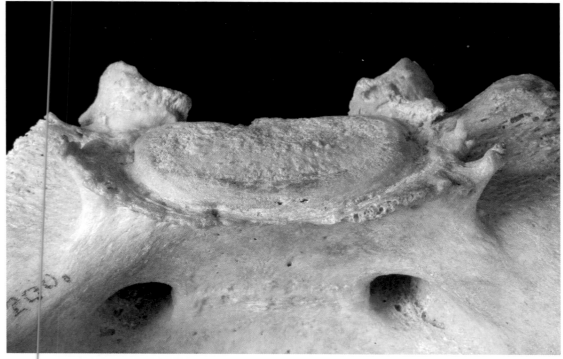

FIGURE 103. Osteophytes (OA/DJD) along the rim (promontory) of the first sacral segment (NMNH-T 200). See Figures 126 and 127 for fusion of sacrum and ilium.

field and has the features of a lumbar vertebra. One way to detect the presence of a transitional, accessory or absent vertebra is to count down the spine beginning with C2, rather than counting up from L5. Using this technique, Hahn et al. (1992) found 15 cases of sacralized L5 and nine cases of lumbarized S1 in 200 patients. Both sacralization and lumbarization represent transitional

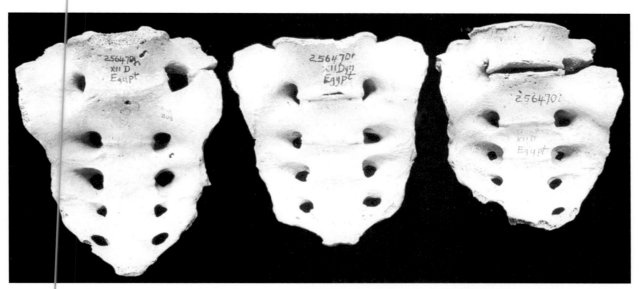

FIGURE 104. Variations of sacral shape, segments and lumbarization (NMNH) (see Plate 109a).

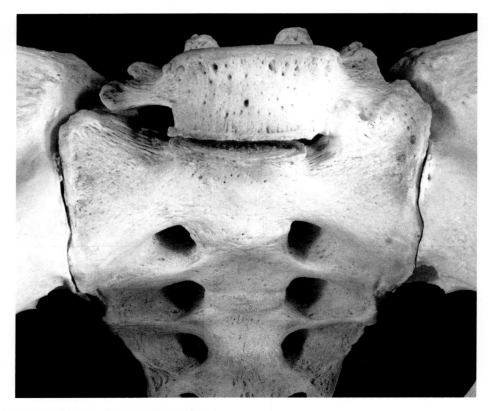

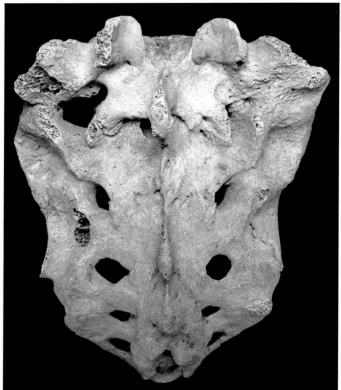

FIGURES 105 (*top*) and 106 (*bottom*) (see Plate 109a).

vertebrae. Normal five segmented sacra (four sacral foramina) can be distinguished from six segmented sacra (five sacral foramina) by counting the number of paired sacral foramina. While approximately 95 percent of people have 7 cervical, 12 thoracic and 5 lumbar vertebrae, about 3 percent have 1 or 2 more ("extra") vertebrae and about 2 percent have one less (Moore and Persaud 2003). When computing stature, allowance must be made if there is an additional vertebra in the spine (Lundy 1988). See also Aufderheide and Rodriguez-Martin (1998), Barnes (1994:108–113, 250–255), Cimen and Elden (1999), Delport et al. (2006), Hahn et al. (1992), Hughes and Saifuddin (2005), Kim and Suk (1997), Konin and Walz (2010), Leboeuf et al. (1989), and Tini et al. (1977).

Spina Bifida

Spina bifida may affect any vertebrae, but is most commonly found in the sacrum, especially in L5 or S1, occurring in approximately 15–18 percent of individuals with a reported incidence of 9 percent in females and 13 percent in males (Cowell and Cowell 1976) (**see Figures 107 and 108 and Plates 94 and 109b**). Two classes are identified, spina bifida occulta and cystica and having two grades, meningocele and meningomylocele. Occulta class is the more common and least severe of the two. Eubanks and Cheruvu (2009) examined 3,100 human sacra and found 355 (12.4%) with spina bifida occulta; the authors reported that the condition was more common in whites and males and that its frequency decreased with age. Cystica is a more rare condition and often results in paralysis due to the spinal cord extruding from the neural arch, and traumatized and rates of expression for cystica have been reported by Fishbien (1963) and Norman (1963) at approximately 3:1000. There may be hip deformity and dislocation which will exacerbate the sacral condition (Canale et al. 1992). Besides congenital hip displasia, spina bifida has also

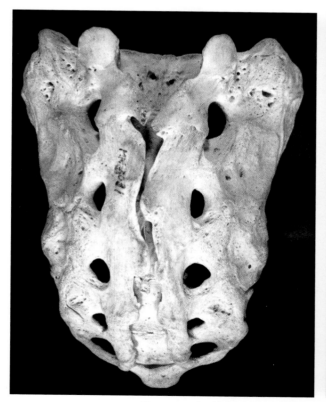 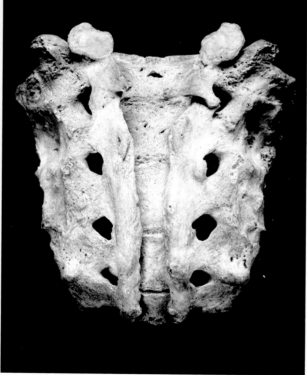

Figures 107 (*left*) and 108 (*right*). Sacral spina bifida (see Plate 109b).

been associated with other developmental problems such as hydrocephalus, rib anomalies and clubfoot (Zimmerman and Kelley 1982:31). It is an uncommon to common finding in populations due to its hereditary properties. See also Barnes (1994:41–49, 259–265), Webb (1995:235–241), and Zimmerman and Kelley (1982:29–31).

Accessory Sacral Facets

Accessory sacral facets (accessory sacroiliac joint, rudimentary sacral zygapophyseal joint) appear as raised contact areas in the ilia are usually located in the most inferior portion posterior to the auricular surface (**Figure 109**). To detect these facets articulate the sacrum and ilium. Ehara et al. (1988) reported that accessory sacral facets are either present at birth (diarthrodial) or develop (fibrocartilaginous) as a result of stress associated with weight-bearing. It is a common

finding. See also Derry (1911), Mahato (2010), and Trotter (1937, 1940).

Sacral Hiatus

A sacral hiatus, a normal feature of nearly all sacra, is a canal that extends up one or two segments from the most inferior segment of the sacrum (**Figure 110**). If the hiatus extends as far as the third sacral segment/third sacral foramen (starting with the most inferior segment), it should be scored as spina bifida. There is no total agreement about the critical criteria among researchers. Research by Sekiguchi et al. (2004) revealed absence of the sacral hiatus in 4 of 94 sacra and that the apex of the hiatus was at the level of S4 vertebrae in 64 percent of sacra. Parashuram (2008), examining 200 dry human sacra, found the shape of the hiatus to be inverted-U in 50 percent of the sacra, inverted-V in 27.5

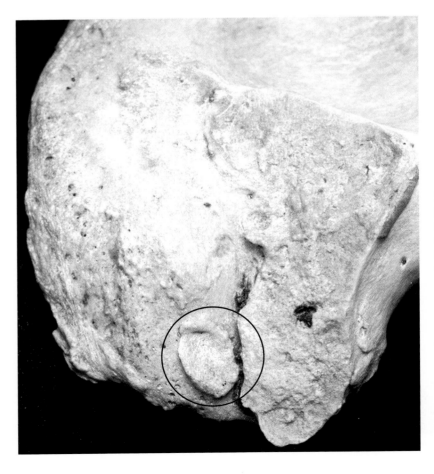

FIGURE 109. Accessory sacral facet (Derry 1911) – frequently, bilateral accessory facets are present at the level of the first or second dorsal sacral foramen for articulation with the ilium (NMNH-T).

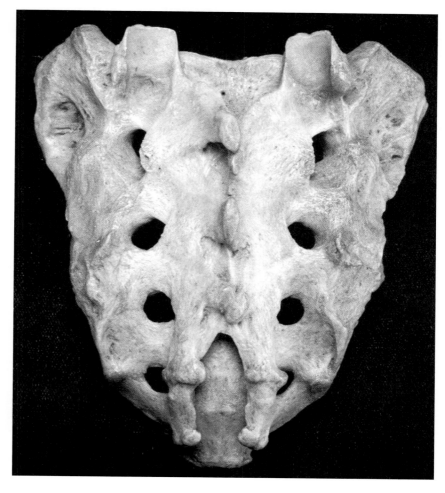

FIGURE 110. Normal sacral hiatus (NMNH-T).

percent, irregular in 15.5 percent and dumbbell in 2 percent and bifid in 1 percent. Complete spina bifida was present in 4 of 200 (2%) sacra and the sacral hiatus was absent in 2 sacra (1%). Sekiguchi et al. (2004) and Parashuram (2008) both stressed normal variation in the sacral hiatus. See doctoral dissertation by Parashuram (2008) for more information.

INNOMINATE

Enthesopathy and Osteophytosis

Macroscopically, enthesophytes (some researchers mistakenly refer to these projections as osteophytes) appear as spike-like projections, spicules, spurs, and ridges or irregular ossification where tendons and ligaments attach (enthesis) to the bone (**Figure 111**). This new bone growth, due to inflammation, may accompany old age, obesity, or repeated acute minor stress related to a particular motion or activity. Frequent sites of involvement are the linea aspera, trochanteric fossa, greater and lesser trochanters of the femur, iliac crest, ischial crest and tuberosity, ischial spine, and obturator foramen in the innominate; attachment of the Achilles tendon in the calcaneus, supinator crest of the ulna, radial tuberosity of the radius, and soleal line (popliteal muscle) of the tibia. In some cases, it is difficult to distinguish enthesopathy from normal skeletal variation

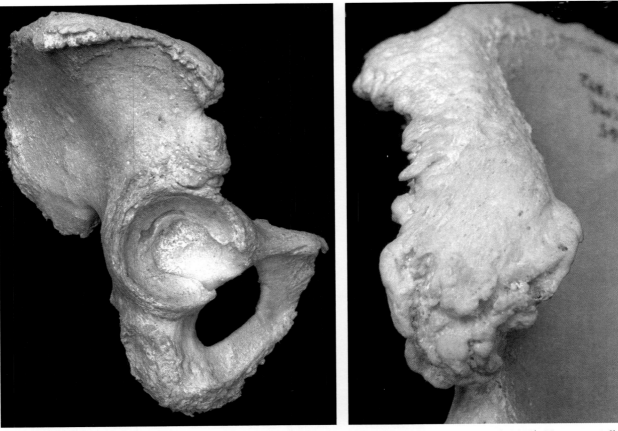

a   b

FIGURE 111. Enthesophytes (Dieppe et al. 1986; Resnick and Niwayama 1988; Taylor and Resnick 2000). Note especially the iliac crest and ischial region. Figure 111b is a close-up of the superior iliac spine (NMNH 387665). See Chapter X for color examples of enthesophytosis.

(robustness). Enthesophytes, possibly reflecting DISH or fluorosis, are common findings in individuals 60 plus years of age, and may be more prominent on one bone or one side of the body as a result of increased pulling stresses at these sites such as handedness or paralysis. Enthesophytes do not reflect osteoarthritis (inflammation of synovial joints; **Figures 113, 114, 115 and 123**) and may accompany leprosy (Carpintero-Benitez et al. 1996). See also Boldsen (2001, 2008), Benjamin et al. (2000), Maas et al. (2002). For assessment of enthesophyte and osteophyte correlations, see Rogers et al. (1997) and Stirland (1991b).

Flange Lesion and Bipartition

The cause and correct name of this rare condition (**Figure 112**) have not been definitely established. With some reservation, Wells (1976) and Knowles (1983) attribute a similar condition to acute, but temporary, dislocation of the femoral head on the rim of the acetabulum, while Stirland (1991b) attributes it to hyperflexion of the hip (**Figure 124**). This "lesion," present in one or both innominates, is approximately 2 to 3 cm in length and appears as an eroded area with either exposed trabeculae or as a smooth, flattened depression. See Hergan et al. (2000), Plotz et al. (2002), Ponseti (1978), and Yochum and Rowe (1996) for discussion of os acetabuli.)

Earlier work by Terry (1933) describes a cotyloid bone as a large separate bone in the pubic region of the acetabulum and resembling a bipartition (clearly not the same condition described by Knowles or Wells). However, there is an example of a cotyloid bone (labeled as such) fitting the

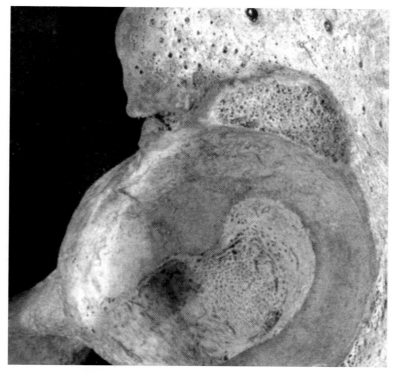

FIGURE 112. Flange lesion (Knowles 1983, Wells 1976), cotyloid bone (Steele and Bramblett 1988) or os acetabuli/marginal acetabular epiphyses (Resnick 2002; Taylor and Resnick 2000).

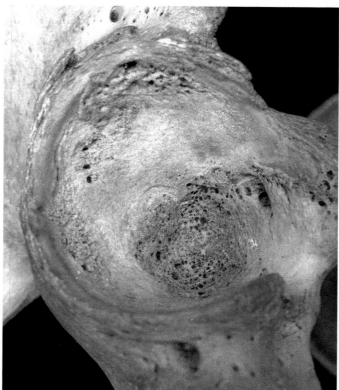

FIGURE 113. Osteoarthritis (pitting and bony buildup) of the right acetabulum accompanying older age. See Plate 100.

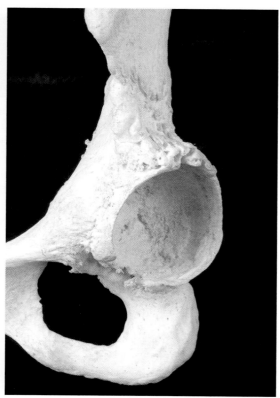

FIGURE 114. Severe osteoarthritis ("collaring") of the left hip socket (NMNH). See Plate 101 for a color example.

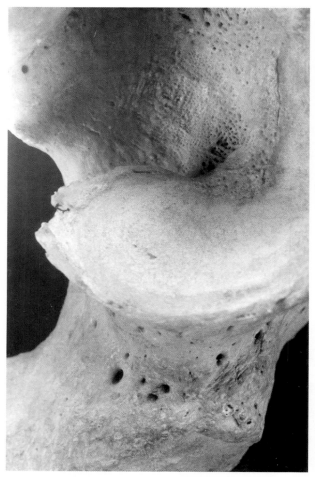

FIGURE 115. Marginal osteophytes ("lipping") on the left acetabulum reflecting degenerative joint disease (DJD). The porosity in the nonarticular portion of the acetabulum is normal.

The authors have only encountered such lesions in two young individuals, both showing the porous form bilaterally most likely associated to bilateral hip displasia. See Aufderheide and Rodriguez-Martin (1998:69–70) for discussion. Care must be taken not to misinterpret normal irregularity in the form of indentations or undulations of the acetabular rim with this condition. Examine the head and neck of the corresponding femur for signs of trauma or alteration reflecting OA. If encountered, it is probably best not to identify the condition by name but rather carefully describe the lesions. If, like the authors believe, Resnick (2002) Taylor and Resnick (2000) are referring to flange lesions, they report that these triangular ossicles adjacent to the superior acetabular rim are present in about 5 percent of the population. Rare finding, however, it is present in most skeletal samples.

A triangular-shaped defect or nearly detached U-shaped "tag" or "ear" of bone located in the superior third of the acetabulum (**Figure 116**). This nonmetric trait may be a remnant of fusion of the bones forming the acetabulum (tri-radiate cartilage). It is a common finding. See also Anderson (1963), and Saunders (1978).

Enlarged Nutrient Foramen

Enlarged nutrient foramina (**Figure 117**) can be the result of many diseases and conditions, including leprosy, tuberculosis and normal variation. Caution: it is better to note the presence of this condition rather than attempt to "diagnose" it.

Tuberculosis and other Lytic Lesions

A concavity or perforation in the inner surface of the ilium (**Figures 118, 119, and 120**) usually resulting from neoplasm, tubercular (tuberculosis), actinomycotic or syphilitic invasion via the psoas muscle from the lumbar vertebrae (Simpson and McIntosh 1927). It is a rare finding in most skeletal populations. However, a skeletal sample with many cases of tuberculosis will show this trait in varying frequencies. See also Fitoz et al. (2001), Franco-Paredes and Blumberg (2001), Ganguili (1963), Hallock and Jones (1954), Hodgson et al. (1969), Micozzi and Kelley (1985), Morse (1961, 1969),

description of the flange lesion in the Hamann-Todd collection in Cleveland, Ohio (Marc Kelley, personal communication 1989). The latter example has the appearance of an accessory bone attached (postmortem) to the rim of the acetabulum with a wire. Regarding etiology, an accessory bone along the rim of the acetabulum as described as Taylor and Resnick (2000) is more plausible, and similar in appearance to other accessory bones, than traumatic dislocation of the femur or trauma associated with activity as suggested by Stirland (1991b). Either flange lesions, cotyloid bones and os acetabuli represent different conditions (traumatic versus developmental), or there is terminological confusion that obscures the topic.

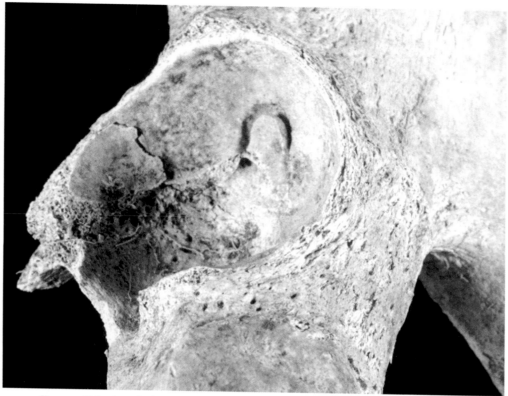

FIGURE 116. Acetabular mark or notch (normal anatomical variant of the skeleton).

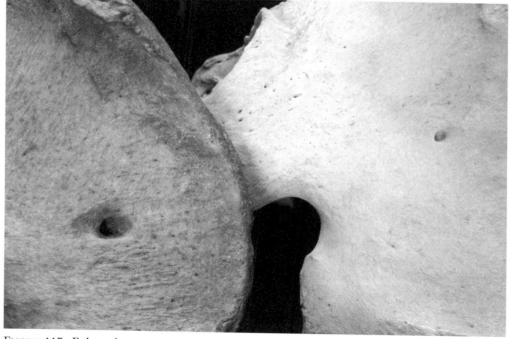

FIGURE 117. Enlarged nutrient foramen (*left*) in comparison to "normal" (*right*) – both are common findings (NMNH-T).

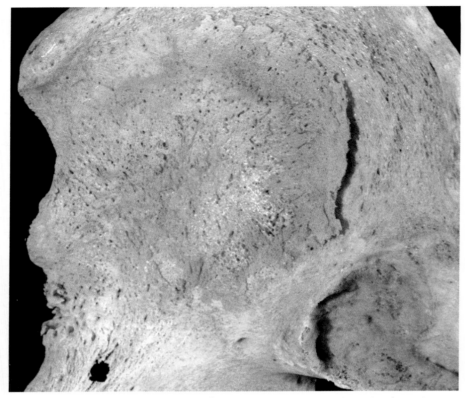

FIGURE 118. Psoas abscess (usually associated with tuberculosis).

Muckley et al. (2003), Pálfi et al. (1999), Steinbock (1976:176, Figure 71), Versfeld and Solomon (1982), and Younes et al. (2002). See Plate 104 for differential diagnosis of metastatic carcinoma.

Preauricular Sulcus

Depending on the shape and elevation of the sacroiliac articulation, a groove of varying width and depth will be found anterior to the auricular surface (**Figure 121 and Plate 108a**). This is generally believed to be a non-metric sex indicator on the female ilium (Bass 1995:208–218).

The preauricular sulcus (Zaaijer 1866) is often attributed to be the result of pulling stresses of the anterior sacroiliac ligaments during birth delivery. Some researchers, however, distinguish two types of preauricular grooves – the groove of pregnancy and the groove for ligamentous attachment (Houghton 1974, 1975; Saunders 1978); both are located anterior to the auricular surface of the ilium. Some researchers attribute

the groove of pregnancy to pulling stresses of the ventral sacroiliac ligament and subsequent inflammation (bleeding) associated with childbirth. Hormonal responses such as relaxin result in a loosening of the sacroiliac joint and localized growth resulting in a larger birth canal. A groove of pregnancy differs from a groove of ligamentous attachment in that the former will generally show discrete or coalesced pits or craters within the groove (which extends more superiorly along the anterior border of the auricular surface of the ilium). The latter will simply be a shallow, short groove present in both sexes (normal imprint of a strong ligament) (Houghton 1975). Both are difficult to interpret.

One study by Spring et al. (1989) designed to determine the relationship of the preauricular sulcus and parity (pregnancy) revealed in 190 women a sulcus in 4 of 41 (10%) nulliparous women and 25 of 149 (17%) women with positive pregnancy histories. A review of 3200 pelvic radiographs (1508 males and 1692 females) by

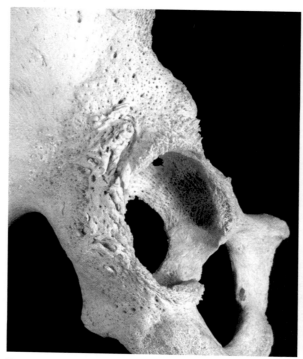

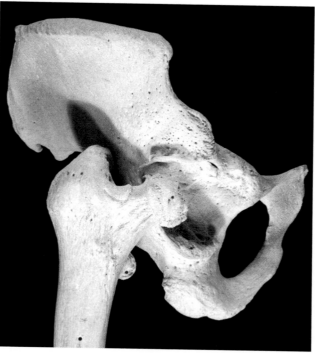

FIGURE 119. Tubercular destruction of the acetablum with reactive bone around the socket (Ganguili 1963; Versfeld and Solomon 1982; Resnick and Niwayama 1988).

FIGURE 120. Tubercular destruction of the right femoral head and hip socket (see Apley and Solomon 1988 for several good radiographic examples – for a close-up image of the proximal femur, see Figure 160).

FIGURE 121. Preauricular groove/preauricular sulcus ("groove of pregnancy," parturition groove) – these grooves do not necessarily indicate having given birth (NMNH-T 1173). See Plate 108a for a color example.

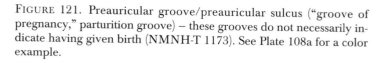

Gulekon and Turgut (2001) revealed preauricular grooves in only 393 of the females (23%) and none in males. The authors concluded that the presence of a deep preauricular sulcus was not necessarily an indication of past pregnancy. It is a common finding. See also Dee (1981), Hatfield (1971), Houghton (1974, 1975), Kelley (1979), Saunders (1978), Spring et al. (1989), and Tague (1990) for discussion of preauricular area in *Macaca mulatta*.

Dorsal Pubic Pits of "Parturition"

Circular or linear depressions or grooves (**Figure 122 and Plate 103a**) can be found on the dorsal surface of the pubic symphyses, usually in females, but can be found in males at approximately a 4 percent frequency (Suchey et al. 1979, 1986). These pits and/or grooves may resorb in old age. Holt (1978) examined 68 female pubic bones with known medical histories and found that nearly 38 percent of the females who had not given birth had small to large pits and scars on the dorsal pubic bones.

Snodgrass and Galloway (2003) examined the pubic bones of 148 modern females and concluded that pressure to soft tissue structures during pregnancy may be a critical factor in the development of dorsal pits. The authors cautioned, however, that further research is needed before drawing any definitive conclusions on the relationship between dorsal pits and number of births. Acute trauma to the pubic symphysis in males from a fall or accident may produce similar dorsal pitting. These findings cast serious doubt on the veracity of using dorsal pitting as an indicator of pregnancy and number of births. It is a common finding in females. See also Holt (1978), Kelley (1979), Stewart (1957, 1970), and Suchey et al. (1979).

Pubic Symphysis Diastasis

In this condition (**Figure 125**) the urinary bladder may open into the external surface of the abdominal wall (exstrophy) resulting in an unusually wide gap (diastasis) between the pubic symphyses (Kajbafzadeh et al. 2010). A

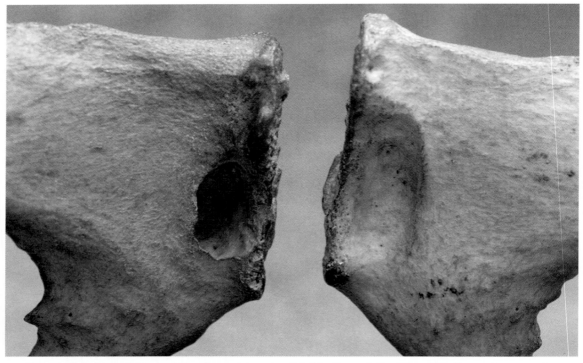

FIGURE 122. Dorsal pits or pitting (parturition pits, "birthing" scars). See Plate 103a for a color example.

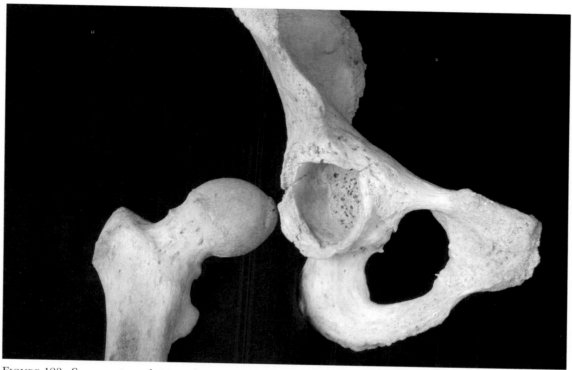

FIGURE 123. Severe osteoarthritis with remodeling of the acetabular rim and femur head (see also Figure 114) (NMNH-T 321R).

rare developmental defect that may be accompanied by other skeletal conditions including acetabular dysplasia, an abnormal lumbosacral spine, increased tibial torsion, patellofemoral instability (clinically) and shortening of the pubic rami. See also Ait-Ameur et al. (2001), Aufderheide and Rodriquez-Martin 1998:70; Kaar et al. (2002), Ortner and Putchar (1985), Sponseller et al. (1995, 2001), and Yazici et al. (1999).

Sacroiliac Joint Fusion

Fusion of the sacrum and the ilium at the sacroiliac joint (**Figures 126 and 127**) can be the cause of a plethora of reasons, from DISH, ankylosing spondylitis, enthesopathy, to infectious diseases (e.g. tuberculosis) and the effects of trauma to the joint from a single occurance to long-term habitual stress. The association or co-occurrence of other skeletal pathological conditions should be noted and described in evaluation of sacroiliac fusion. It is a common finding in modern populations.

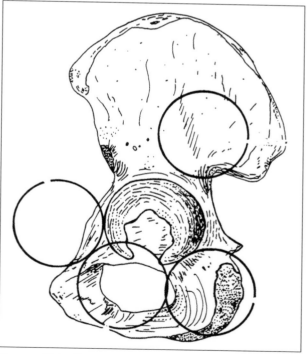

FIGURE 124. Common positions of the dislocated femur head. See Plate 102 for a color example.

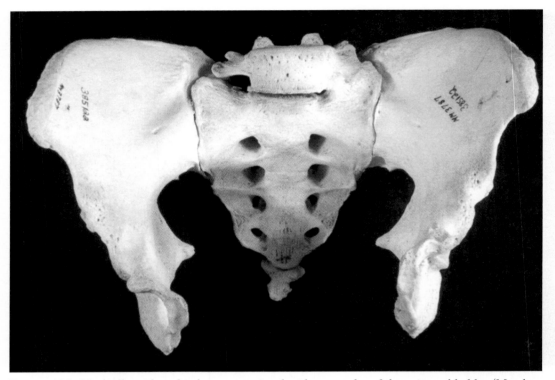

FIGURE 125. Markedly wide pubic bones associated with exstrophy of the urinary bladder (Meschan 1984) – also note unilateral sacralization of L5 and lumbarization of S1 (AFIP MM3787).

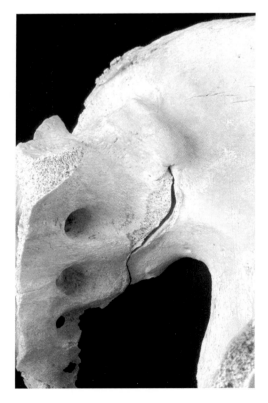

FIGURE 126. Fused left sacroiliac joint showing a large bridging osteophyte. Regular sacrolitis may be unilateral or bilateral and is indicative of several conditions, including ankylosing spondylitis or degenerative hypertrophic spondylitis (Ortner 2003:571, Figures 22-8, 22-10) – a fused sacroiliac joint is inconsistent with DISH. See Plates 106 and 107 for color examples.

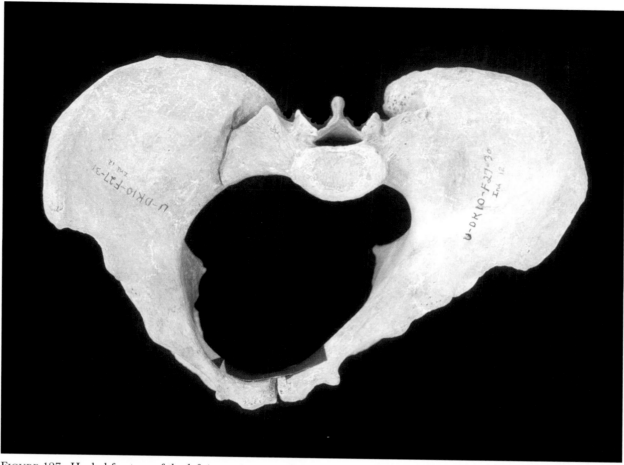

FIGURE 127. Healed fracture of the left innominate resulting in a deformed pelvic girdle with traumatic fusion of the left sacroiliac joint (sacralitis).

MANUBRIUM AND STERNUM

Sternal Foramen (Sternal Perforation)

A sternal foramen or perforation is an oval or circular developmental defect that is often misidentified as a healed bullet wound or perforating wound of the sternum (**Figures 128 and 129 and Plate 71**). This condition, which is considered a minor anatomical variant, results from the lack of complete fusion of the lower two or three sternal segments, frequently between segments three and four, as they ossify separately from left and right centers (Barnes 1994:227–230; Grant 1972); fusion begins cranially and progresses caudally, finishing last with the xiphoid. Yekeler et al. (2006), in a CT study of 1000 patients consisting of 582 men and 418 women aged 20–92 years, found sternal foramina in 45 (4.5%) subjects. The size of the foramina ranged from 2 to 16 mm and none was found in the manubrium. See Barnes (1994:210–230) for a comprehensive discussion of the multiple variants of morphology in the sternum with numerous citations of the paleopathological literature. It is an uncommon variant but present in most all populations. See also Ashley (1956), Cooper et al. (1988), Fokin (2000), Krogman (1940), McCormick (1981), O'Neal et al. (1998), Resnick (2002), Saunders (1978), and Yekeler et al. (2006).

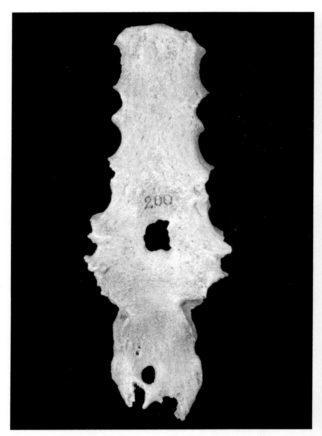

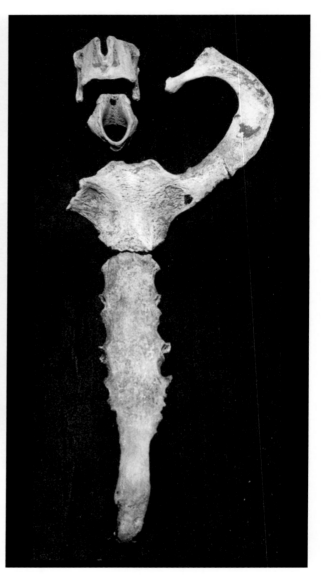

FIGURE 128 (*above*). Sternal and xiphoid perforation/fora-mina, aperture (cleft sternum) (Knight and Morley 1936–37) (NMNH–T 200). See Plate 71 for a color example.

FIGURE 129 (*right*). Ossification and elongation of the xiphoid process, and ossification of costal cartilage attaching to the left first rib (also in the photo are the thyroid [Adam's apple] and cricoid cartilage of the throat). All features older age related calcification of the cartilages (see also calcified xyphoid carti-lage in Figure 128 and Plates 70 and 71b) (NMNH 390022).

CLAVICLE

Rhomboid Fossa

The classic rhomboid fossa (**Figures 130 and 131**) appears as an irregularly shaped crater, groove, depression, pit, or excavation along the inferior surface of the clavicle. This fossa, a normal anatomical variant, associates with the attachment of the costoclavicular ligament (also known as the rhomboid ligament because of its shape) between the sternal end of the clavicle and superior surface of the first rib, connecting the clavicle and first rib (Depalma 1963). Other catchall varieties of this normal anatomical variant appear as smooth, raised eminences or depressions, crescent-shaped ridges and crests. Right and left clavicles from the same individual may differ both in overall size of the bone and type of costoclavicular attachments (e.g., raised in left clavicle and a deep rhomboid fossa in the right clavicle). In most people, however, there is

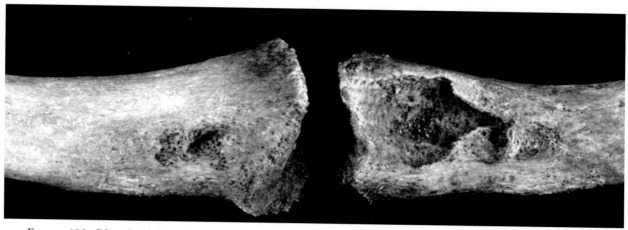

FIGURE 130. Rhomboid fossa for attachment of the costoclavicular ligament (Depalma 1963; Rogers et al. 2000).

no evidence of this trait visible in the clavicle or bony alteration at the attachment site of the costoclavicular ligament. Parsons (1916) noted rhomboid fossae in 10 percent of 183 clavicles. An examination of 10,000 chest fluorograms of individuals between 8 and 70 years revealed this trait in 5 percent of the sample; the youngest individual in the sample with a rhomboid fossa was 11 years old (Khazhinskaia and Ginzburg 1975; Shauffer and Collins 1966).

A preliminary study (contemporary autopsy sample of 350 adults by one of the authors [RWM]) revealed that rhomboid fossae longer than 15 mm in length were usually found in males. Although females commonly showed this trait, it was usually much smaller than those in males. Research by Rogers et al. (2000) and Prado et al. (2009) revealed similar findings, concluding that rhomboid fossae were more commonly found in males and could be used to estimate sex (the largest were in 20–30-year-old males) and that younger individuals rather than older individuals tended to have fossae. Thus, remodeling in older ages reduces the expression of this fossa. Paraskevas et al. (2009) found a high frequency of the excavated type of rhomboid fossa on the dominant-hand side, suggesting a mechanical etiology. See Flournoy et al. (2000) for a discussion of the rhomboid fossa as an indicator of sex and age.

While the etiology of the rhomboid fossa (commonly diagnosed as a lesion on radiographs) (Gerscovich et al. 1991) is unknown, it may be aggravated by strenuous activity of the pectoral girdle or dominant hand/arm. The present authors believe that this crater/groove, like those often found in the proximal diaphysis of the humerus (cortical excavations) and the distal femoral cortical excavation (DFCE), is a normal variant that may be especially pronounced in subadults, and is associated with muscle or tendinous insertion.* See Resnick (2002) for more detailed information and radiographic illustrations. Rhomboid fossa is often bilateral, but usually not symmetrical in size and shape. See also Cave (1961), Hagberg and Wegman (1987), Jit and Kaur (1986), Longia et al. (1982), Prado et al. (2009), Rogers et al. (2000), Saunders (1978), Shauffer and Collins (1966), Silverman and Kuhn (1993), Stirland (1991b), Taylor and Resnick (2000), and Williams and Warwick (1980).

Conoid Process (Joint/Tubercle)

Conoid joint/process (Cockshott 1958), coracoclavicular joint or bar (Depalma 1963; Taylor and Resnick 2000) or conoid tubercle of clavicle (Yochum and Rowe 1996) on the bottom illustration is a rare finding in most skeletal populations

---

*For an excellent overview and discussion of skeletal markers of occupational stress, refer to authors Kennedy, Wilczak, Hawkey, Steen and Lane, Stirland, Robb, Peterson, and Churchill and Morris in the 1998 issue of *International Journal of Osteoarchaeology,* 8(5):303–411.

(**Figures 131 and 132**). This raised plateau-like bony extension is a normal variant with no clinical significance that articulates with the superior surface of the coracoid process of the scapula with a reported incidence of 1 to 1.2 percent of individuals. Look for a corresponding facet on the coracoid (**Figure 137**). This is most oftenly expressed bilaterally.

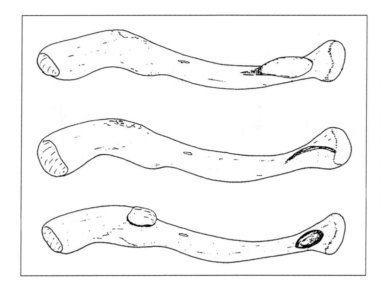

FIGURE 131. Variations of the rhomboid fossa and conoid facet (bottom figure). *Upper drawing – Raised* plateau-like attachment site for the costoclavicular ligament. It is an uncommon finding. *Middle drawing – Raised* ridge-like attachment site. It is an uncommon finding. *Bottom drawing – Depressed* (excavated) crater-like fossa – the typical shape is oval or oblong. The fossa will show sharply defined cortical margins and a porous center of exposed trabecula. It is a common finding.

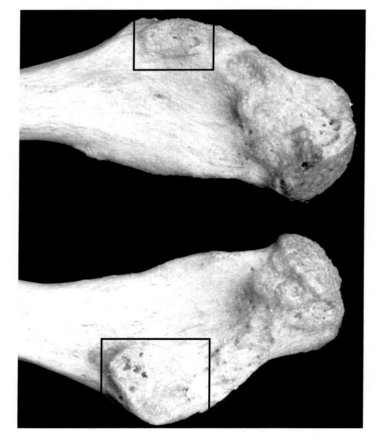

FIGURE 132. Conoid joint of the clavicle (see conoid joint of the scapula, Figure 137).

SCAPULA

Os Acromiale

Os acromiale (separate acrominal process), a non-metric trait (**Figure 133**) representing an un-united ossification center of the lateral end of the acromion. There are three separate centers of ossification for the acromion process, which will coalesce and reach normal final fusion between 16 and 25 years (Mudge et al. 1984; Park et al. 1994: Rockwood and Matsen 1998). In a radiographic study of 1800 shoulders, Liberson (1937) found the trait bilateral in 62 percent of cases and an incidence of 1.4 percent. Mudge et al. (1984) and Park et al. (1994) further reported that a correlation exists between rotator cuff tears and os acromiale. Sammarco (2000) observed from 1198 individuals from the Hamman-Todd Collection (Cleveland) that 8 percent of the population showed os acromiale with 33 percent being bilateral. Sammarco also found that os acromiale was more prevalent in blacks (13.2%) while it was much less frequent in whites (5.8%) and more frequent in males (8.5%) than females (4.9%). A complete survey of the Terry Collection by one of the authors (Hunt and Bullen 2007) resulted in the overall incidence of os acromiale at 8.3 percent and group frequencies ranking greatest in black males (12.5%), then black females (9.2%), white males (6.8%) and the lowest occurrence in white females (3.2%) (the combined sex frequencies, 11% for blacks and 5% for whites). The authors would derive from these results that there is some genetic component to the occurrence of os acromiale, with the significant difference of expression between American blacks and whites. Angel et al. (1987) reported a high incidence of os acromiale in the First African Baptist Church cemetery in Philadelphia (8.3%) and patterned special distribution within the cemetery, which they suggest as hereditary inheritance. In contrast, Stirland (1984) found a high frequency (12.5%) of os acromiale in the crew of the tudor ship *Mary Rose* and hypothesized that the condition was the result of pulling stresses on the unfused final element of the acromion related to long-term use of heavy longbows. However, Stirland's sample was composed primarily of young men below the age of 25 years and only

the continued unfused epiphyses in individuals over 25 years of age should be scored as os acromiale (Saunders 1978). See also Angel et al. (1987), Chung and Nissenbaum (1975), Edelson et al. (1993), Kalantari, et al. (2007), Kurtz et al. (2006), McKern and Stewart (1957), Resnick (2002), Saunders (1978), Stirland (1984, 1991b), and Symington (1900).

Osteoarthritis of the Glenoid Joint/Fossa

Raised, sharp margins along the glenoid fossa (or any other bony joint) can be detected either visually or by using a fingernail (**Figures 134 and 135**). Common finding accompanying middle age and older as well as individuals functionally stressing their shoulders in various activities (OA). See also Graves (1922). See Plate 76 for humeral head changes.

Small to large pits (macroporosity) in the articular surface of the glenoid fossa. The normal fossa should be smooth and slightly concave or have an undulating surface. It is a common finding.

Glenohumeral Joint Dislocation and Rotator Cuff Syndrome

This condition (subluxation/luxation) may result in flattening, erosion, porosity, and eburnation of the glenoid fossa (scapular deformity) (**Figure 136**). Although the condition of dislocation (clinically) may be fairly common in many populations, it is uncommon to rare to find discernible bony changes related to dislocation in either the humeral head or glenoid fossa or rotator cuff syndrome. See also Edelson et al. (1993), Ortner (2003:160, Figures 8-62, 8-63, 8-66, and 8-67), Park et al. (1994), Resnick and Niwayama (1988) and Steinbock (1967:40).

Costoclavicular Articulation

In some individuals the inferior surface of the distal clavicle (**Figures 131, 132 and 137**) will exhibit a raised, circular plateau/table of bone for articulation with the superior portion of the coracoid process of the scapula. This is an uncommon finding.

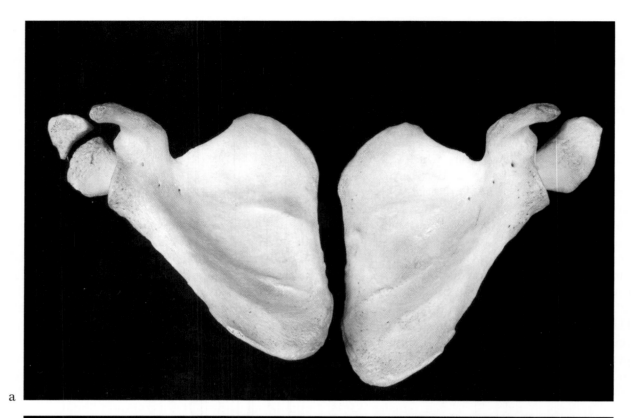

a

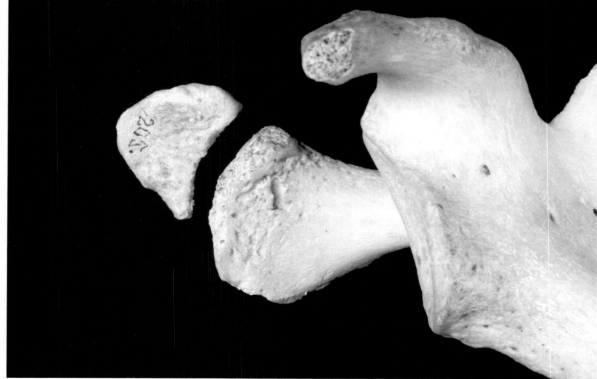

b

FIGURE 133. Os acromiale of the right scapula (NMNH).

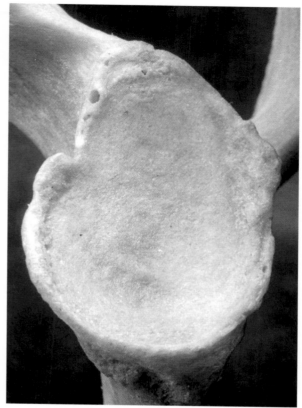

FIGURE 134 (*right*). Marginal osteophytes and pitting (OA) of the glenoid fossa (NMNH-T).

FIGURE 135 (*below*). Surface porosity and moderate-severe degenerative joint disease (OA) on right, compared to normal glenoid fossa on left (NMNH-H).

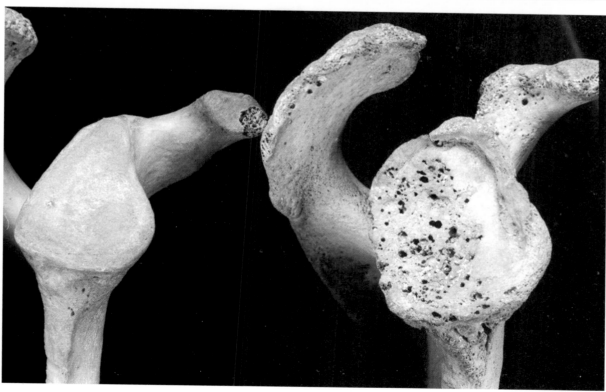

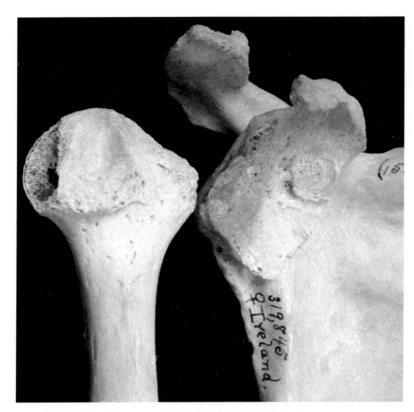

FIGURE 136 (*left*). Chronic unreduced dislocation (anterior) of the humeral head (glenohumeral joint) (NMNH-H 319845).

FIGURE 137 (*below*). Conoid joint or costoclavicular joint (DePalma 1963) (NMNH-T).

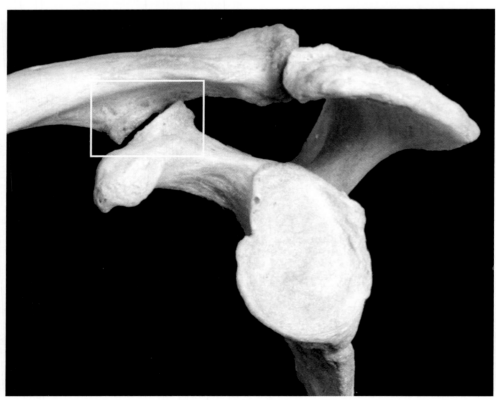

HUMERUS

Cortical Defect (Depression)

A porous groove (cortical defect) (Brower 1977; Caffey 1972; Keats 1988) upper humeral notches (Taylor and Resnick 2000) for attachment of the pectoralis, latissimus, or teres major muscles (**Figure 138**). In subadults this groove is probably a normal anatomical variant that remodels and later fills in, often leaving a shallow depression in the adult. The pectoralis groove is lateral to the smaller teres groove and both are oriented parallel to the long axis of the diaphysis. Cortical defects are common in children, adolescents (10–16 years of age) and young adults, especially individuals who are unusually physically active, but rare in adults (Mann and Murphy 1989; Murphy and Mann 1990; Ann Stirland, personal communication 1989). (See Cervical Fossa of Allen for a possible relationship between this and cortical excavations elsewhere in the skeleton.) See also Yochum and Rowe (1996).

The teres major cortical defect is a porous groove similar in appearance to the pectoralis defect noted above. This groove is a normal feature frequently seen in adolescents and young that normally remodels and fills in by adulthood, but may be accentuated (depened) due to acute or chronic trauma to the shoulder (e.g., prolonged tennis activity). This is an uncommon to rare find in adults. See also Mann and Murphy (1989) and Ann Stirland (personal communication 1989).*

Supracondyloid Process (Supratrochlear Spur)

A small roughly triangular and hook-shaped exostosis (**Figure 139**) projects 5 to 7 centimeters above the medial epicondyle and varies in length from 2 to 20 millimeters (Genner 1959; Fowler et al. 1959).

This exostosis serves as an accessory ligamentous attachment for origin of the pronator teres muscle. Through the tunnel formed by this fibrous band (Struther's ligament) passes the median nerve and brachial artery and may result in neurovascular impingement. Congenital trait reportedly found in 7 of 1000 living subjects by Schaeffer (1942) and in approximately 1 percent of people of European ancestry (Barnard and McCoy 1946; Gray 1948; Terry 1921, 1926, 1930). This trait has a high rate of heritability and has been found in embryos (Adams 1934), children of all ages, and adults. It is an uncommon finding in most archaeological samples. See also Cady (1921), Dwight (1904), Genner (1959), Hrdlička (1923), Marquis et al. (1957), Parkinson (1954), Pikula (1994), Rau and Sivasubrahmanian (1931), and Witt (1950).

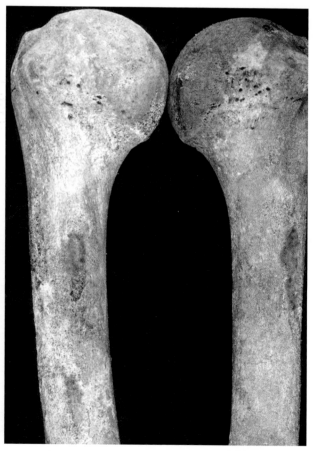

FIGURE 138. Cortical defect (teres muscle).

---

*For an excellent overview and discussion of skeletal markers of occupational stress, refer to authors Kennnedy, Wilczak, Hawkey, Steen and Lane, Stirland, Robb, Peterson, and Churchill and Morris in the 1998 issue of *International Journal of Osteoarchaeology,* *8*(5):303–411.

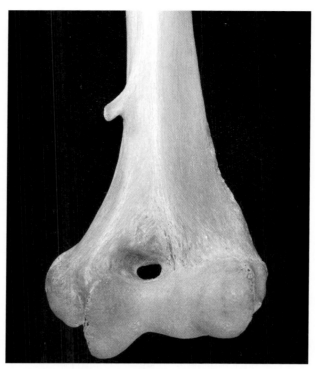

FIGURE 139. Supratrochlear spur. Supra-condyloid process (Struthers 1873), supracondyloid process (Schaeffer 1942; Fowler et al. 1959), supratrochlear spur (Saunders 1978), supracondylar spur (Kessel and Rang 1966), supracondylar process (Yochum and Rowe 1996) (NMNH-H).

Septal Aperture (Olecranon Perforation)

Septal aperture, also known as olecranon perforation or supratrochlear foramen, is a hole in the olecranon fossa that may range from the size of a pinpoint to the width of a pencil (**Figures 139 and 140**). Although the etiology of this perforation, which connects the olecranon and coronoid fossae, is uncertain (e.g., congenital, developmental/mechanical or hereditary), it is often found more frequently in females than males and occurs in 4 to 13 percent of individuals (Bergman et al. 1988). It is a common finding. See also Bass (1995:54), Finnegan (1978), Hrdlička (1932), Lamb (1890), Mays (2008), Ming-Tzu (1935), Singhal and Rao (2007), and Trotter (1934).

Osteoarthritis of the Distal Humerus

Severe OA of the posterior distal left humerus (**Figure 141 and Plate 78a**). Note the large osteophytes along the medial margin of the articular surfaces ("mushrooming") and extension into the olecranon fossa. Severe forms of OA of the distal humerus may also result in bony deposits (hyperplasia resembling small mounds), large

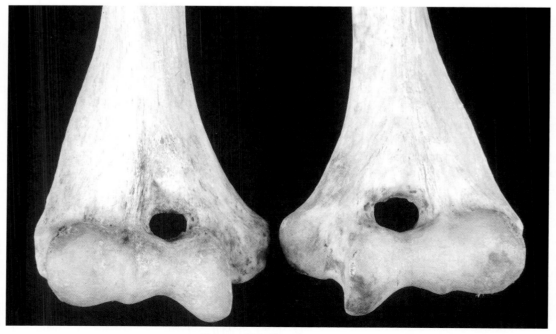

FIGURE 140. Septal aperture.

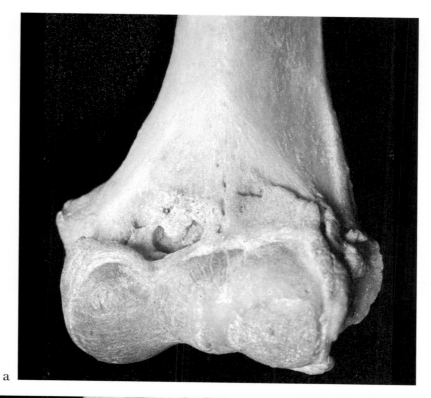

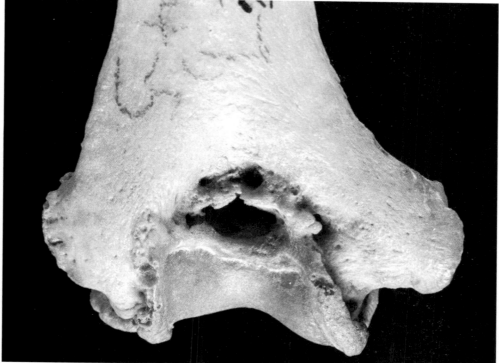

FIGURE 141. Surface osteophytes of the dorsal surface, filling of the coronoid and olecranon fossae of the distal humerus and moderate-severe DJD with distortion of the articular margin and eburnation of the capitulum (NMNH-H).

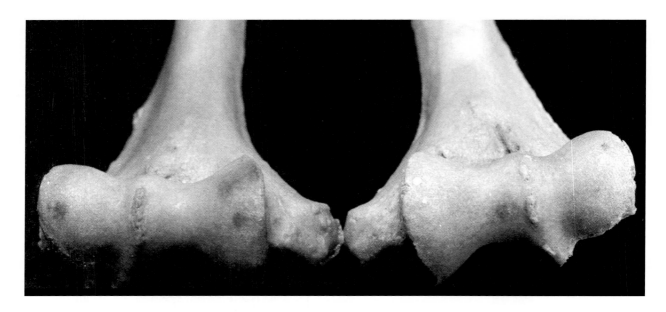

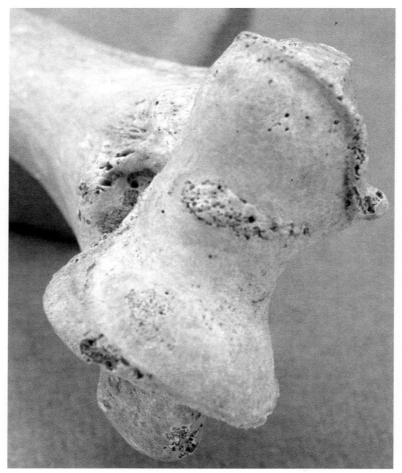

FIGURES 142 (*top*) and 143 (*bottom*). Early stage osteoarthritis of the distal humerus with a ridge-like buildup of bone.

pits, and distortion of the articular surface. Although similar severe changes may be the result of local acute trauma, most instances are associated with old age. Look for fractures of the proximal ulna and distal humerus. The severe form is uncommon.

Porosity may be interspersed in or along this thin, raised ridge of new bone formed between the trochlea and capitulum, responding to inflammation at the radioulnar articulation (**Figures 142 and 143**, facing page). It is a common finding. See also Ortner (1968).

Amputation

Evidence of an amputation (**Figure 144 and Plate 11b**) is apparent in that the normally open/hollow medullary cavity at the site of amputation has closed/remodeled and the distal end of the diaphysis is blunt and rounded (Zimmerman and Kelley 1982:54, Figure 37). See also Aufderheide and Rodreguez-Martin (1998:29–30), Steinbock (1976), and Zaki et al. (2010). For references on archeological specimens see Brothwell and Sandison (1967:640–641).

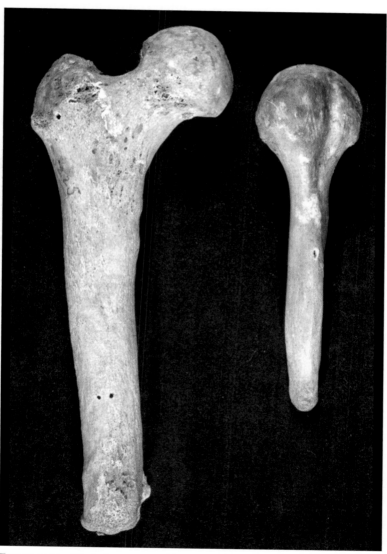

FIGURE 144. Healed amputated humerus (and femur) at about midshaft (see Plate 116).

RADIUS AND ULNA

Osteoarthritis of the Proximal Radius

Pitting within and enthesophytic development (**Figures 145 and 146**) along the radial tubercle (especially along the lateral border) is the result of inflammation of the biceps muscle insertion from strenuous activity. This feature is a common finding in all populations and will increase in size and severity with age of the proximal radius (see **Figure 147**).

Colle's Fracture and Smith's Fracture

Plate 80a–b illustrates Colle's fracture of the distal radius. Colle's fracture is a general term for fracture of the distal radius in the region of the metaphysis – fracture can be extraarticular, comminuted, articular, or any combination thereof. Note whether the angulation of the displaced bone is dorsally (Colles' type fracture) or volarly

(Smith's type fracture). See Duncan and Weiland (2004), Fernandez and Jupiter (2002), or the many other textbooks on wrist fractures for a description of Colles' fracture and its pathogenesis. See Plate 77 for healed fracture of humerus and ulna.

Pseudarthrosis (Pseudoarthrosis or "False" Joint)

Continued motion between two broken ends of a complete fracture from unrestricted motion often can produce a pseudarthrosis (**Figure 148 and Plate 81**). Restriction of motion at the site should allow callus development to reunite the bone, unless a cartilage barrier has been formed, resulting in a permanent pseudarthrosis. See also **Figure 98** for pseudarthrosis of a rib. This is an uncommon to rare finding. See also Ortner (2003:151–153) and Steinbock (1976:21–22).

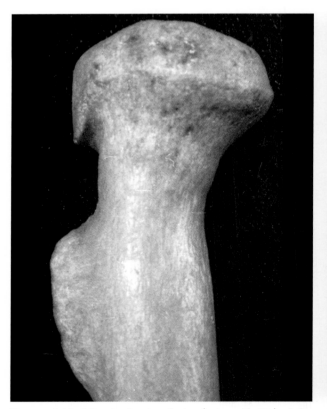

FIGURE 145. Marginal osteophytes (osteoarthritis) of the proximal radius.

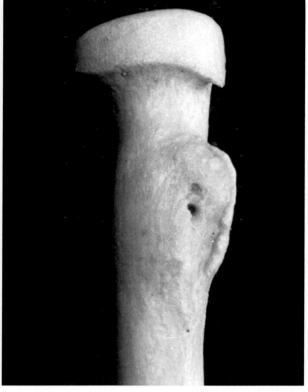

FIGURE 146. Enthesophyte and pitting of the radial tuberosity.

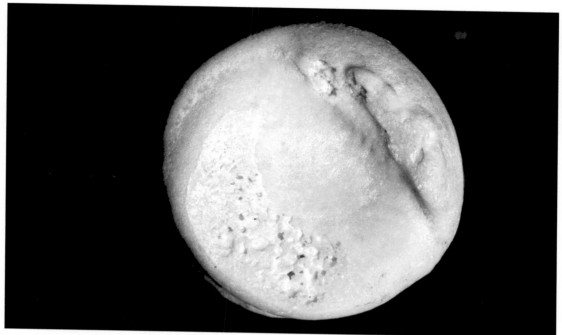

FIGURE 147. Pitting form (moderate to severe) of osteoarthritis of the proximal radius. Note the crescent shaped defect reflecting complete loss of cartilage and bony destruction and eburnation due to contact with the distal humerus.

Parry Fracture

Parry fracture (**Figure 149**) is generally associated with trauma when a person raises their arm to block a blow to the face or head, the ulna taking the brunt of the blow resulting in fracture. Fracture at the distal end of the ulna may also be associated with trauma induced in breaking a fall. Check distal radius for possible associating fracture (Cone's) before assuming this is a defense fracture (Nielsen et al. 2001). Cf. Warwick et al. (1993).

Interpersonal violence is not always the cause of a Parry fracture, this is emphasized by Grauer and Roberts (1996), and can be the result of accidental and osteoporotic complications (Krishnan 2002; Lindau et al. 1999). Incidence is generally uncommon but is dependent on the amount of warfare or domestic violence in a population. See also Knowles (1983) and Ortner (2003:137–138).

Periostitis

Figure 150 provides an example of severe active periostitis of radius. In this example, the periosteum has been irritated and lifted from the bone surface by extensive infection. The reactive bone is poorly formed due to the restriction of normal bone matrix formation by the infectious pus.

Radial Styloid Erosion

Erosion of the radial styloid process is one of the criteria associated with rheumatoid arthritis (Resnick and Niwayama 1988). However, osteoarthritis can also deteriorate the styloid process and may resemble a rheumatoid condition. Consult a rheumatologist to evaluate erosion with little or no bony proliferation around the distal ulnar and radial articulation (Leak et al. 2003). Fracture of the styloid process is also common in Colle's fracture. See also Nielsen et al. (2001) and Dieppe (1986).

Osteoarthritis of the Ulna

Figures 151 and 152 represent arthritic changes of the proximal and distal ulna.

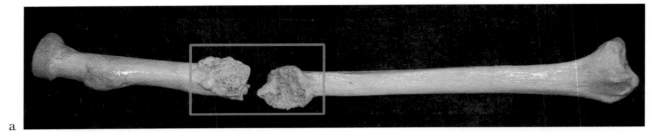

a

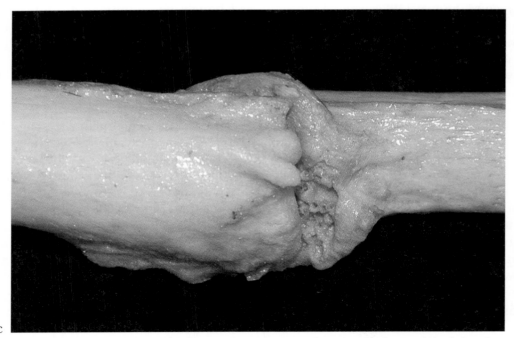

b

c

FIGURE 148. Pseudarthrosis ("false joint" and malunion) of the diaphysis of the left radius (KKU 186).

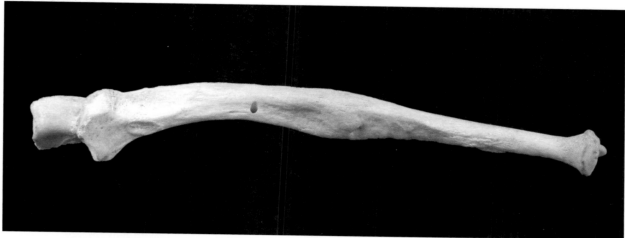

FIGURE 149. Parry fracture ("nightstick" defense fracture of the forearm) of the ulna (NMNH-H).

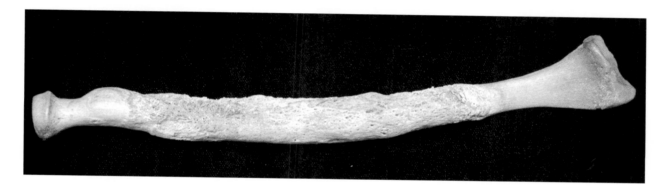

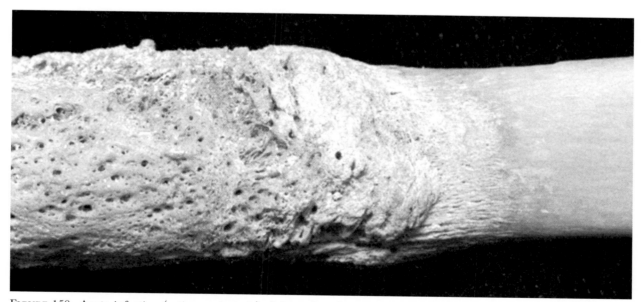

FIGURE 150. Acute infection (active periostosis) of the radius showing line of demarcation between reactive new bone (diseased) and normal bone. (See Aufderheide and Rodriguez-Martin (1998:179–181) and Ortner (2003:206–214) for other periostosis examples. See also **Figures 183 and 184** for periostosis on tibia (NMNH-H 319963).

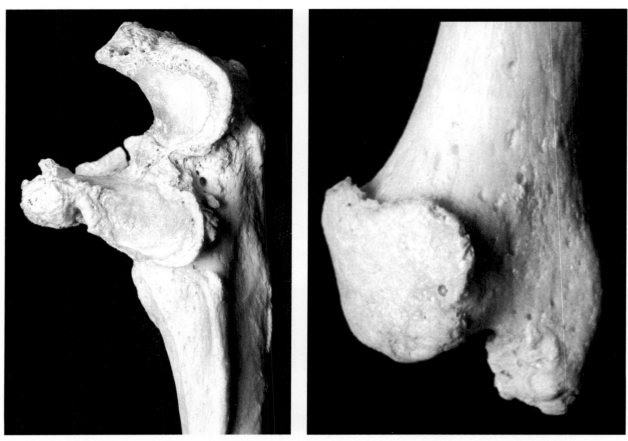

FIGURES 151 (*left*) and 152 (*right*). Moderate osteoarthritis of the proximal and distal left ulna (NMNH).

HAND

Boutinnere deformity (Swan-neck Deformity)

Although this condition is often associated with rheumatoid arthritis, localized trauma and infection must be considered as part of the differential diagnosis when only one joint is involved. Boutinnere, buttonhole, or "swan-neck" deformity (**Figure 153**) results from flexion at the proximal interphalangeal joint and hyperextension at the distal interphalangeal joints. Other skeletal indicators of rheumatoid arthritis include periarticular cysts, loss of joint space, and angulation of joints. See also Boyer and Gelberman (1999), Coons and Green (1995), Klepinger (1979), Resnick and Niwayama (1988), and Sambrook (2000). See Plates 82 and 83a for other hand deformities.

Enlarged Nutrient Foramen in the Phalanx

This condition, possibly the result of increased vascularity or velocity, can result from a variety of diseases and conditions, including leprosy, but can also be a normal variant (**Figure 154**). In this particular specimen, the enlargement is most likely associated with osteoarthritis as is evident by extension of the distal margin.

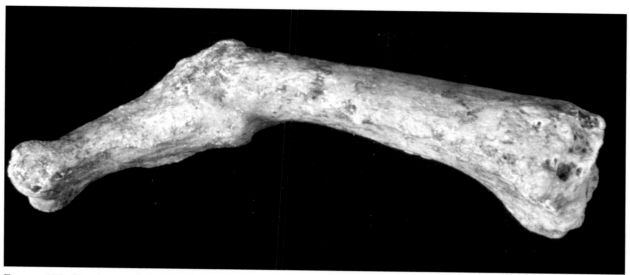

FIGURE 153. Boutinnere deformity (destruction and, possibly, bony fusion of the proximal interphalangeal joint [PIP]) of the hand.

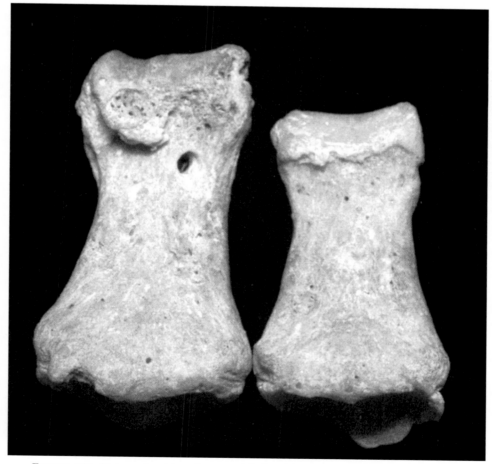

FIGURE 154. Enlarged nutrient foramen (phalanx) (NMNH-H). See also Figure 117.

FEMUR

Osteoarthritis and Eburnation on the Proximal Femur

Osteoarthritis commonly affects the knee and the distal femur in the form of marginal osteophytes bordering the articular surface of the condyles and, to a lesser degree, pitting of the articular surface. In severe cases, ivory-like eburnation with flattened and grooved areas may be produced in the articular surface from loss of the cartilage and subsequent "bone on bone" wear. Eburnated bone can range in size from extremely small (early stage) to areas involving the entire articular surface. Eburnation has a smooth/polished shiny appearance that reflects light and is frequently yellow, resembling old piano keys. It is an uncommon to rare finding associated with the elderly. See Rogers and Waldron (1995:38, Figure 4.4 and Figure 4.10) for examples. It is also good to perform radiographic investigation

to evaluate extent of lipping and cystic changes in the femoral head from necrosis.

Note the following bony changes:

A. *Periarticular bone (hypertrophic bone, periosteal osteophytes, "mushrooming").* Loss of bone height in the femur head has resulted in new bone growth near and encircling the neck (OA) (**Figure 155**). Most believe that osteoarthritis is the result of changes in the abnormal mechanical forces in and around the affected joint.

B. *Flattened areas of bone where there has been degeneration and resorption of most of the femur head (OA).* (See **Plates 110 and 111**).

C. *Bone cyst (subchondral cyst) below the articular surface.* Common radiographic finding in elderly individuals with advanced osteoarthritis and avascular/ischemic necrosis (dead bone usually resulting from trauma and loss of blood supply to the femur head; especially common in very old individuals) (**Figure 161**). See also Claffey (1960),

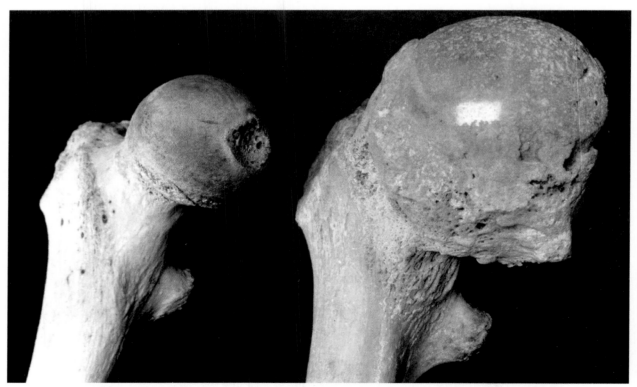

FIGURE 155. Osteoarthritis of the femur – note the large periarticular osteophytes, "mushrooming" and eburnation (shiny area) in the specimen on the right (NMNH).

Dieppe et al. (1986), Jaffe (1969, 1975), and Tanamas et al. (2010).

If degenerated femoral heads and deformed acetabula are found in children, this pathological condition may be the result of Legg-Calve-Perthes disease (Caterall 1982; Karpinski et al. 1986; Schoenecker 1986; Schwarz 1986) or osteochondritis of the femoral head, or slipped capital femoral epiphysis (**Plate 111**) (Busch and Morrissay 1987; Schoenecker 1985, 1986; Schiesinger and Waugh 1987), but histological investigation should be considered to confirm the bone cell changes which are associated with this disease.

Osteochondritis Dissecans

Stirland (1991b), in examining the 179 crewmen of the British ship *Mary Rose* that sank in AD 1545, reported finding six femora (two left and four right) with "unusual pits" (OD) in the femoral heads, superior to the fovea capitis (**Figure 156**). See also Aufderheide and Rodriguez-Martin (1998:81–83).

Osteochrontritis dissecans (OD) is also seen as a scooped-out lesion, usually in the medial condyle of the distal femur. First described in 1870 by James Paget, OD is usually attributed to avascular necrosis that begins with acute trauma to the joint (not to be confused with a periarticular cyst that, as the name implies, is located along the articular margin, not on it; **Figure 157**). However, numerous other causes have been suggested including abnormal ossification of the epiphyseal cartilage, contact with the tibial spines, hereditary influences (Lee et al. 2009), and generalized disorders (Smillie 1960; Wells 1964, 1974). Lee et al. (2009), for example, reported on three males in the same family that exhibited bilateral idiopathic osteochondritis dissecans of the femoral head, indicating a familial linkage.

Most researchers believe the lesion begins with the separation of a portion of cartilage and underlying bone fragment resulting in a "loose body" in the joint ("joint mice") (**Figure 206**). In time the bone and cartilage may resorb, creating a crater-like defect in the mature skeleton (Bradley and Dandy 1989). OD can affect any joint but has

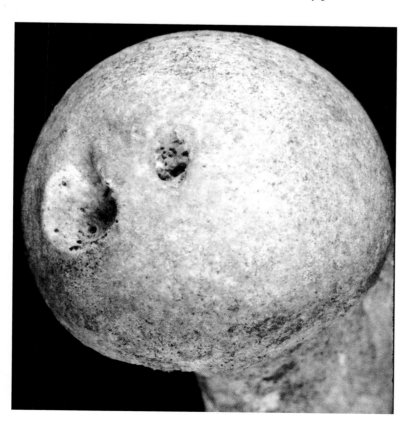

FIGURE 156. Osteochondritis dissecans (osteochondritic pit) in the femoral head.

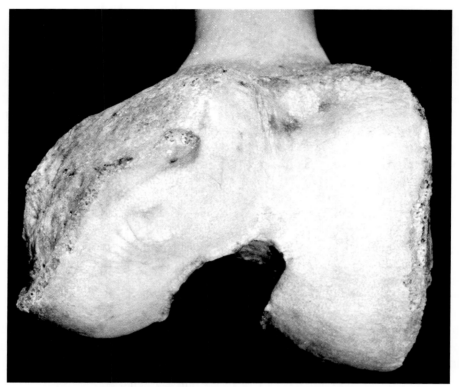

FIGURE 157. Periarticular cyst of the distal femur (similar cysts may be found at any joint – sometimes referred to as synovial cysts or pseudocysts of osteoarthritis).

a predilection for the medial condyle of the femur and appears during the second decade of life. Hughston et al. (1984) found 78 defects in the medial femoral condyle and 17 defects in the lateral femoral condyle in a sample of 83 patients with OD. Clinically, OD is seen in 15 to 21 cases per 100,000 in the femur, although many cases may go undetected unless accompanied by pain and detected on radiograph. It is an uncommon to common finding in most skeletal samples. See also Aufderheide and Rodriguez-Martin (1998: 81–83, Figure 5.5), Barrie (1987), Bowen et al. (1986), Clanton and DeLee (1982), DeSmet et al. (1997), Griffiths (1981), Lee et al. (2009), Ortner (2003: 353, Figure 13.8), Stirland (1991b), Manchester (1983), and Zimmerman and Kelley (1982:72, Figure 5.5).

Bradley and Dandy (1989:518), in contrast, examined lesions of the femoral condyles in 5000 knee arthroscopies and found that OD develops as an expanding concentric lesion that forms in an otherwise normal epiphysis. They reported that OD develops in the second decade of life and progresses to a steep-sided defect in the mature skeleton. They further hypothesized that OD and acute osteochondral fractures with or without loose bodies, were different conditions and should not be classified or referred to as OD.

A scooped-out concavity, pit or depression situated along the articular margin of a true joint. If the concavity "cyst" is located on the articular surface, especially near the middle of the joint of the distal femur or proximal tibia, an osteochondritic pit should be considered in the differential diagnosis. Similar pits may result from a variety of reasons including acute or repeated trauma to a joint and infection. See also Hill et al. (2003), Resnick (2002), and Resnick and Niwayama (1988).

Enthesopathy (See Chapter X)

Features of this specimen (**Figures 158 and 159**):
Surface osteophytes. Raised, isolated bony growths on the articular surface.

Marginal osteophytes. Raised lip of bone along the junction of the head and neck of the femur or fovea capitis (epiarticular osteophytes) (Jeffery 1975).

Surface porosity. Small to large pits in the articular surface.

Enthesophytes. New bone at the insertions of tendon and ligaments. Enthesophytes appear as irregularly raised areas (excrescences), roughened attachment sites, bony spicules or "whiskers," spurs or projections commonly found on the greater and lesser trochanters, trochanteric fossa, and linea aspera. This condition is due to stresses (inflammation) at tendon and ligament attachments (any) to bone. Some researchers state that enthesophytes occur only at the site of tendon and ligament insertions (Dutour 1986) because more stress is placed on the insertions, rather than the origins, due to the smaller area of attachment fibers into the bone. Enthesophytes are strongly correlated with old age (Plate 19). Care must be exercised not to confuse enthesophytes with myositis ossificans or periostitis (see Plate 118). See also Dutour (1986), Mariotti et al. (2004, 2007), Merbs (1983), Pecina and Bojanic (2004), and Resnick and Niwayama (1988).

Irregular spikes and crests of bone at muscle attachment sites along the linea aspera. Enthesophytes, also sometimes known as traction spurs, reflect pulling stresses and inflammation at ligament and tendon insertions.

Surface and marginal osteophytes are common findings in the elderly.

Subchondral Bone Cyst and Aseptic Necrosis

Subchondral bone cysts. Radiographs will reveal radiolucent (dark) areas indicating fluid-filled

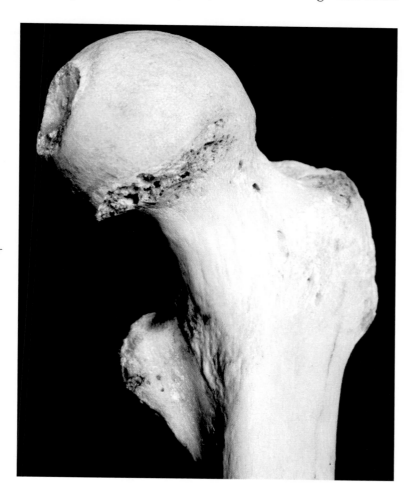

FIGURE 158. Osteoarthritis and enthesopathy of the proximal femur (anterior view).

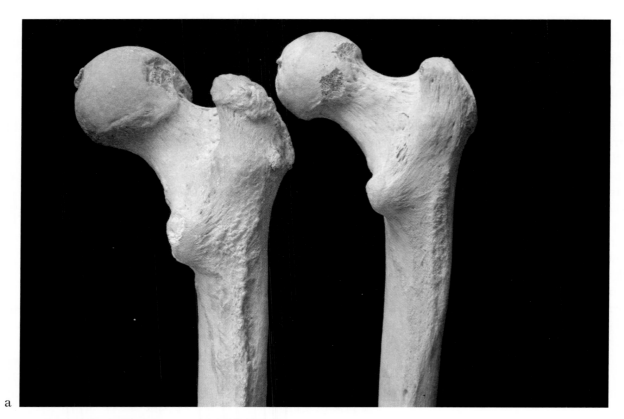

a

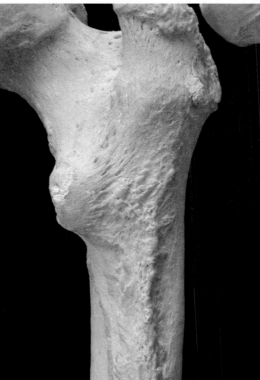

b

FIGURE 159. Enthesophytes (bony spikes and projections) along the linea aspera.

cysts/pockets beneath the articular cartilage (**Figure 161**). Common finding associated with degenerative articular changes accompanying osteoarthritis, rheumatoid arthritis, and aseptic necrosis. The cyctic changes shown can reflect acute or long-term repeated trauma, infection (**Figure 160**), or avascular necrosis, among others (Hough and Sokoloff 1989). See also Resnick and Niwayama (1988), and Schajowicz et al. (1979).

Anterior Acetabular Fossa, Allen's Fossa and Poirier's Facet

This facet (**Figure 162**) is most likely the result of contact of the tendon of rectus femoris on the neck of the femur (Kate 1963) and not compression of the antero-superior surface of the femoral neck by the acetabulum, the so-called anterior acetabular imprint (Meyer 1924). It is worthy of note that similar fossae have not been noted in habitually squatting nonhuman primates. Kate suggested that this fossa is the evolutionary result of the erect posture and is not due to squatting as others have stated (Molleson et al. 1998). Allen's (1882) original description of the femur neck stated that it "is marked in front near the articular surface by a faint depression, which is often cribiform in appearance and may receive the name cervical fossa." Findings by subsequent researchers have since made it difficult to distinguish one facet from another with the addition of such terms as the imprint of Bertaux (1891), Testut (1911), Poirier (1911), and cribra femoris (Molleson et al. 1998). There are a variety of

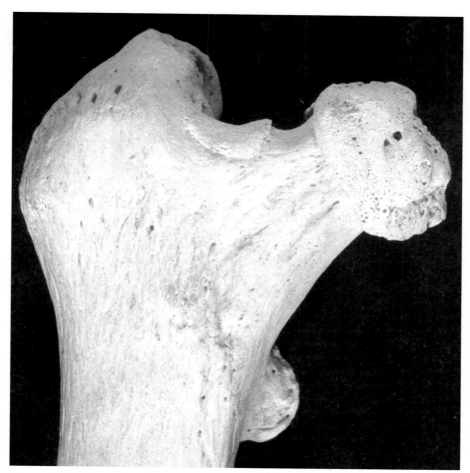

FIGURE 160. Tuberculous destruction of the femur head with a "saddle-shaped" plaque formation on the neck from articulation with the superior border of the acetabulum.

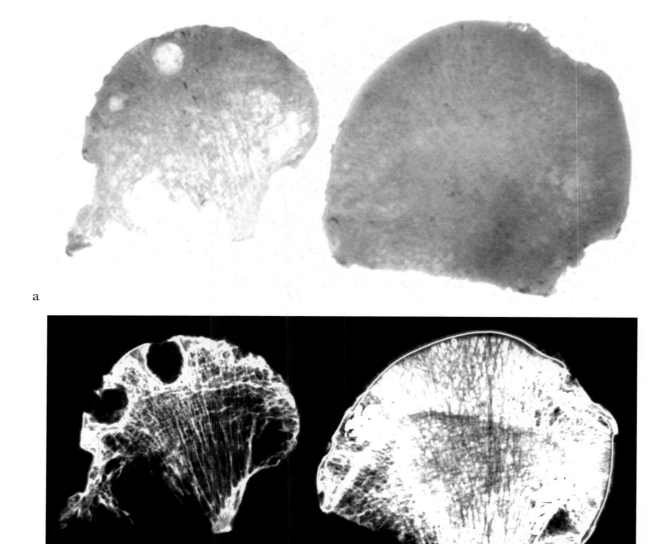

a

b

FIGURE 161. Cross-section and radiograph of femoral heads illustrating common indicators of osteoarthritis (*left*) normal comparative specimen on right. The rounded cavities that open through the articular surface (*left*) are subchondral cysts.

depressions, raised plaque-like areas and extensions of the articular surface (plaque and Poirier's facet) (**Figure 163**), and cribiform (porous/trabecular "sieve-like" [cervical fossa of Allen, or Allen's fossa]) lesions that may be present at various locations on the femur neck, possibly reflecting the bursa for crossing of the psoas tendon across the femoral neck. Allen's fossa can be distinguished from Poirier's facet and plaque (resembles Poirier's facet but with a raised rim) in that the former appears as if a "window" of the cortex has been cut out of the femur, revealing the underlying trabeculae. Look for other skeletal "excavations" at the attachment sites of the latissimus dorsi, teres and pectoralis in the proximal humerus, medial head of the

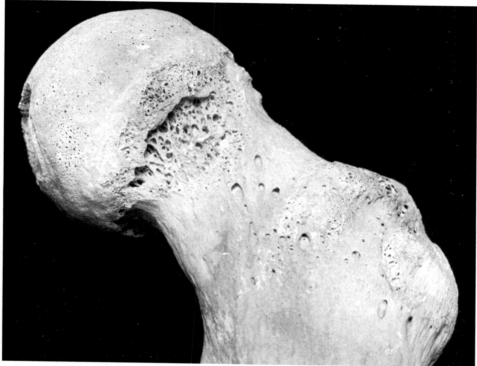

FIGURE 162. Anterior view of left proximal femur showing a cervical fossa of Allen (note porous and cribriform nature of the defect and exposed underlying trabeculae).

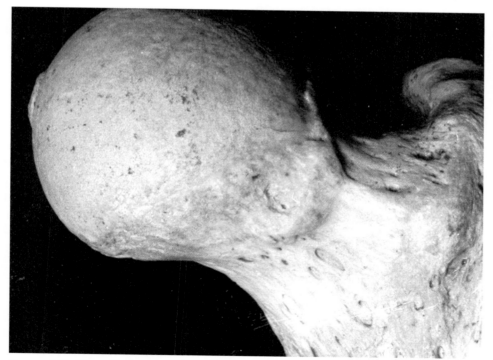

FIGURE 163. Kostick's facet (1963) or Poirier's facet (1911) or plaque – note the raised rim or mound of bone extending onto the neck.

gastrocnemius in the distal femur, and the costo-clavicular ligament in the sternal end of the clavicle. These excavations are often found together in adolescents and young adults engaged in strenuous activities (such as soldiers) and may result from the same biomechanical or growth stress. Horackova and Vargova (2004) examined 560 femurs from Moravian and Egyptian archaeological sites (and 10 cadavers) and found Allen's fossa in 18.1 percent of Czech femurs and 16.3 percent in Egyptian femurs. Poirier's facet was found in 10.6 percent of Czech and 8.7 percent of Egyptian femurs, respectively. It is a common finding, especially in adolescent and young adult males. See also Finnegan and Faust (1974), Kostick (1963), Meyer (1924), and Saunders (1978).

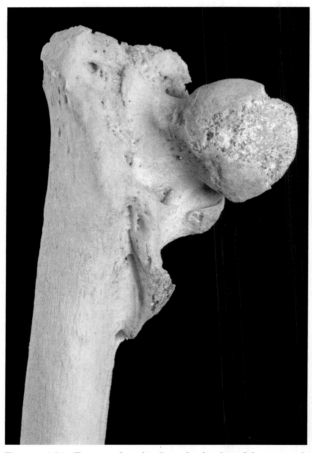

FIGURE 164. Fractured and inferiorly displaced femur neck with reactive (hypertrophic) bone growth in the region (NMNH-H).

Poirier's Facet

An area of smooth or slightly raised bone, often about the size and shape of a thumbprint, along the anterior margin of the femur in the area of the bursa for the psoas tendon. This facet often appears as an extension of the articular surface of the femur head (**Figure 163**) and may present either as a smooth extension or an extension (Poirier's) with a raised rim (plaque).

Femoral Neck Fracture

This is a common occurrence in the elderly ("broken hip"), especially postmenopausal and senescent females (**Figure 164 and Plates 112a and 117**). Fractures of the femoral neck can result from acute trauma, such as from a fall, or from osteoporosis in the aged, such as sitting down or lifting oneself from a chair. Fractured femur necks often lead to avascular necrosis (bone death). Death of the individual can be the ultimate result of a femoral neck fracture by disability and complications by immobility (e.g., pneumonia). In contemporary populations, 10–20 percent of elderly people who sustain hip fractures die within six months of injury (Genant 1989). See also Cummings et al. (1985), and Gardner (1965).

Myositis Ossificans (Heterotopic Ossification)

Note the exuberant bone formation (often referred to as an exostosis), which would have extended into the muscles of the upper thigh. Heterotopic bone such as this (i.e., myositis ossificans) results from trauma to the muscle and subsequent ossification (**Figures 165, 166, and 167 and Plates 112, 113, 114, 127, and 128**). See also Resnick and Niwayama (1988), and Schajowicz (1981). (See **Figure 180 and Plate 75** for comparison.)

Arthritic Changes to the Distal Femur

Marginal osteophytes. Raised, sharp lipping indicative of OA that may be the result of acute

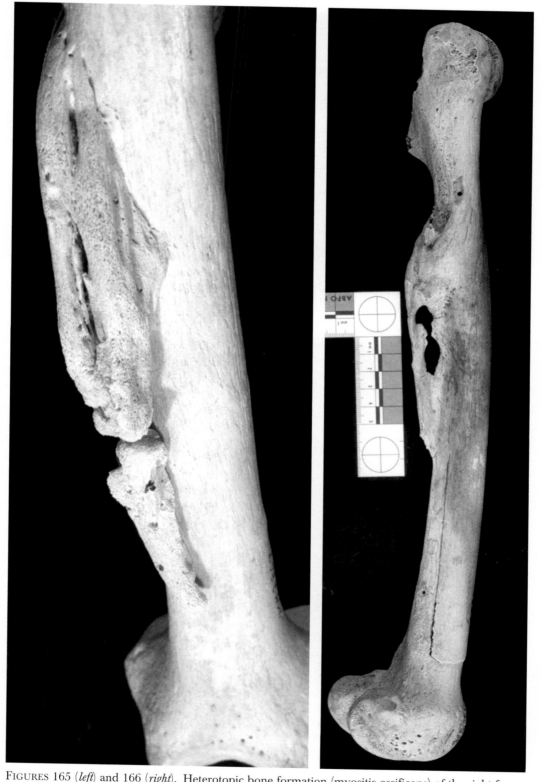

FIGURES 165 (*left*) and 166 (*right*). Heterotopic bone formation (myositis ossificans) of the right femur (AFIP).

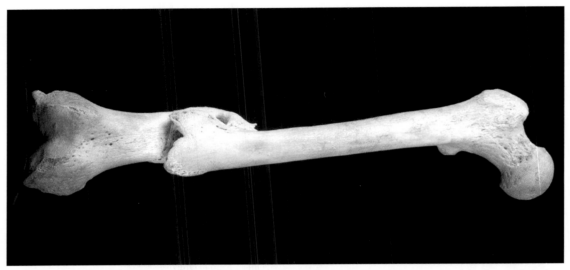

FIGURE 167. Healed fracture of the femoral diaphysis. Note angulation and shortening of the femoral shaft as a result of misalignment and overriding of the ends of the broken bones and heterotopic ossification (see Plates 114 and 115 for another example) (NMNH-H 321053).

FIGURE 168. Severe OA/DJD of the distal femur showing marginal osteophytes and pitting of the articular surface with grooving and eburnation (polishing) of the patellar surface – eburnation often resembles old ivory (NMNH).

trauma (e.g., knee injury) or associated with old age (**Figures 168, 181, and 182 and Plate 125a**).

Articular surface porosity. Small to large single or coalesced pits accompanying old age.

Articular surface osteophyte. Raised areas of bone on the articular surface.

Eburnation, resulting from the loss of cartilage and polishing ("shiny") of the articular surface from "bone on bone" wear. Eburnation of the femur is often accompanied by similar DJD of the patella and, possibly, the proximal tibia.

Osteophytes and macroporosity are common findings in the elderly. Eburnation is less common and more often associated with more acute trauma to a joint where necrosis has occurred. See also Aufderheide and Rodriguez-Martin (1998:94–96), Rogers and Waldron (1995:38), and Steinbock (1976:282–284).

Long Bone Growth – Porosity, "Cut back" and Metaphyseal Plate Morphology

These areas of rapidly growing and remodeling bone (Figures **169, 170, and 171**) are often misinterpreted as "diseased." See discussion concerning the "cut back" zone in normal bone growth of children presented by Ortner (2003:15–16).

Harris Lines (Lines of Growth Arrest)

Harris lines or growth arrest lines (**Figures 172 and 173**) appear as thickened lines in dry bone and opaque/sclerotic lines (radiographically) extending across the diaphysis and/or metaphysis of long bones (and teeth as well) as result of additional osseous deposition due to the failure of continuous growth at the metaphyseal cartilage. Note the wavy horizontal lines visible in the metaphysis in **Figure 172** and multiple lines in the distal ends of the humerus and femur and proximal end of the humerus in **Figure 173**. Controversy exists over the etiology of Harris lines as individuals never experiencing fever or other serious childhood illnesses or physiological disturbances have been shown to have these lines (Zimmerman and Kelley 1982). Gindhart (1969) reports that only 25 percent of severe diseases in immature subjects result in the formation of Harris lines. Harris lines, which may be visible in any of the long bones, may resorb over time leaving no evidence

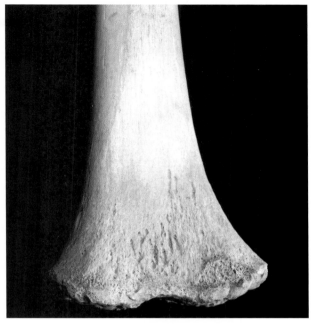

FIGURE 169. Normal growth areas (pitting, striations and discoloration) in the dorsal surface of a child's distal femur (NMNH-H).

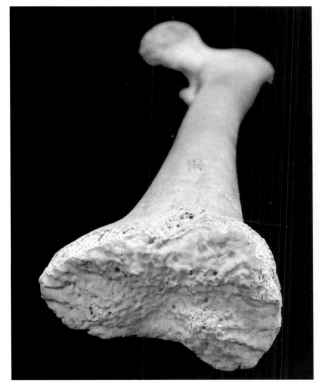

FIGURE 170. Normal roughened metaphyseal plate in the distal femur of a child (NMNH-H).

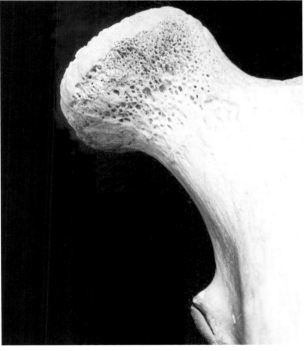

FIGURE 171. Normal porosity and exposed trabeculae in the femoral neck of a child (the femoral head was still unfused to the neck and the lesser trochanter epiphysis is still only partially fused) (NMNH-H).

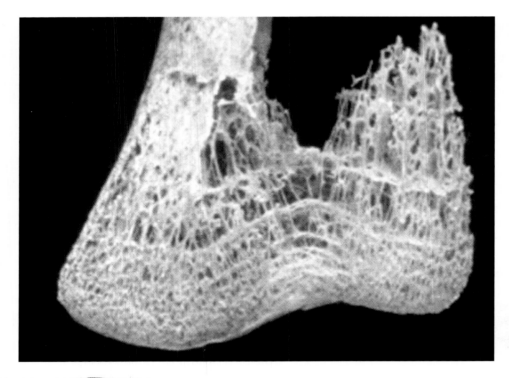

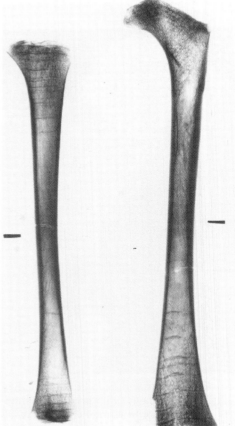

FIGURES 172 (*above*) and 173 (*left*). Harris lines (growth arrest lines, transverse lines, Park lines, Park-Harris lines, stress lines) indicative of multiple bouts of arrested growth (stress) during the period of growth as a subadult. (Areas of missing bone in the present examples are due to post-mortem breakage.)

of these transverse lines of arrested growth. See also Aufderheide and Rodriguez-Martin (1998: 422–424), Byers (1991), Goodman and Clark (1981), Grolleau-Raoux et al. (1997), Harris (1926, 1931, 1933), Lee and Mehlman (2003), Nowak and Piontek (2002a-b), Park (1964), Arnay-de-la-Rosa et al. (1994), and Steinbock (1976:43–55).

Distal Femoral Cortical Excavation (Charles's Facet)

A benign crater-like defect, sometimes referred to as Charles's facet (Kostick 1963) or distal femoral cortical excavation (DFCE), defect, or irregularity (Resnick 2002; Yamazaki et al. 1995) that can measure more than 2 cm in diameter, located above the medial condyle at the attachment of the medial head of the gastrocnemius muscle (**Figure 174**). The floor and margin of these defects often reveal sharp bony spicules and exposed trabeculae as though a plug of cortical bone has been forcibly avulsed. While the pathogenesis of this usually bilateral defect is unclear, most researchers believe it is due to the repeated pulling stresses (traction) or acute trauma of the gastrocnemius muscle, hyperemia (increased blood velocity), and localized bone resorption brought on by the traumatic event (Donnelly et al. 1999; Posch and Puckett 1998; Resnick 2002, p. 4563). It is a common finding of limited existence in children and adolescents that is likely related to rapid bone remodeling during youth. Hrdlička (1914) noted its presence in the skeletons of ancient Peruvian adolescents but not in children. The present authors have seen this feature in many subadults and adolescents, especially ancient Egyptians, Native Americans and Civil War and War of 1812 soldiers.

Effects of Paralysis on the Knee Joint

Narrow and seemingly elevated lateral condyle relative to the medial condyle (**Figure 175**). The lateral condyle serves to guide and hold the quadriceps tendon in place.

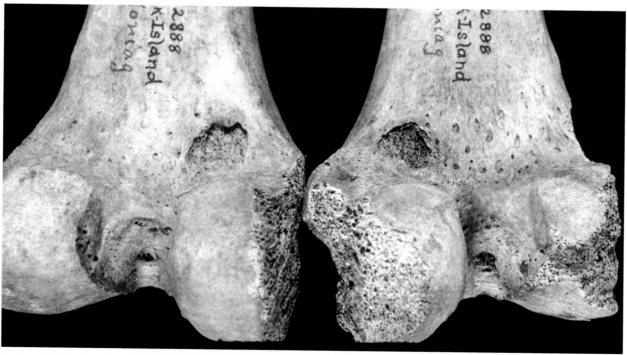

FIGURE 174. Distal femoral cortical excavation (DFCE) (Resnick and Greenway 1982), (subperiosteal cortical defect, cortical defect, cortical desmoid), ("tendon lesion," Hrdlička 1914), avulsive cortical irregularity (Posch and Puckett 1998), fibrous metaphyseal defects (Ritschl et al. 1988, 1989) at the site of attachment of the medial head of the gastrocnemius. This crater, also known as Charles's facet (Kostick 1963), typically fills in but may also result in a raised mound (NMNH 372888).

a

b

FIGURE 175a-b. Left femur and patella (from below) showing bony changes due to prolonged flexion contractures of the lower leg (often due to poliomyelitis). Deepening of the trochlear fossa may be an indicator or prolonged immobility or, possibly, a bent-knee gait (NMNH-T 781).

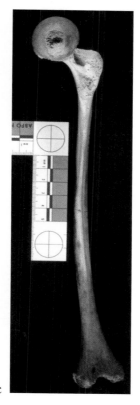

c

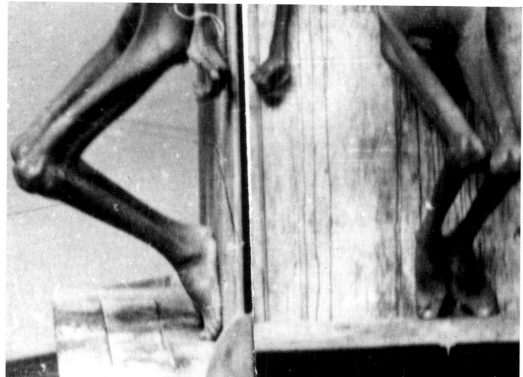

d

FIGURE 175c-d. Note wasting and thinning of the femoral shaft (c) and "knobby" knees (d).

Deep trochlear fossa (patellar surface) due to contact pressure erosion of the quadriceps tendon. The average depth of most adult trochleae is 5.2 mm, but some may be as deep as 10 mm (Casscells 1979). Trochleae deeper than 10 mm may represent a pathological condition in humans reflecting long-term flexion contractures of the legs (knees drawn up towards the chest in a fetal position) such as being bedridden due to disease or injury (e.g., paralysis; Mann et al. 1990) or bipedalism, as quadrupeds have deep fossae while those in bipeds are shallow. Condition may be unilateral or bilateral (Conner 1970). See also Abraham et al. (1977), and Micheli et al. (1986).

Note the following bony changes:

- *Wide intercondylar fossa.*
- *Large, raised, and widely spaced lateral articular facet.*
- *Small, raised, and widely spaced medial articular facet.*

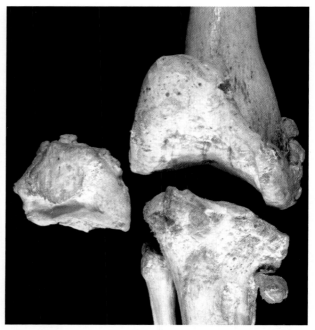

FIGURE 176. Neuropathic joint (Charcot's joint, anterior view of knee) (AFIP).

If changes of this type are noted in either the femur or patella, it is almost certain that the individual endured many years with the lower legs in flexion (e.g., legs drawn toward the chest), possibly due to paralysis (associated with the condition noted in B above). Increased contact pressure of the quadriceps tendon against the distal femur results in erosion of the trochlear surfaces. A number of conditions, including poliomyelitis, spinal cord trauma, and spina bifida, can cause flexion contractures of the legs. The finding of deep trochlear fossae or widely spaced patellar facets suggests cultural implications of caring for the individual who might have been unable to properly care for him/herself. Rare finding in most archaeological samples and uncommon finding in contemporary populations.

Charcot's Joint

Neuropathy may affect any joint but shows a predilection for the large weight-bearing joints such as the knee (**Figure 176**). Although the specific etiology is debated, it is likely that Charcot's joints (severe osteoarthritis) develop after sensory loss (underlying neurological defect) to a joint

and continued joint stress (Hellman 1989). This example depicts a hypertrophic Charcot joint of the knee involving the femur, tibia, and patella. Note the massive exuberant new bone and misalignment of the femur and tibia (the midline axis has changed).

Conditions predisposing the development of a Charcot joint include alcoholism, diabetes mellitus, leprosy, congenital insensitivity to pain, and syringomyelia (Sella and Barrette 1999). It is a rare finding. See also Aufderheide and Rodriguez-Martin (1998:107–108), Brower and Allman (1981), Bryceson and Pfaltzgraff (1990), Eichenholtz (1966), Jaffe (1975), MacAusland and Mayo (1965), Ortner and Putschar (1985), and Raina et al. (2007).

Rickets

Rickets (**Figure 177**) is caused by the lack of calcium and phosphorous in the diet, or the inability to metabolize or absorb these minerals by intestinal maladies such as chronic diarrhea, metabolic insufficiency, renal failure, etc. See further discussion in Auferheide and Rodriguez-Martin

(1998:305–310). See also Ortner (2003:393–404), and Zimmerman and Kelley (1982:60).

Hypotrochanteric Fossa

This fossa, located along the postero-lateral surface of the femur and inferior to the lesser trochanter, appears as an elongated groove or shallow depression running parallel to the long axis of the femur is often found in American Indians and Asians (**Figure 178**).

Osteomyelitis and Cloacae

Cloacae or draining sinuses (for pus) are frequent findings in chronic suppurative osteomyelitis (**Figures 179 and 189 and Plate 120**). The light island of bone visible within the cloaca (in

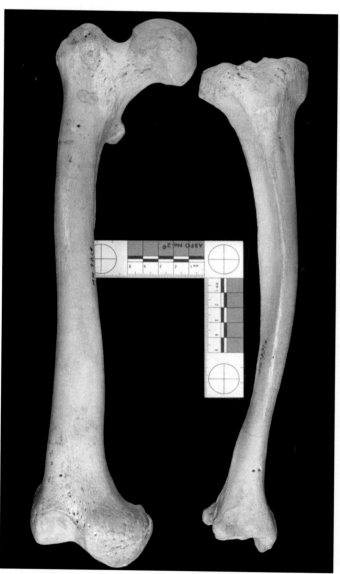

FIGURE 177. Rickets of the femur and tibia. See also Aufderheide and Rodriguez-Martin (1998:305–310), Mankin (1974a-b, 1990, Ortner (2003:393–404), Steinbock (1976:262–273), and Zimmerman and Kelley (1982:58–60) AFIP MM3913).

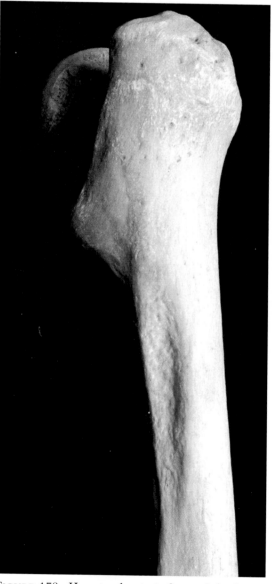

FIGURE 178. Hypotrochanteric fossa of the femur (normal variant more commonly found in American Indians.

Figure 189) is a sequestrum, that is, and isolated dead bone fragment that became walled off by exudate, granulation tissue, or scar. The outer envelope of irregular bone is an involucrum (i.e., sheath of newly formed bone). See Ortner (2003:179–206) for numerous illustrations of osteomyelitic effects to bone (Figures 9-1 through 9-32) and Steinbock (1976:60–85). As stated by Carney and Wilson (1975), "The special pathologic tissue reactions which are characteristic of suppurative osteomyelitis are the lysis of bone, the formation of new bone, and the presence of dead bone." If the fragment(s) is small enough it will gradually be resorbed/absorbed). In dry bone the sequestrum may be freely mobile and rattles when shaken. It is an uncommon to rare

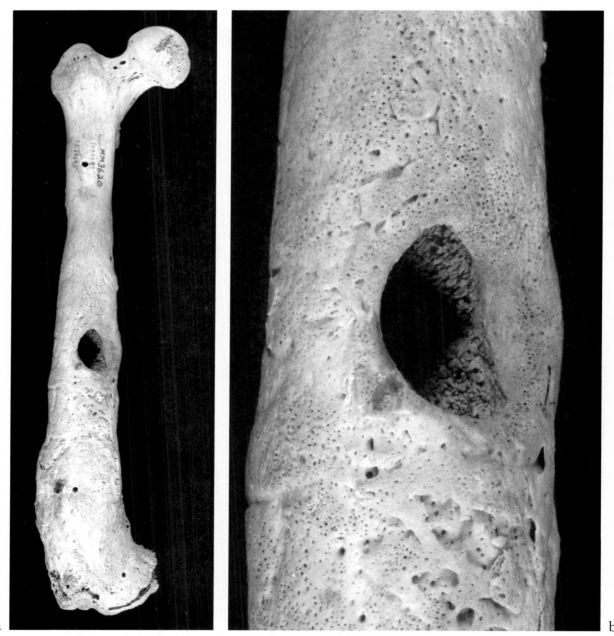

a b

FIGURE 179. Osteomyelitis of the femur with a cloaca (draining sinus) (AFIP MM3620). See Plate 120 for another example.

finding in archaeological specimens, but it may be common in hospital and battle-related skeletal samples where surgical intervention prolonged and often exacerbated infection (e.g., Civil War soldiers). See also Aufderheide and Rodriguez-Martin (1998:174–177) (especially a complete sclerotic involucrum of a tibia at the Mütter Museum) (Figure 7.45). See Plate 121 for lytic "holes" that are from carcinoma, not from infection.

TIBIA

Periostitis

Periostosis results from inflammation stimulus from infectious activity or as the result of direct trauma such as a blow to the bone where the haematoma lifts the periosteum from the bone, stimulating bone deposition (**Figures 183 and 184**). One possible result of the latter is an ossified periosteal hematoma (Day 1960) (see **Figure 180a-b**). Other causes for periostosis can be

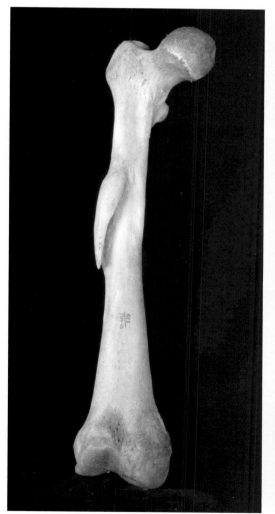
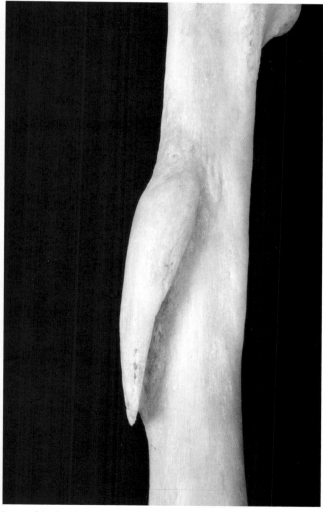

a   b

FIGURE 180. Healed subperiosteal hematoma of the femur (similar lesions may be found in any bone). See also Day (1960), Kullman and Wouters (1972), and Steinbock (1976) (NMNH-H).

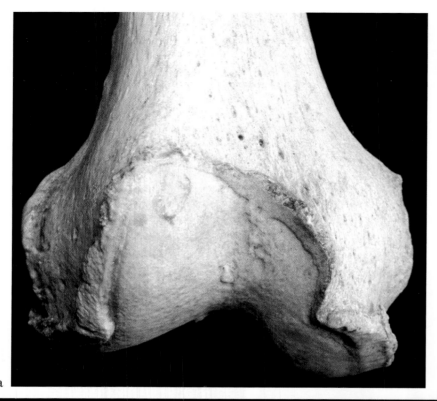

a

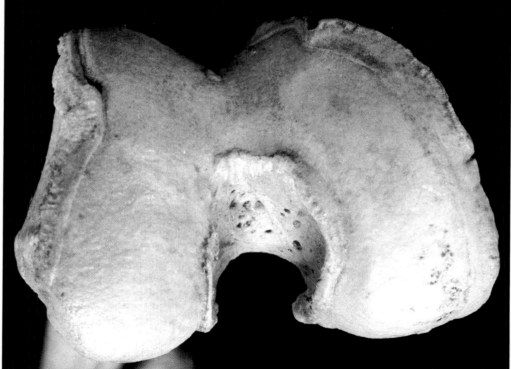

b

FIGURE 181. Marginal and surface osteophytes (osteoarthritis) of the distal femur. See Plate 125 for a color example.

FIGURE 182. Moderate osteoarthritis of the femoral head and fovea capitis. The degree of severity in this example is commonly found in modern groups, but less so in ancient groups. The floor of the fovea capitis often becomes rough and porous (degenerative joint disease/OA) and its rim raised and "over-hanging" in elderly individuals.

blood-borne infection (syphilis, tuberculosis), venous insufficiency as in varicose veins (Daniels and Nashel 1983; Gensburg et al. 1988), scurvy, or any other host of factors (Aufderheide and Rodriguez-Martin 1998:310–311). Common usage of the term *periostitis* generally refers to bone formed only on the outer cortex while osteitis refers to changes within the cortex, and osteomyelitis to changes that affect both the marrow and bone (Caraveo et al. 1977). Many times it is difficult, if not impossible, to ascertain the cause of the periosteal remodeling. Periostosis, in the active stage, is easily recognized by its color, texture, and raised appearance. Active periostosis (A) is often discolored from the underlying bone, pitted due to the increased vascularity, striated, and has defined, raised margins (B). Active periostosis has the look of small, loosely attached sheets of tree bark or thin layers of sponge. See also Levine et al. (2003), Ortner (2003:206–215). See **Plate 131a-b** for a color example of periostitis and its resemblance to tree bark.

In healed cases, the margins of the defect are less distinct, is less porous/pitted than in active cases, and "blend" into the surrounding bone. The rough striated or spongy morphology has begun to be remodeled, making a more smoothed appearance. It should be remembered that the periosteum in fetuses, newborns, and children is only loosely adhering to the cortex and can be easily lifted resulting in massive periosteal remodeling (e.g., congenital scurvy or syphilis; Toohey 1985). See also Ortner et al. (2001).

Arthritis of the Proximal Tibia

The tibial spines (**Figure 185 and Plate 130a**), although not ligamentous insertions but articular surfaces, will be slightly elongated and sharp to the touch. Common finding in the elderly.

Raised islands of bone on the articular surface (small arrow) or along the articular margin are bony responses to the destruction and degeneration of the joint capsule and cartilage covering the joint (large arrows). Marginal osteophytes

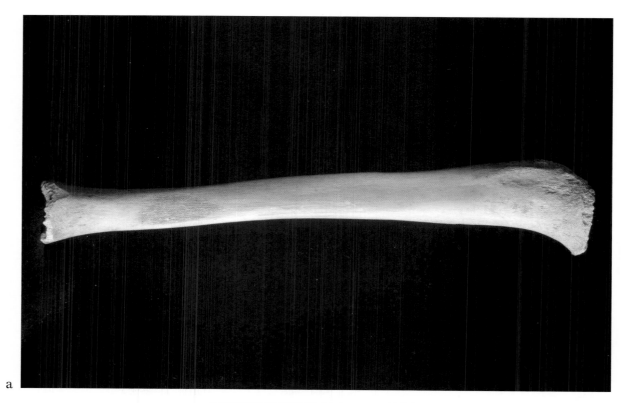

a

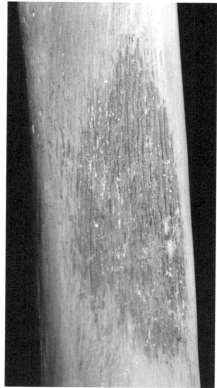

b

FIGURE 183. Active periostosis (lateral view of right tibia) (NMNH-H). See Plate 131a-b for a color example.

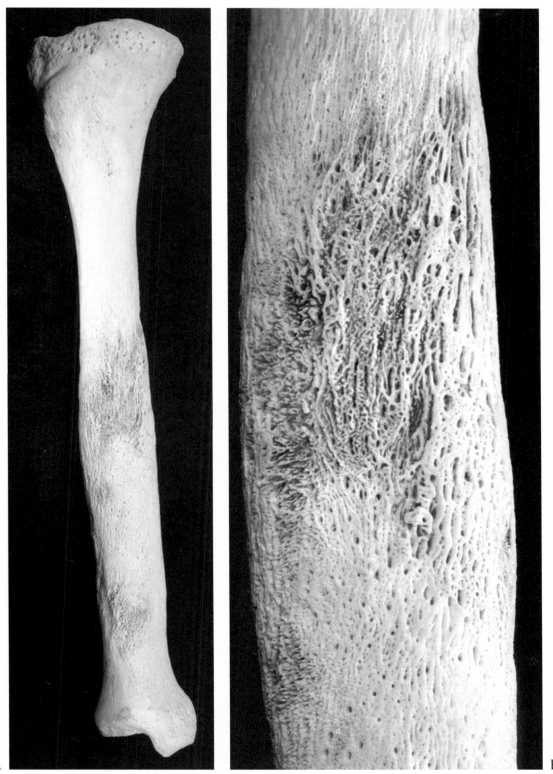

a b

FIGURE 184. Healed periostitis of the tibial diaphysis (mild forms are common in many skeletal groups and along any bone) (NMNH-H). See Plate 118 for a color example.

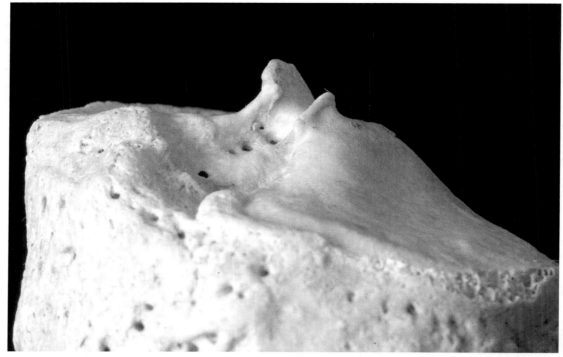

FIGURE 185. Spiking/peaking of the tibial spines – indicator of osteoarthritis (Alexander 1990) (NMNH).

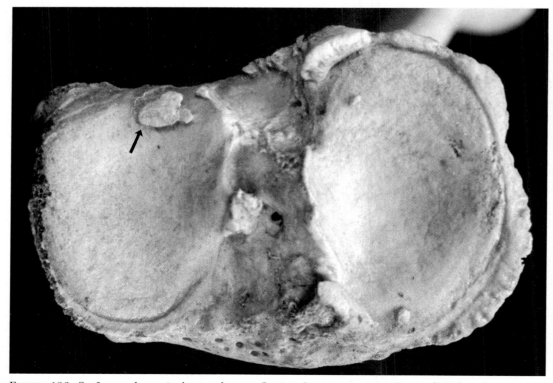

FIGURE 186. Surface and marginal osteophytes reflecting degenerative joint disease (DJD) on the proximal tibia. Mild to moderate forms of DJD are common findings in the elderly.

appear as irregular, raised bony margins resulting from endochondral ossification that obliterate the articular margin (**Figure 186**). This is unlike syndesmophytes which are inflammatory spurs, and enthesophytes which are traction spurs at the insertion site of tendons and ligaments. This is a common finding associated with increasing age. See also Rogers et al. (1997).

Osseous Tumors and Exostoses

Rounded, smooth or irregular bony projections, usually extending perpendicular to the long axis of the diaphysis. These growths may reflect trauma to muscle or tendinous attachment or may be formed by benign ossous tumor (see **Figure 187 and Plates 125b–c and 126** for normal bone and **Figure 188** for abnormal exostosis). Frequent sites of involvement are the metaphyseal areas of the shoulder, knee, and ankle (Spjut et al. 1971). It is an uncommon finding in most archaeological samples. For more concise identification as to the cause of the formation, seek the advice of a radiologist and/or perform histological assessment. See also Lange et al. (1984), Ortner (2003:509–510, Figure 20-8), and Steinbock (1976:319–336).

Septic or Pyogenic Arthritis and Joint Fusion

Many diseases (particularly infectious) as well as congenital factors or trauma (especially fracture) and subsequent remodeling may result in bony union of two or more bones (**Figures 190 and 212**). The bones fuse through the proliferation of callus, osteophytes, or exostoses.

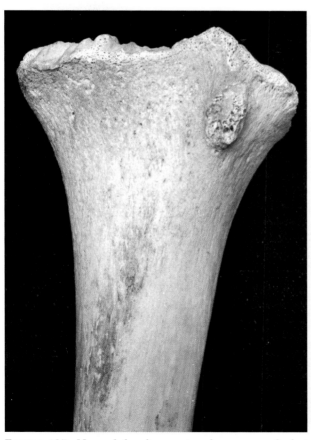

FIGURE 187. Normal development and presence of what some may mistake as a bone spur or osteochondroma in the proximal tibia of a child (NMNH-H).

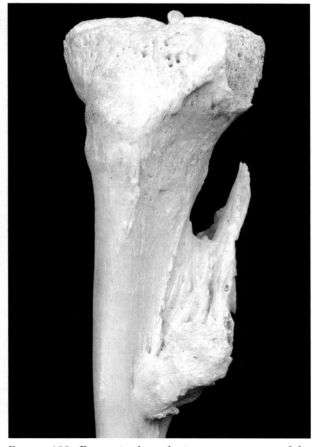

FIGURE 188. Exostosis along the interosseous crest of the tibia (NMNH-H). Cf. Figure 180.

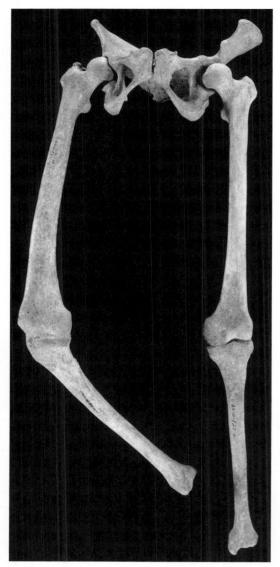

FIGURE 190. Bony ankylosis (fusion), in this example, of the right knee.

Septic or pyogenic arthritis may fuse joints together, and illustration of involved knees are presented in Ortner (2003:255), Webb (1995:167) and in Zimmerman and Kelley (1982:95). Bony ankylosis of the joints of major long bones is uncommon to rare. When ankylosis of a joint is encountered, seek the advice of a specialist for determination of the cause. Ankylosis of the ribs to the vertebrae (AS) and bones of the hands and feet, especially the middle and distal phalanges of the feet, is an uncommon to common finding.

FIGURE 189. Osteomyelitis of a subadult tibia with sequestrum of bone and cloaca. See Ortner (2003:200–203, Figure 9-29) for in-depth discussion of this specimen (NMNH 378243).

Bony ankylosis of joints is rarely the result of osteoarthritis (Lasater and Groer 2000).

Tibial Bowing

Tibiae affected by acquired or congenital syphilis (**Figures 191 and 192**) or some of the other treponematoses (e.g., yaws in tropical areas) may exhibit periostosis (A) along the anterior and middle portion of the shaft, resulting in an anterior enlargement ("bowing" or sabre-shin) of the bone. Sabre-shin tibiae were noted in 9 of 100 (Caribbean) and 11 of 271 (native-born Americans in Boston) patients with late congenital syphilis (Fiumara and Lessell 1983, 1970. Pavithran (1990) has identified sabre-shin features in leprosy. See also Andersen (1991), Aufderheide and Rodriguez-Martin (1998:159–160), Goff (1967), Hackett (1975, 1976), Krogman (1940), Mays (1998), Mays et al. (2003), Ortner (2003:294–296), Palmer and Reeder (2001), Rasool and Govender (1989), and Steinbock (1976:98–158).

Note the presence of new bone (radio-opaque [whitish] area) along the proximal diaphysis of the tibia giving it the characteristic "sabre-shin" or "boomerang" shape.

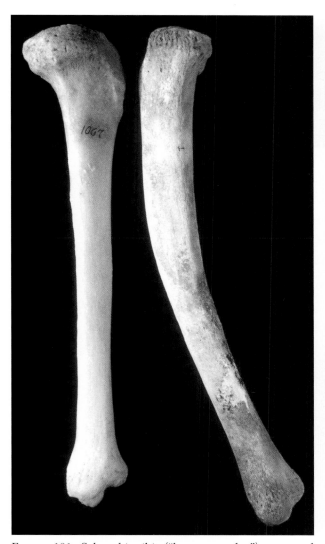

FIGURE 191. Sabre-shin tibia ("boomerang leg") compared to normal tibia.

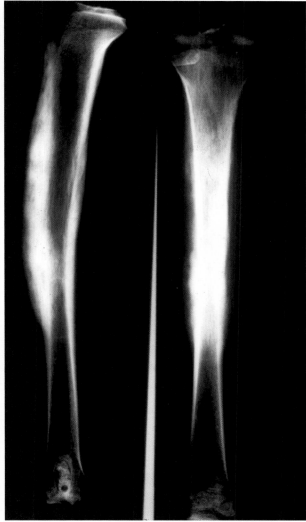

FIGURE 192. Radiograph of syphilitic ("sabre-shin") tibia.

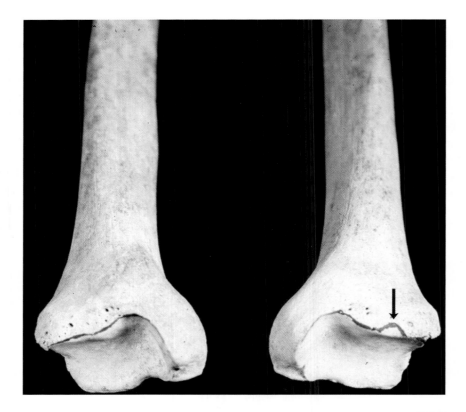

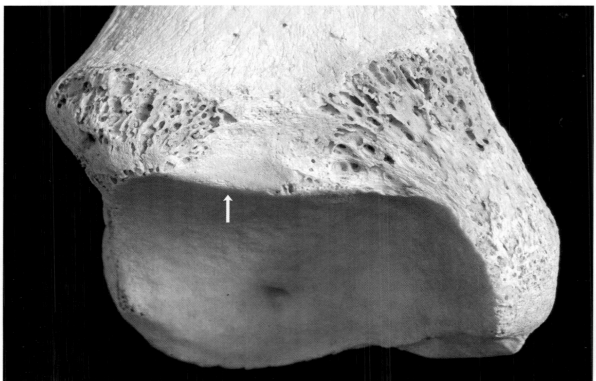

FIGURES 193 (*top*) and 194 (*bottom*). Medial squatting facets of the distal tibia (Figure 194-NMNH 293624).

Squatting Facet

A squatting facet (**Figures 193 and 194**, facing page) is scored as present whenever there is a break in the continuity of the anterior margin of the distal tibia, which should be a straight line. The anterior surface of the tibia abuts the dorsal portion of the talar neck when the knee and foot are hyperdorsiflexed and is usually attributed to the squatting position (Boulle 2001; Jeyasingh et al. 1979; Oygucu et al. 1998; Pandey and Singh 1990). Massada (1991) found a high incidence of squatting facets in soccer players. There may be one or two facets (medial and lateral) and are common findings in most if not all populations. The authors agree with many other researchers that not all squatting facets develop as a direct result of squatting but, possibly, other physical activities such as running and climbing that result in dorsiflexion (upward bending) of the foot toward the knee. Squatting facets may also be present on the talus, calcaneus, distal metatarsals, and proximal tibia. See also Barnett (1954), Baykara et al. (2010), Kate and Robert (1965), Lima (1928), Oygucu et al. (1998), and Ubelaker (1999:103).

Enthesophytes of the Tibial Tuberosity (Osgood-Schlatter's Disease)

Bony spurs are formed by the ossification of the patellar tendon, most often from trauma (Resnick 2002:3589). This outgrowth (**Figure 196**) may also be associated with Osgood-Schlatter's disease, but diagnosis (which is not a radiographic determination) can be tentative (D'Ambrosia and MacDonald 1975). These bony growths are initially formed in subadults when the epiphyses of the tibial tuberosities are fusing, but stress trauma or other injury cause localized inflammation (**Figure 195** – normal metaphysic of the proximal tibial diaphysis). See also Aufderheide and Rodriguez-Martin (1998:85), and Ortner (2003:353–354). See Figure 13-9 for illustration of Osgood-Schlatter's with radiograph.

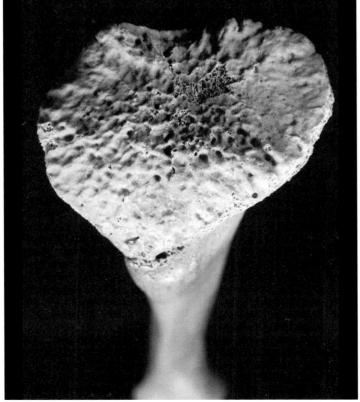

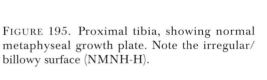

FIGURE 195. Proximal tibia, showing normal metaphyseal growth plate. Note the irregular/ billowy surface (NMNH-H).

FIGURE 196. Enthesophytes of the tibial tuberosity. (Mild cases are commonly encountered in most skeletal groups studied.) (NMNH-H).

FIBULA

Pott's Fracture and Distal Interosseous Enthesophytes

Pott's fracture (not to be confused with Pott's disease), first described by Percival Pott in 1769, is loosely used to refer to fractures and dislocations of the distal tibia and/or fibula occurring 2 to 3 inches above the ankle joint (Apley and Solomon 1988). The type of fracture (**Figure 197**) is dependent on the type and direction of force (e.g., lateral impact) exerted to the lower leg. For example, eversion and external rotation of the lower leg can result in fracture of the lateral or medial malleolus, with rupture of the interosseous ligaments. This tramatic rotation of the ankle (especially by the talus) will separate the fibula from the tibia, tearing the interosseous ligaments, resulting in hemorrhage which can lead to subsequent ossification of these ligaments. In most groups, common finding in adults and rare in children. See Sammarco and Cooper (1998) for fracture classification types using the Lauge-Hansen or Weber system. See also Adelaar (1999), Brown et al. (2003), Hamilton (1984), Jahss (1991), Ortner (2003:158), Resnick and Niwayama (1988), Sisk (1987), and Wilson (1975).

PATELLA

Bipartite Patella

Bipartite patella is a condition in which the patella appears as to have a notch taken out of the superolateral corner (most frequently observed location) at the insertion of the vastus lateralis muscle (**Figure 198**). Halpern and Hewitt (1978) reported a possibly unique case of bilateral medial bipartite patellae. The notch is usually located at the site for attachment of a smaller accessory

FIGURE 197 (*right*). Pott's fracture (Dupuytren fracture) of the distal fibula with exostosis of the interosseous ligaments.

FIGURE 198 (*below*). Bipartite patella (emarginate patella, patellar subdivision, accessory patella, multipartite patella) as though a bite was taken out of it (the ossicles were not recovered with this individual).

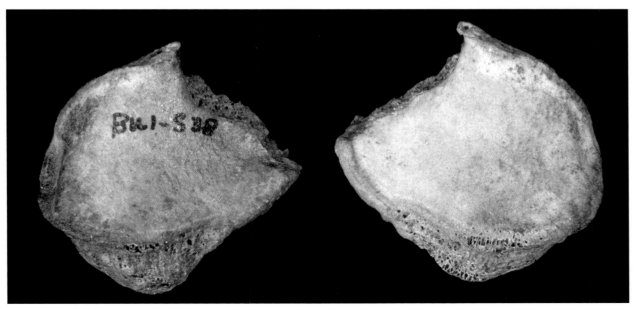

ossicle, the so-called "patellula" in the older literature (Oetteking 1922).

Although the etiology of this condition is unknown, it is commonly believed to represent an accessory center of ossification at the superolateral pole and not the result of direct or indirect trauma to the patella (Green 1975; Halpern and Hewitt 1978; Miles 1975; Ogden et al. 1982). Bipartite patellae occur in approximately 2 percent of the population (George 1935), are more common in males by 9 to 1 (Ogden et al. 1982), and are unilateral in 57 percent of the cases (Weaver 1977). Pain may or may not have been associated with this condition. See Yochum and Rowe (1996) for a good radiographic example of this feature. See also Stanitski (1993).

A review of the literature revealed that some researchers make no distinction between a bipartite patella and vastus notch (E) (see Figure 204). Some early anatomists (Kempson 1902; Todd and McCally 1921) viewed defects of the superolateral border as manifestations of a single anomalous condition (emargination) with a vastus notch representing the mildest form, and a bipartite

patella the most severe form. Most researchers, however, now distinguish between a vastus notch and bipartite patella. A bipartite patella has a porous, roughened central area surrounded by smooth-bordered cortical bone for attachment of an accessory (bipartite) bone. A vastus notch, on the other hand, is a smooth surfaced, depressed or flattened area with no accompanying porosity or accessory bone. Bipartite patella, as opposed to the much rarer tripartite patella, is a common clinical, but uncommon archaeological finding. Vastus notches are very common. See also Adams and Leonard (1925), Callahan (1948), Giles (1928), Jones and Hedrick (1942), Miles (1975), Pecina and Bojanic (2004), Resnick (2002), Saunders (1978), Wright (1903), and Zimmerman and Kelley (1982, p. 36, Figure 19).

Osteoarthritis and Eburnation

The posterior surface of the patella may exhibit a raised rim at any point along its margin (**Figures 199, 200, and 201**). Other indicators of osteoarthritis (OA) are surface osteophytes or

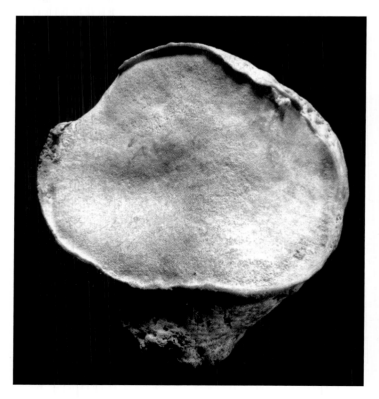

FIGURE 199. Marginal osteoarthritic "lipping" of the patellar facets (NMNH-H).

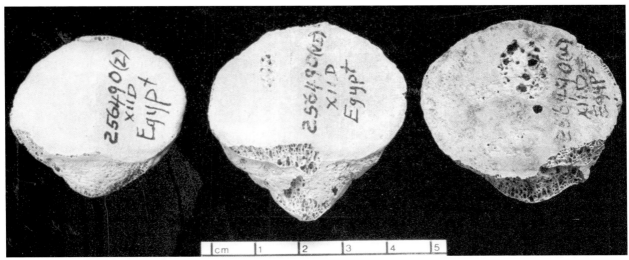

FIGURE 200. "Pitting" type of osteoarthritis on the patellar facets (normal = left, middle = mild/moderate, right = moderate/severe) (NMNH 256490).

FIGURE 201. Eburnated and grooved articular surface due to destruction and loss of articular cartilage at the knee (NMNH-H).

surface porosity and pitting, usually present in the medial facet. OA of the patella is a common finding in the elderly, but is usually first evident in the late thirties (Todd and McCally 1921). OA in young individuals may reflect faulty/soft cartilage known as chondromalacia or shearing stress and cartilage/osseous degeneration due to acute trauma particularly with the knee flexed (Edwards and Bentley 1977). Normal patellae exhibit smooth, sometimes undulating facets for articulation with the distal femur. Large, widely separated and raised oval facets may indicate paralysis and flexion contractures (bent knees) of the legs (**Figure 175**). This feature is uncommon.

Osteoarthritis (OA) can be distinguished from osteochondritic pits and other normal variants, in that pitting reflecting arthritis is usually characterized by irregular, jagged margins.

Note the vertical orientation of grooves (**Figure 201**) from contact with the distal femur (in this case, the femur exhibited similar grooves). Also note the presence of marginal osteophytes (a mild to moderate degree of severity inconsistent to the severity of the eburnation) encircling the patella (**Figure 202**).

Enthesophytes (see Chapter X)

Elongation and projection of the inferior border of the patella resulting from pulling stresses at the inferior ossification center of the immature patella by the patellar tendon (Smillie 1962). It is an uncommon finding. Exuberant bone seemingly attached to and flowing across the anterior surface of the patella (**Figures 202 and 203 and Plates 131a–b and 132**). Small bony projections (enthesophytes) on the superoanterior face of the patella are common findings in most populations. However, development (See **Figures 202 and 203**) is more likely to be seen in the elderly, typically beyond 60 years of age and likely reflects DISH. Usually bilateral and symmetrical in appearance. If there is spiking of only the superoanterior surface of the patella the condition reflects enthesopathy, not ossification of the quadriceps tendon or osteoarthritis. See also Dieppe et al. (1986), and Resnick and Niwayama (1988).

Vastus Notch

This non-metric variant appears as notch or depression in either the superolateral (commonly) or superomedial corner of the patella (**Figure 204**). The vastus notch can be distinguished from the bipartite patella in that the surface of the vastus notch is smooth and nonporous. The shape of the notch varies from nearly flat to concave. It is a common finding. See also Finnegan and Faust (1974).

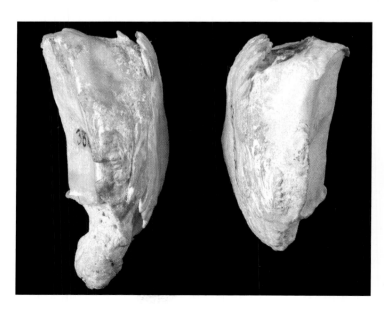

FIGURE 202. Elongated inferior pole (*left*) and enthesophytes (ossification) at the insertion of the quadriceps femoris tendon.

bone fragment of a calcaneus secundarius (**Figure 210**) is rarely recovered with archaeological remains and often misidentified as a sesamoid bone of the hand or foot, the notch at the anterior aspect of the calcaneous will be visible. The notch will be concave, roughened, and porous. The presence of porosity serves to distinguish a calcaneus secundarius from normal variation in the shape of the anterior calcaneal facet. Research by one of the authors (RWM) revealed that this trait is probably present in all populations with a frequency of between 1.4 and 6.0 percent (Mann 1989, 1990). Care must be exercised not to confuse this normal detached ossicle with a fracture and loose bone (Hodge 1999). See also Anderson, (1988), Ceroni et al. (2006), and Kalantari et al. (2007).

Talocalcaneal Coalition

Osseous or cartilaginous coalition of the talus and calcaneus (**Figure 211**), usually of the middle facet of the sustentacumlum tali, is an inherited embryonic anomaly resulting from a failure of primitive mesenchyme to segment in the fetus (Kawashima and Uhthoff 1990; Richardson 2000). Talocalcaneal middle facet and calcaneo-navicular coalitions are the most common of the tarsal coalitions (Percy and Mann 1988). Tarsal coalition has been likened to a "spot weld" of a joint and is found more often in males. This coalition is often associated with talar beaking and is found bilaterally in 50 percent of cases. There appears to be no population specificity and is asymptomatic in most adults, or accompanied by minor pain and stiffness. See also Beckly et al. (1975), Bhalaik et al. (2002), Bohne (2001), and Solomon et al. (2003).

Ankylosing Spondylitis and Spondyloarthropathies

This condition (**Figure 213**) results from a combination of genetics, repetitive biomechanical

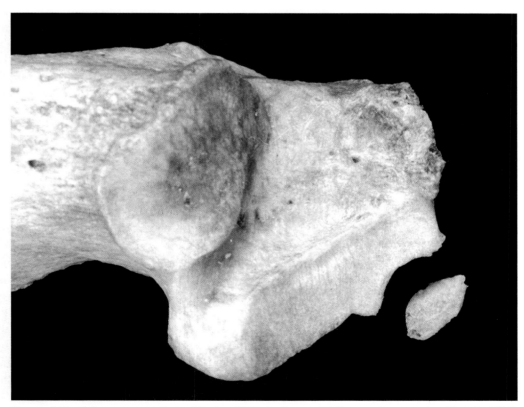

FIGURE 210. Calcaneus secundarius (Gruber 1871; Anderson 1988), secondary os calcis (Dwight 1906).

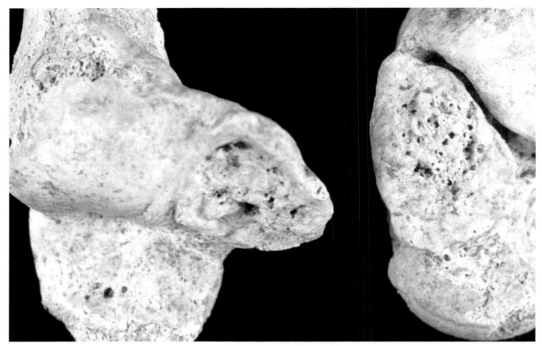

FIGURE 211. Talocalcaneal coalition.

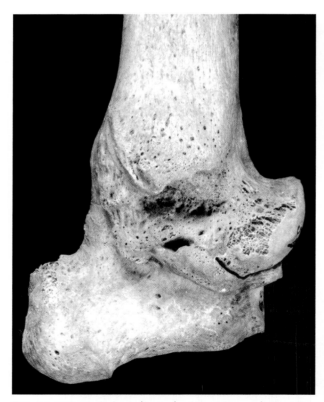

FIGURE 212. Ankylosis (fusion) of the tibia with the talus and calcaneus (AFIP).

stress, and an immune response to bacterial products leading to inflammatory changes at specific anatomical sites and joints. The spondyloarthropathies include psoriatic arthritis and ankylosing spondylitis, among others. See also Bernhard (2002), Deesomchok and Tumrasvin (1993), Luong and Salonen (2000), and Steinbock (1976:304–312).

Leprosy

Note the destruction (acroosteolysis) of the metatarsals and phalanges in leprosy. Many of the foot bones become "penciled," cupped and pegged (metatarsals and phalanges), deformed, and misaligned (**Figure 214 and Plate 140**). The nutrient foramina, especially those of the phalanges, may be enlarged. Frostbite and rheumatoid arthritis, however, can result in similar deformities and disintegration. The tibia and fibula may also become involved and exhibit bilateral symmetrical inflammatory changes including pitting and marked periosteal new bone formation (Manchester 1989). If this condition is encountered, seek the advice of a specialist. See

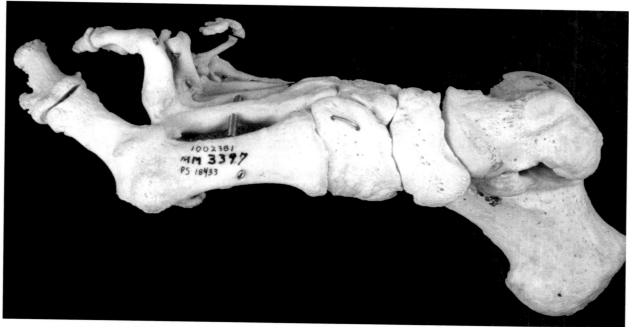

FIGURE 213. Seronegative spondyloarthropy of the foot (Mann et al. 1990) (AFIP MM 3397).

also Andersen (1969, 1991), Anderson (1982), Aufderheide and Rodriguez-Martin (1998:141–154) (epidemiology and geographic distribution as well as lesions and their affects on the bone are discussed), Boldsen (2001, 2008), Burgener and Kormano (1991), Harverson and Warren (1979), Kulkarni and Mehta (1983), Møller-Christensen (1953, 1961, 1983. See Ortner (2003:263–271) for an overview of skeletal lesions and paleopathological review. See also Paterson (1965), Paterson and Job (1964), Patil (2000), Steinbock (1976:192–212), and Wastie (1975).

Foot Binding

Note severe loss of bone (osteoporosis/osteopenia) and angle of the foot as a result of many years of being wrapped/bound ("foot binding") by pulling the fore foot toward the hind foot (**Figure 215 and Plates 134 and 135**). This process, practiced by the ancient Chinese from the tenth century until 1911, often resulted in damage to the foot, gangrene, and sometimes death. The gradual process of foot binding, which greatly altered the woman's gait and activities,

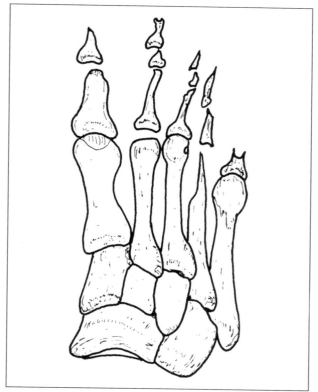

FIGURE 214. Leprosy of the feet (the talus and calcaneus not shown). See Plate 140 for a photographic example and Ortner (2008).

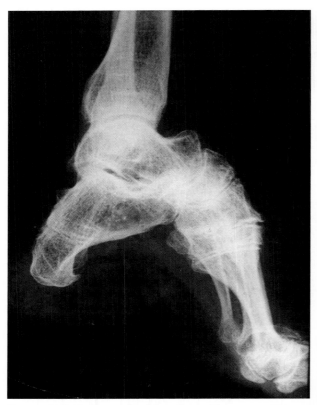

FIGURE 215. Radiograph of the effects of long-term Chinese foot binding (Mann et al. 1990). See Plates 134 and 135 for examples.

was begun at a very young age and progressed until the soles of the feet were highly arched (pes cavus, concave), the calcaneus nearly vertical and the feet were shortened, usually by about age three. As a result, the highly decorated shoes ("lotus" – see **Plate 135** for a soft tissue example) for such bound feet were very small (2–3 inches long). Throughout the process the feet were gradually made smaller by tighter bindings and the shoes were decreased in size until achieving the desired size. Richardson (2009) reported on a 99-year-old Chinese woman who sought medical care due to a fall. Although she had her feet bound as a child, they were unbound at the time of the fall and examination, they exhibited marked pes cavus, nearly vertical calcaneus, foreshortening of the feet and osteoarthritis. Not unexpectedly, binding had permanently altered the shape of her foot bones and her feet maintained the "bound" appearance, despite being

unbound for many years. See also Mann et al. (1990). See Ortner (2003:164, Figure 8-69) for articulated foot deformed by foot binding.

Enlarged Nutrient Foramen

Enlarged nutrient foramina (**Figure 216**) are often seen on adjoining metatarsals and may be symmetrical. Fink et al. (1984) reported that normal foramina can attain a size of 1 mm. Although the etiology of enlarged foramina in the foot is unknown, such a condition in the phalanges of the hand has been found in association with thalassemia major (Poznanski 1974), Gaucher's disease (Fink et al. 1984) and leprosy (Møller-Christensen 1967). It is possible that increased blood flow to the extremities is responsible for such enlargement.

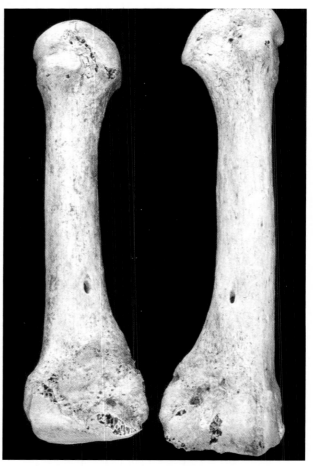

FIGURE 216. Enlarged nutrient foramina in metatarsals.

Ankylosis of Joint

Bony ankylosis of any joint (**Figures 212 and 217**) may be the result of many conditions and variables, including: infectious disease, developmental or congenital birth defects, endocrinological disturbances or trauma. An accurate interpretation for the most probable cause for the ankylosis must be based on the overall evaluation of the associating skeleton, since it may be a combination of causes to produce the fusion. As an example of indirect causation, Balen and Helms (2001) report ankylosis of the foot as the result of thermal and electrical burns. One of the most common forms of osseous fusion is distal symphalangism of the toes (pedal) resulting in fusion of the middle and distal digits. A study of 488 Japanese feet revealed symphalangism of the fifth toe to be 72.5 percent and 11.9 percent in the fourth toe, with no significant side or sex differences (Nakashima et al. 1995).

Talotibial Osteophytes

Numerous studies have shown that overuse or overstressing activities such as soccer, American football, dancing, and kicking may result in the formation of talotibial osteophytes longer than 1 mm. Tol et al. (2002) investigated hyperplantar flexion in soccer players, measured the magnitude of impact force with the soccer ball, and noted the location of osteophytes in 15 elite soccer players. The authors concluded that ankle impingement syndrome is related to recurrent ball impact, a repetitive microtrauma to the anteromedial aspect of the ankle (Eisele and

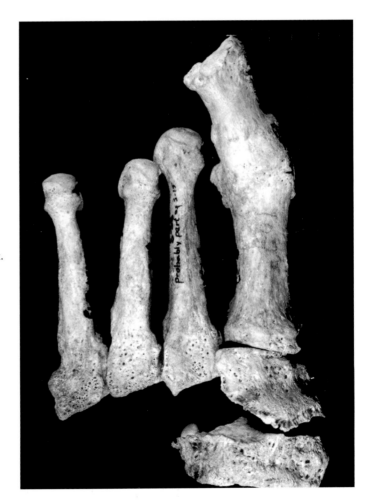

FIGURE 217. Fused left first metatarsal and phalanx.

FIGURE 218. Osteochondritis dissecans of the talar dome (Canale and Belding 1980; Pecina and Bojanic 2004).

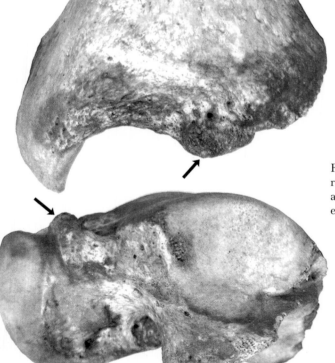

FIGURE 219. Anterior ankle impingement syndrome resulting in exostoses on the talar neck (*arrow*) and/or anterior tibia (*arrow*) (also known as "soccer player's exostosis" or "footballer's ankle").

Sammarco 1993). Other authors have attributed osseous remodeling where the talus abuts with the tibia (**Figure 219**, facing page) resulting in the formation of either a squatting facet or osteophyte (along the talus and/or tibia) (Massada 1991). Berberian et al. (2001) found that anterior talotibial osteophytes have a specific pattern of formation and location such that the talar spur (osteophyte) occurs on the medial aspect of the talar neck, while the tibial spur forms lateral to the midline. The best way to discern osteophytes along the anterior rim of the tibia and/or talar neck is to articulate the two bones such that they abut, and look for alterations in the two surfaces where they come into contact with one another. An os trigonum may also be present in individuals with footballer's ankle (talotibial osteophytes) (Masciocchi et al. 1998). See also Haverstock (2001), Hawkins (1988), Lepow and Cafiero (1980), Myers (1987), Parkes et al. (1980), Spitz and Newberg (2002), Stoller et al. (1984), Umans

(2002), Van Dijk et al. (2002), Vincelette et al. (1972), and Zhang et al. (2002).

Osteophytes and Resorption in the Distal Phalanx

Distal phalanx osteophytes are common in older ages (**Figure 220**). Tuft resorption in either hand or foot phalanges may reflect psoriatic arthritis (Battistone et al. 1999; Burgener and Kormano 1991). Although rare in most skeletal samples, "spade-like" terminal hand phalanges may be associated with acromegaly (Taylor and Resnick 2000).

Steida's Process (Variant of Os trigonum)

If this process (**Figure 221**) is completely separated, the variant is referred to as os trigonum (**Figure 207**). The corresponding irregularly shaped loose bone is commonly overlooked, misidentified, or not recovered with the remains.

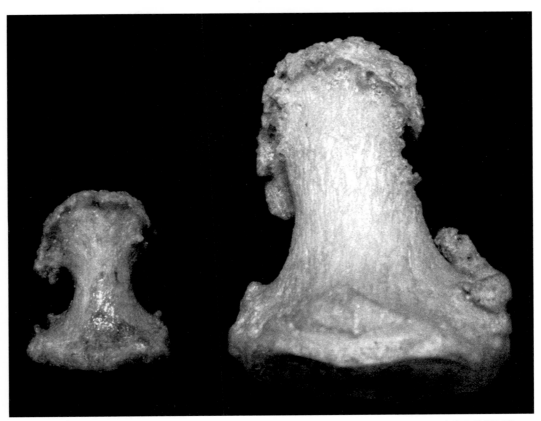

FIGURE 220. Tufting and osteoarthritis of the distal phalanges (in this case, the foot) (NMNH-H).

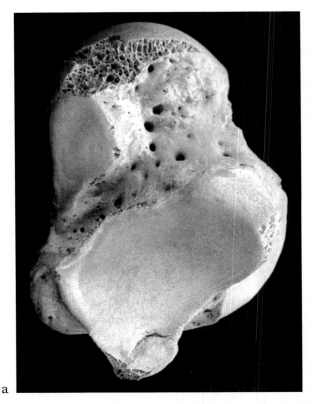

a

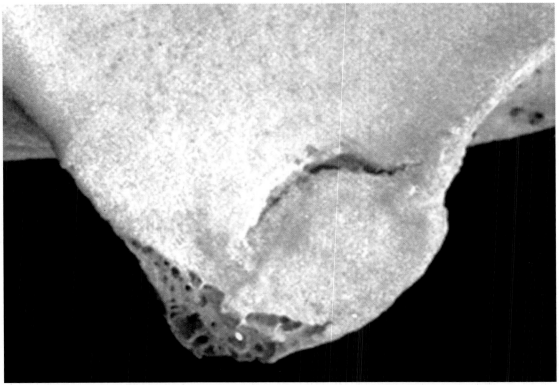

b

FIGURE 221. Steida's process of the talus (incomplete separation of the posterior talar process).

Chapter VI

FUNGAL INFECTIONS

Fungi are plant-like organisms that, for the most part, are not harmful to man. In fact, they play an important role in the decay of organic matter. Only a small group of fungi and fungi-like organisms are pathogenic (disease causing) to humans. The crucial factor is the patient's immune response.

Bone lesions in many fungal infections are similar in appearance (e.g., blastomycosis, coccidioidomycosis, and cryptococcosis). Most infections invade the bone by direct extension from a soft tissue lesion. The fungi then burrow into the bone producing cratered or scooped-out lesions that appear as small (0.3 to 1.5 cm), round or ovoid depressions with thin, sharp margins. The lesions tend to be clustered and confluent with minimal periosteal reactive bone. In older lesions the bone may exhibit marked thickening, small spicular bony projections, and shallow craters. See also Aufderheide and Rodriguez-Martin (1998), Colonna and Gucker (1944), Jones and Martin (1941), and Reeves and Pederson (1954).

It is extremely difficult and often impossible to distinguish one form of mycosis from another without soft tissue and a patient's history. Differential diagnosis in mycotic infection depends largely on the patterning of skeletal involvement and fungi associated with the geographic region where the skeleton was recovered (**Figure 222 and Table 1**).

The following are a few of the more commonly encountered fungal organisms or organisms that produce a skeletal response similar to fungi:

Actinomycosis (lumpy jaw, leptothricosis, streptotricosis) – a chronic granulomatous infection caused by the fungal-like ("higher" bacterial) *Actinomyces israelii.* Actinomycosis has a worldwide distribution. There are three anatomical sites of primary infection: cervicofacial (most common), thoracic, and abdominal. The skeletal area affected depends on where the organism enters the body. The lesion will almost always start as a periosteal reaction (periostosis) and new bone formation with eventual erosion of the underlying cortex. Although any bone is susceptible, the mandible and vertebrae are most commonly affected. Actinomycosis of the spine has often been incorrectly identified as tuberculosis (Simpson and McIntosh 1927). However, unlike tuberculosis, actinomycosis frequently results in erosion of the articular facets, laminae, transverse and spinous processes. See also Goldsand (1989), and Ortner and Putschar (1985).

Blastomycosis (North American blastomycosis, Gilchrist's disease, Chicago disease). A chronic granulomatous fungal infection that starts as a pulmonary or cutaneous inflammation that often disseminates to a systemic disease. Blastomycosis is as soilborne fungus that grows on decaying material and is most commonly found in the southeastern and central United States (Mississippi and Ohio River basins). Males have a higher incidence than females with the average age of onset between 20 and 50 years (Jaffe 1975). Twenty to 50 percent of people with blastomycosis develop skeletal lesions, and systemic blastomycosis is fatal 90 percent of the time (Colonna and Gucker 1944; Reeves and Pederson 1954). While all bones are susceptible, the most common sites are the vertebrae, skull, ribs, tibiae, tarsals, and carpals (Jones and Martin 1941). The differential

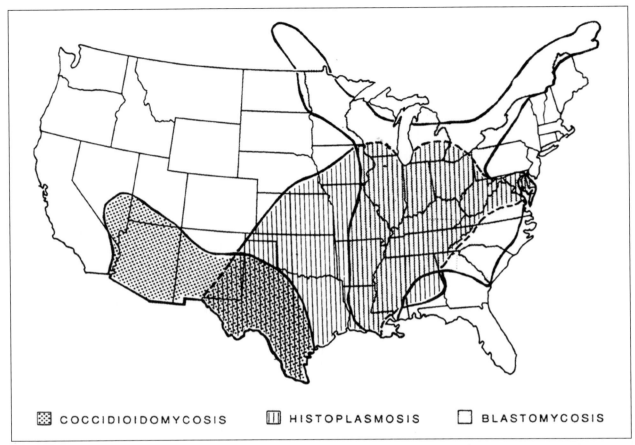

FIGURE 222. Geographic distribution of fungal infection in the continental United States (from Rippon 1988; Youmans et al. 1980).

diagnosis between blastomycosis and tuberculosis is that blastomycosis is more diffuse in the skeleton, will often involve several vertebrae and will affect the dorsal as well as the ventral portions.

Table 1
SELECTED DISEASES THAT PRODUCE BONY
REACTIONS SIMILAR TO MYCOTIC AND
ACTINOMYCOTIC INFECTION

Bones Involved	Differential Diagnosis
Vertebrae	Tuberculosis, brucellosis, nonspecific osteomyelitis, metastatic carcinoma
Long bones	Nonspecific osteomyelitis, tuberculosis, treponematosis, various neoplasms
Skull	Treponematosis, tuberculosis, multiple myeloma, metastatic carcinoma, eosinophilic granuloma, histiocytosis X, osteomyelitis

See also Aufderheide and Rodriguez-Martin (1998:214–215), Colonna and Gucker (1944), Kelley and Eisenberg (1987), Martin and Smith (1939), Moore and Green (1982), Ortner (2003: 326), Reeves and Pederson (1954), and Zimmerman and Kelley (1982:89).

Coccidioidomycosis (coccidiomycosis, Posadas disease, valley fever, desert rheumatism, San Joaquin fever). Caused by the inhalation of soil-borne fungal spores, this disease is generally a benign inflammation of the respiratory tract. Only a small number of people (usually immune deficient) will contract the disease (Fiese 1958). Coccidioidomycosis is endemic to a well-defined area of southwestern Texas, Mexico, and Central and South America (Binford and Conner 1976). All bones are potential sites with the ribs, vertebrae, skull, and bones of the extremities favored (Fiese 1958).

Periosteal new bone formation and sequestra are uncommon. The lesions tend to be lytic and favor the cancellous areas of bone. See also Aufderheide and Rodriguez-Martin (1998:215–217), Bried and Galgiani (1986), Hoeprich (1989a–b), Ortner (2003:326–327), Zeppa et al. (1996), and Zimmerman and Kelley (1982:89–90).

Histoplasmosis (Darling's disease, cave disease, Ohio Valley disease, Tingo Maria fever). A common pulmonary infection with worldwide distribution (Binford and Conner 1976). The fungus grows in soil (particularly soil rich in bird feces) and enters the body through the inhalation of spores.

Histoplasmosis is endemic in the eastern and central United States. See also Aufderheide and Rodriguez-Martin (1998:217–218), Ortner (2003:327), and Youmans et al. (1980).

For further information refer to Attah and Cerruti (1979), Aufderheide and Rodriguez-Martin (1998:212–222), Brook et al. (1977), Caraveo et al. (1977), Dalinka et al. (1971), Daveny and Ross (1969), Echols et al. (1979), Greer (1962), Ortner (2003:325–332), Poswall (1976), Procknow and Loosli (1958), Reeves and Pederson (1954), Seligsohn et al. (1977), Wheat (1989), and Zimmerman and Kelley (1989:88–91).

Chapter VII

TREPONEMATOSIS (SYPHILIS)

Treponematosis is a chronic infection caused by a corkscrew-shaped organism (spirochete) of the genus Treponema (meaning "twisted thread"), most commonly called syphilis. Actually, due to the clinical and geographical variation, the infection is divided into four types: pinta, yaws, endemic syphilis, and venereal (and congenital) syphilis (lues or leutic disease) (Table 2). With the variations of the disease and the wide range of its expression in the skeleton, if treponematosis is suspected in an individual, seek the advice of at least one paleopathologist familiar with the disease. Diagnosing bone as syphilitic (from the other forms) is difficult due to the similarity in the skeletal response to the infectious agent as is seen in nonspecific osteomyelitis and other allied infectious diseases. Syphilis is often called the "great imitator" (Thomas 1985). The information about treponematoses offered below should be considered only as an introduction to this variable disease and is not meant to serve as a specific tool for diagnosis, especially in atypical cases.

Pinta (*Treponema carateum*) is the least destructive and most benign form of treponematosis. Pinta (Spanish for "spot," "dot" or "mark") is endemic only to American tropics and is most frequently found in Mexico and parts of Central and South America. Pinta is spread through direct nonsexual contact that initially appears as a skin rash, undergoes pigment changes, and spreads (secondary pinta). The tertiary (final) stage is usually marked by a general depigmentation of the lesions. Pinta is the only treponemal infection that does not cause bone lesions.

Yaws (*Treponema pertenue*), like pinta, is a nonvenereal endemic juvenile disease (Hudson 1940) that is spread through close contact with an infected sore (e.g., children playing). Yaws is found worldwide in hot and humid tropical regions, usually begins in childhood, and starts as a painful localized lesion/sore on an exposed extremity (Binford and Connor 1976). If the sore is located near bone it will induce periostosis (polydactylitis is a frequent finding in early stage yaws) (Hoeprich 1989). Bone lesions in yaws are rare (ca. 15%) and generally considered indistinguishable from other treponemal infections (Hackett 1963, 1976, 1983). One of the most consistent findings in yaws is anterior bone hypertrophy of the tibia ("boomerang leg"); only rarely does the fibula show this feature. In the late stages, gummatous periostosis and osteomyelitis develop which are almost identical (although not as severe) to tertiary syphilis. Synonyms are bouba, frambesia, tropica, parangi, and pian (Thomas 1985).

Endemic syphilis (treponarid, bejel) is a nonvenereal infection found in warm, arid to semi-arid environment (e.g., Africa, the Middle East, and Asia). Endemic syphilis is spread through close contact with infected persons (e.g., kissing) or contaminated objects (e.g., drinking utensils). Binford and Connor (1976) propose that the clinical symptoms of endemic syphilis and yaws are so similar that the infectious agent should be classified as Treponema pertenue (yaws) instead of *Treponema pallidum* (venereal and endemic syphilis). Steinbock (1976) suggests that the endemic form is an intermediate manifestation between yaws and venereal syphilis. Hudson (1958) writes "the standard of living and the level of hygiene in a given community determines whether its syphilis will be venereal or non-venereal."

Table 2
OSSEOUS CHANGES PRESENT IN PATIENTS
WITH LATE CONGENITAL SYPHILIS
(Fiumara and Lessell 1970, 1983)

Stigmata	*Boston (1960–1969) n/%*	*Caribbean (1975–1981) n/%*
Hutchinson teeth	171/63.1*	54/100**
Mulberry molars	176/64.9*	71/100**
Frontal bossae	235/86.7	96/96
Short maxillae	227/83.3	100/100
Sternoclavicular thickening	107/39.4	81/81
Relative protuberance of the mandible ("Bull-dog-Jaw")	70/25.8	84/84
High palatal arch	207/76.4	79/79
Saddle nose	199/73.4	92/92
Sabre shin	11/4.1	9/9
Scaphoid scapulae	2/0/7	3/3

*Not a true picture of the incidence because many patients had had their teeth extracted or were edentulous.
**The remainder were edentulous.

Regardless, bone lesions are uncommon and are difficult to distinguish from yaws and venereal syphilis. Research by Donald J. Ortner suggests the development of joint lesions in some cases of yaws that are not seen in syphilis. (See Ortner 2003:273–319 for a review of the literature, discussion of its evolutionary history, and the paleopathological evidence of treponematosis.).

Venereal syphilis (*Treponema pallidum*, "pale twisted thread") is transmitted through sexual contact. Venereal syphilis is found worldwide and is the only treponemal infection that can be passed from mother to fetus (congenital syphilis). The venereal form, also known as acquired syphilis, uncommonly involves bone (10 to 20%), usually in the tertiary stage of the disease, and two to ten years after the initial infection (Ortner and Putschar 1985; Rudolph 1989; Steinbock

1976). The bones most commonly affected are the tibia and skull, although any bone(s) may be involved (Jaffe 1975).

In the skull (**Figure 223**), syphilis almost always first attacks the outer table, unlike tuberculosis and some neoplasms that start in the diplöe and work outwards (**Figures 184, 191 and 192;** see **Plate 5 for a color example**). The most characteristic cranial lesion is the pattern of scarring seen on the frontal and parietals (occasionally affects the occipital and temporals), called caries sicca (pronounced "sick-uh"). There are three types of lesions that together form caries sicca: stellate scarring, nodes, and cavitations (Hackett 1976). Cavitations appear as areas of depressed bone that usually do not perforate the inner table. The walls and rims of the depressions are comprised of smooth sclerotic bone (nodes) and stellate (star-like) scars, also known as radial scars. These scars appear as smooth-rimmed furrows that radiate from a central point. Caries sicca represents a healed stage of the disease and will remain visible throughout life. Active areas of early inflammation may be difficult to differentiate from other infections such as tuberculosis, soteomyelitis, neoplasm, and others. See Webb (1995) for treponemal infections in Australian aboriginal populations. Zimmerman and Kelley (1982:101–102) present an example of tertiary syphilis from the Mutter Museum.

For further information refer to Baker and Armelagos (1988) (historical overview), Hackett (1975, 1976, 1981), Hudson (1940, 1958), Murray et al. (1956), Ortner (2003:273–319) (for historical and literature as well as paleopathological references), Ortner and Putschar (1985), Stewart and Spoehr (1952), Steinbock (1976:86–169) (for review and paleopathological references), Suzuki (1984), Turk (1995), and Zimmerman and Kelley (1982:96–102).

a

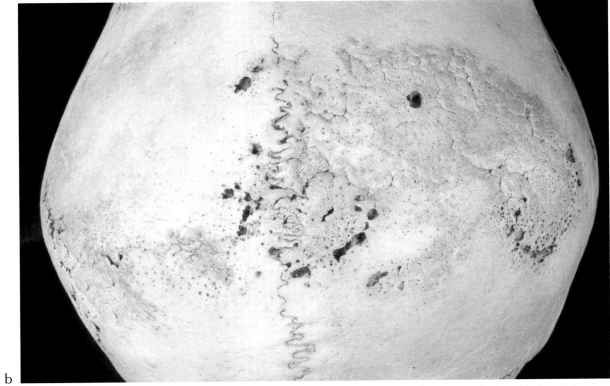

b

FIGURE 223. Probable early stage lesions reflecting treponematosis (syphilis?) in the skull of an Alaskan Eskimo.

Chapter VIII

TUMORS

A tumor, or neoplasm, is new growth of tissue that is uncoordinated with the normal tissue and may interfere with normal physiology (**Tables 3–4**) (Madewell et al. 1981a-b). Tumors may be either benign or malignant. Benign tumors are usually slow growing, localized, noninvasive, and usually present no serious threat to the life (e.g., osteoma) (Bushan et al. 1987). Malignant tumors, on the other hand, grow rapidly, spread (metastasize) to other tissue, and often eventually result in death of the individual (e.g., osteosarcoma). Invasion of other tissue by metastasis can occur either through the lymphatic system, bloodstream, or by direct extension of the tumor. Not all tumors fall clearly into these two categories. Giant cell tumors, for example, are generally considered benign but up to 30 percent develop malignant tendencies (Hutter et al. 1962; Madewell et al. 1981a-b).

Table 3
TUMOR AND TUMOR-LIKE CONDITIONS OF BONE BY REGION

	Benign	Malignant	Blastic	Lytic	Location
LONG BONES					
Adamantinoma		X		X	D*
Aneurysmal Bone Cyst	X			X	M
Chondroblastoma	X			X	E/M
Chondrama (enchondroma)	X			X	D/M
Chondrosarcoma		X	X	X	M
Eosinophilic Granuloma	X		X	X	M/D
Ewing's Sarcoma		X	X	X	D/M
Fibrosarcoma		X		X	M
Fribrous Dysplasia	X		X	X	D/M
Fibrous Cortical Defect	X			X	M
Giant Cell Tumor	X			X	E/M
Metastatic Carcinoma		X	X	X	M/D
Multiple Myeloma		X		X	M/D
Non-ossifying Fibroma	X			X	M
Osteoid Osteoma	X		X	X	D
Osteoblastoma	X			X	D/M
Osteochondroma	X		X		M
Osteosarcoma		X	X	X	M
Solitary Bone Cyst	X			X	M

continued

Table 3 – *continued*

	Benign	Malignant	Blastic	Lytic	Location
RIB/STERNUM					
Aneurysmal Bone Cyst	X			X	Both
Chondroblastoma	X			X	Ribs
Chondrosarcoma		X	X	X	Both
Fibrous Dysplasia	X		X	X	Ribs
Metastatic Carcinoma		X	X	X	Both
Multiple Myeloma	X	X	Both		
Osteoid Osteoma	X		X	X	Ribs
Osteosarcoma		X	X	X	Both
PELVIC/PECTORAL					
Aneurysmal Bone Cyst	X			X	All
Chondroblastoma	X			X	All
Chondrosarcoma		X	X	X	P/S
Metastatic Carcinoma		X	X	X	All
Multiple Myeloma		X		X	P
Osteoid Osteoma	X		X	X	P/S
Osteosarcoma		X	X	X	All
SPINE					
Aneurysmal Bone Cyst	X			X	All
Hemangioma	X			X	T/L
Metastatic Carcinoma		X	X	X	All
Multiple Myeloma		X		X	All
Osteoblastoma	X			X	All
SKULL					
Eosinophilic Granuloma	X		X	X	Calv.
Fibrous Dysplasia	X		X	X	All
Hemangioma	X			X	Calv.
Metastatic Carcinoma		X	X	X	All
Multiple Myeloma		X		X	Calv.
Osteoblastoma	X			X	All
Osteoma	X		X		Calv.
HAND/FOOT					
Chondroblastoma	X			X	Tars.
Chondroma (enchondroma)	X			X	D
Metastatic Carcinoma		X	X	X	All
Osteoblastoma	X			X	All
Osteochondroma	X		X		All
Osteoid Osteoma	X		X	X	All

*Diaphysis; E = epiphysis; M = metaphysis; P = pelvis; S = scapula; T = thoracic; L = lumbar.

Table 4
TYPICAL AGE RANGE OF THE OCCURRENCE OF BONE TUMORS

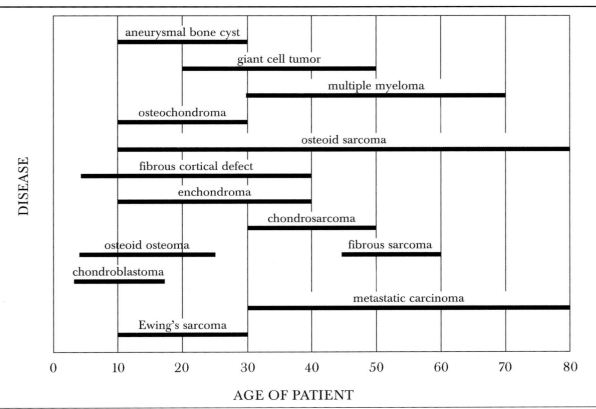

(Reprinted with permission from Griffiths, *Basic Bone Radiology*, Appleton-Century-Crofts, Norwalk, CT, 1981, as modified.)

Tumors are classified according to the tissue of origin or differentiation such as bone or cartilage. The prefix usually indicates the tissue of origin (e.g., osteo = bone, chondro = cartilage). However, not all tumors that affect bone arise from bone and cartilage. For example, liposarcoma (from fat), leiomysarcoma (smooth muscle), fibroma (fibrous tissue), hemangioma (vascular tissue), and neurofibroma (nerve tissue) to name a few. See also Ly et al. (2003).

It should be remembered that differentiating one type of tumor from another disease condition is extremely difficult (**Figure 224**). This task is made even more difficult when attempting a diagnosis in dry bone without benefit of patient history, soft tissues, or the entire skeleton. For example, was there pain or soft-tissue swelling associated with the tumor? Did the individual sustain trauma at the tumor site? Which bones of

the skeleton were affected? All of these considerations must be taken into account if a correct interpretation is to be made. Even with a well-documented patient history an accurate diagnosis is often only narrowed down to a few possibilities that best fit the clinical "picture." If a neoplasm is suspected, first radiograph the bone(s) (refer to **Table 3 and 4 and Figure 224**) and seek the assistance of a qualified paleopathologist, orthopaedic radiologist, or orthopaedic pathologist for the final interpretation.

For other discussion on clinical and paleopathological cases, see Aufderheide and Rodriguez-Martin (1998:371–392), Brothwell and Sandison (1967:320–345), Coley (1949), Dahlin (1967), Grupe (1988), Hutter et al. (1962), Jaffe (1958), Manchester (1983), Ortner (2003), Steinbock (1976:316–401), Suzuki (1987), and Tkocz and Bierring (1984).

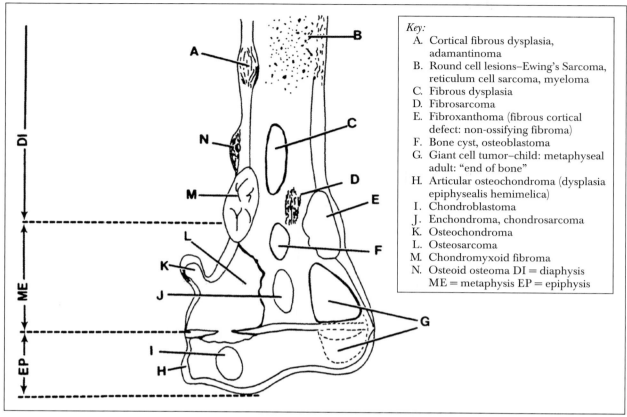

FIGURE 224. Composite diagram illustrating frequent sites of bone tumors. The diagram depicts the end of a long bone that has been divided into the epiphysis, metaphysis, and diaphysis. The typical sites of common primary bone tumors are labeled. *Adapted from Madewell, J.E. et al.: Radiologic and Pathologic Analysis of Solitary Bone Lesions: Part I Internal Margins. *The Radiologic Clinics of North America, 19*(4), W. B. Saunders Company, Philadelphia, Pennsylvania, 1981.

Key:
A. Cortical fibrous dysplasia, adamantinoma
B. Round cell lesions–Ewing's Sarcoma, reticulum cell sarcoma, myeloma
C. Fibrous dysplasia
D. Fibrosarcoma
E. Fibroxanthoma (fibrous cortical defect: non-ossifying fibroma)
F. Bone cyst, osteoblastoma
G. Giant cell tumor–child: metaphyseal adult: "end of bone"
H. Articular osteochondroma (dysplasia epiphysealis hemimelica)
I. Chondroblastoma
J. Enchondroma, chondrosarcoma
K. Osteochondroma
L. Osteosarcoma
M. Chondromyxoid fibroma
N. Osteoid osteoma DI = diaphysis ME = metaphysis EP = epiphysis

Chapter IX

PERIMORTEM VERSUS
POSTMORTEM FRACTURES

Forensic pathologists refer to soft tissue injuries usually as antemortem (inflicted during life) (**Figures 227, 228, and 229**), perimortem (at or near the time of death), or postmortem (occurring after death). The degree of tissue reaction and bleeding are the primary indicators. Occasionally wounds fall in between, inflicted while the individual was in the act of dying, and this is termed *perimortem.* Ascertaining/estimating whether an injury occurred before or after death has significant medicolegal importance and highlights the difficulty of making such a determination. For soft tissues, this period is rather short and precise; in dry bones this term has a slightly different meaning and covers a time that is indistinguishable from antemortem and may persist until weeks after death. The reason for this is twofold. First, many times there is no soft tissue left to aid ante-, peri-, or postmortem determination. Second, living bone is a viscoelastic material with special fracture properties that do not disappear until the elastic (collagen) component deteriorates after death, and this may take weeks or months. Thus, in the absence of soft tissue evidence, the perimortem period loses its precision and must include recent antemortem events and an obscure postmortem period during which the bone retains its "fresh" visco-elastic properties.

Perimortem fractures (**Figures 225 and 226**) usually leave sharp, smooth, often beveled fracture lines. Most of the bone at the site of fracture will be present. Fractures of fresh bone often have associated radiating and/or concentric fracture lines at the site of trauma. The fractured ends are usually the same color and as discolored as the adjacent surface bone (old, dirty appearing). Displaced, curved but still adhering bone fragments or splinters or incomplete fractures with bending of the bone (so-called "greenstick" fractures), and dirt within the breaks are also associated with perimortem trauma. But, it is not always possible to distinguish perimortem fractures from postmortem. Further, because the breakage may have occurred weeks to months after the person's death, but while the bone was still fresh, statements establishing relationship to the cause of death (e.g., gunshot) must be carefully considered and caveated – overstepping one's "precision" in this arena can prove disastrous to many, including the anthropologist. As documented by the authors' research on Civil War soldiers who survived their injuries long enough for remodeling to occur, fracture repair may be evident in as few as 7 to 10 days and, in all cases, by two weeks. See also Sledzik (personal communication 1990) and Barbian and Sledzik (2008).

Transverse (A). Fracture in which the bone is broken perpendicular to its long axis (see Greenstick fracture). If the bone is broken in two (as illustrated in A) the fracture is termed complete.

Comminuted (B). Fracture in which the bone is broken into many pieces or fragments. Common finding in the elderly.

Oblique and displaced (C). An oblique fracture is one in which the bone is completely broken at an angle diagonal to its long axis (**Figure 229**). Displaced fractures may result in misalignment and overriding of the broken ends. Improper or inadequate immobilization of the fracture may result in a pseudarthritis or loss in bone length.

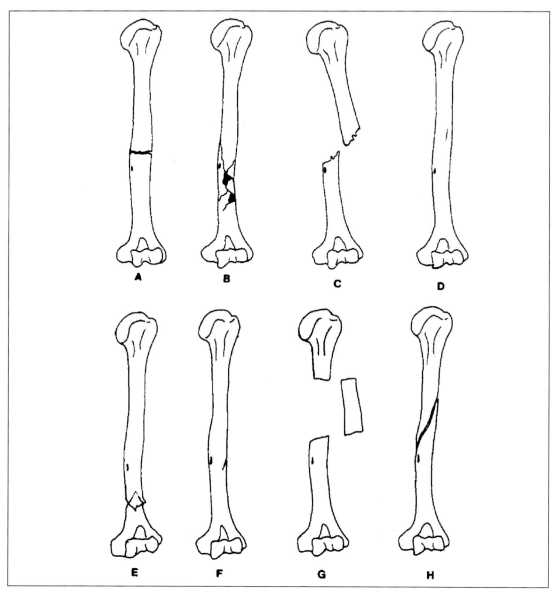

FIGURE 225. Types of long bone fractures.

FIGURE 226. Perimortem fracture of the fibula resulting in two "butterfly" fractures reflecting the type and direction of force (air crash). Note the absence of healing or bone callus (which would indicate that the injury was sustained days before death). (Refer to radiology or orthopaedic texts for further information.)

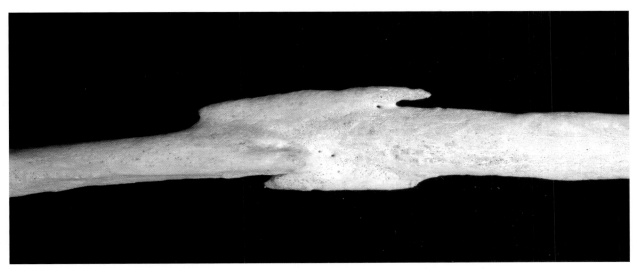

FIGURE 227. Healed "butterfly" fracture of the fibula (NMNH-H).

Hairline (D). Minor fracture in which the bone fragments remain in perfect alignment (e.g., "march" fracture).

Impacted (E). Fracturing and subsequent wedging of one bone end into the interior of another. Impaction results in a reduction in the normal length of the bone.

Incomplete (F). Fracture more severe than a hairline but less severe than a complete with no separation of bone fragments (e.g., sword wound).

Segmental (G). Fracture in which a significant portion (intact segment) of the bone is displaced. Again, loss of bone or improper immobilization may result in a pseudarthrosis or loss of bone length.

Spiral (H). Oblique fracture. Common finding in the elderly and usually resulting from osteoporosis and brittle bones. Spiral fractures are commonly associated with perimortem or "fresh" bone (see **Figure 228**).

Stellate (not shown). Fracture in which lines radiate from the central point of impact (e.g., penetrating wound such as a gunshot); see DiMaio (1999) for information on gunshot wounds (see **Figure 62**).

Undisplaced (not shown). Fracture in which the bone fragment(s) and host bone remain in approximate anatomical position.

Greenstick (not shown). The thick periosteum (and inherent flexibility of a child's bone) surrounding subadult long bones may result in the bending

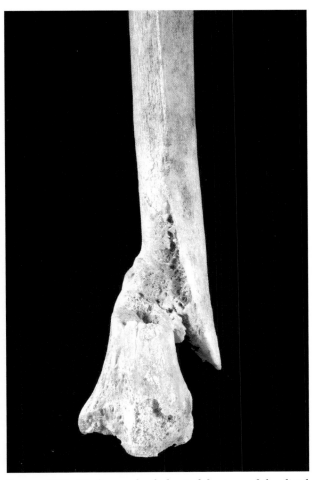

FIGURE 228. Healing or healed spiral fracture of the distal tibia (NMNH-H).

(actually, a slight fracture will be histologically visible) of one side of the cortex and fracture of the other. The appearance is similar to a broken "green stick" (Roser and Clawson 1970; Wilkins 1980).

Butterfly (see **Figures 226 and 227**).

Although the metacarpal (**Figure 229**) was chosen as an example of a healed displaced long bone fracture, the fracture terminology applies to any tubular bone in the body. In this example, the fracture occurred diagonally at the midshaft and resulted in shortening and overriding of the two broken ends. Evidence of healing can be seen by fusion of the proximal shaft to the distal shaft. Extensive callus formation may accompany a fracture. If the bone appears "swollen" and radiographs reveal large areas of sclerosis (especially if there is a cloaca or draining sinus), it is possible that the fracture site was infected (osteomyelitis) (Resnick and Niwayama 1981). See also Rockwood and Green (1975), and Rogers (1982). See also Plate 129 for an example of a healed tibia and fibula.

Although there is disagreement as to how long it takes for the earliest signs of periosteal remodeling (callus formation) to become visible on the surface of the cortex, it is probably safe to say that any active remodeling visible in dry bone suggests at least 10 days of healing and, depending on the age of the individual, that complete remodeling may occur in as little as a few months. Other than nasal bones and vertebrae, healed fractures are relatively uncommon findings in most skeletal samples

Postmortem fracture/breakage usually refer to events that occur after death and clearly indicate that the bone has lost its elastic properties (**Figure 230**). Varied conditions of decomposition impact the timing of the loss of bony elasticity. When bone is exposed to the environment for a long period of time it becomes dry, weathered, subject to distortion, breaking, checking, and cracking. Although it is extremely difficult to ascertain what may have caused a particular break, it is often possible to discern postmortem

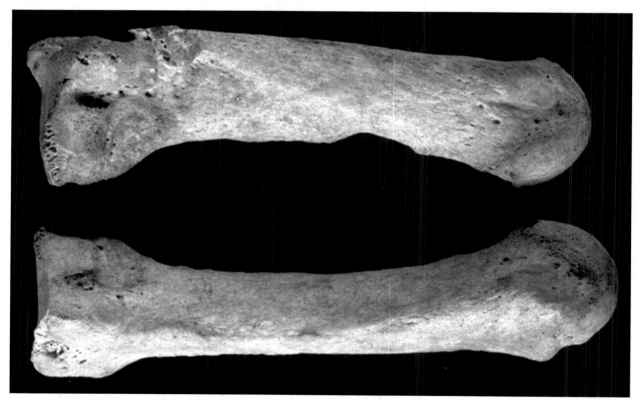

FIGURE 229. Healed fracture of a metacarpal (top) compared with normal bone.

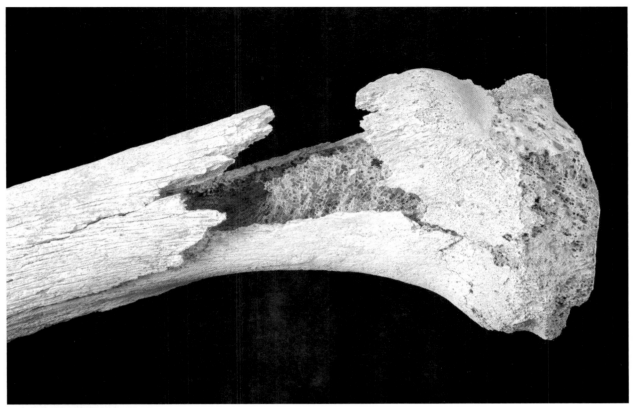

FIGURE 230. Postmortem fractures (more accurately, "breakage") of a tibia – note the jagged and "torn paper" appearance of the broken edges.

features. Postmortem characteristics reflect its viscous or brittle nature, fracture edges are lighter (sometimes whitish) than surrounding bony areas, have an irregular or jagged "torn paper" appearance, but with blunt/dull edges and little or no beveling, few or no tiny radiating fractures, and small areas of missing bone that become "dust" upon breakage.

See Galloway (1999), Wilson (1930), Berryman and Symes (1998) and Sauer (1998) for a more definitive coverage of fractures and trauma to the skeleton. Also see documentation of medieval battle fractures from the Battle of Towton by Fiorato et al. (2001) and medieval periods of Great Britian by Stirland (1996, 1999). See also Mihran and Tachdjian (1994), Morrissy et al. (2000), Resnick (2002), and Resnick and Niwayama (1988). Also refer to any of the radiology or orthopaedic books and journals such as the *American Journal of Sports Medicine, American Journal of Orthopedics, British Journal of Sports Medicine, Clinical Orthopaedics and Related Research, Journal of Sports Science and Medicine, International Journal of Sports Medicine, Seminars in Musculoskeletal Radiology,* and the *Journal of Bone and Joint Surgery* (American or British) for case reports and types of activity that result in particular fractures, trauma, and rates of healing.

Chapter X

ENTHESEAL CHANGE

Matthew P. Rhode

ENTHESEAL CHANGE

This chapter offers a brief overview of the skeletal manifestations variously known as activity-induced pathologies, evidence for occupation, markers of occupational stress, musculoskeletal stress markers, activity-induced stress markers, muscle markers, musculoskeletal attachment sites, enthesopathies, or simply entheseal changes (Merbs 1983; Dutour 1986; Kelley and Angel 1987; Hawkey 1988; Hawkey and Street 1992; Hawkey and Merbs 1995; Kennedy 1989, 1998; Wilczak and Kennedy 1998; Capasso et al. 1998; Weiss 2003a-b, 2007, 2009; Rhode 2006; Cope 2007; Cardoso 2008; Cardoso and Henderson 2010; Villotte et al. 2010; Jurmain and Villotte 2010; Mariotti et al. 2004, 2007). The subject matter are the rugosities, tuberosities, crests, spurs, pits, grooves, and furrows observed at sites of muscle tendon and ligament origin and insertion (**Table 5**), whose distribution and patterning are commonly interpreted as resulting from the performance of strenuous, habitual, or repetitive activities (Hawkey and Merbs 1995). The sites of bony attachment for muscle tendons and ligaments are known as entheses, which act as levers for and strain "dissipaters" during muscle contraction (Cooper and Misol 1970; Resnick and Niwayama 1983; Francois et al. 2001; Benjamin et al.. 2002, 2006; Zumwalt 2006; Benjamin and McGonagle 2009). The medical literature distinguishes two types of entheses, both of which are examined in entheseal change research, fibrous and fibrocartilaginous

(Hems and Tillman 2000; Benjamin et al. 1986, 2006; Benjamin and McGonagle 2001). Fibrous entheses are further classified as bony or periosteal. Bony fibrous entheses characterize muscles that attach directly to bone. Periosteal fibrous entheses characterize muscle tendons and ligaments that attach indirectly to bone through the periosteum. Fibrocartilaginous entheses characterize muscle tendons or ligaments possessing a transitional layer of fibrocartilage that separates the muscle tendon or ligament from the bone.

Specific understanding of what initiates entheseal changes remains unclear. However, they are commonly associated with both degenerative pathological conditions and overuse injuries, with inflammation of the enthesis known as enthesitis and any pathological change to the enthesis known as an enthesopathy (Niepl and Sit'aj 1979; Resnick and Niwayama 1983; Adirim and Cheng 2003; Shaibani et al. 1993; Rogers and Dieppe 1997; Olivieri et al. 1998; Francois et al. 2001; Benjamin et al. 2002, 2006; D'Agostino 2006; Mariotti et al. 2004, 2007). Among the pathological conditions associated with entheseal changes most often occur in older adults and include diffuse idiopathic skeletal hyperostosis (DISH), ankylosing spondylosis (AS) and osteoarthritis (OA) (Ragsdale et al. 1981; Resnick and Niwayama 1983; Rogers and Waldron 1989; Olivieri et al. 1998; Benjamin and McGonagle 2001, 2009; Benjamin et al. 2002, 2006; Henderson 2008, 2009). Additionally, the range of possible

Table 5
COMMONLY STUDIED UPPER/LOWER BODY MUSCLES AND LIGAMENTS.[1]

Bone	Muscle (M) – Ligament (L)			Attachment	Enthesis Type[2]
Upper Body					
Clavicle	Deltoid	M		Origin	Fibrous
	Costoclavicular	L		Insertion	Fibrocartilaginous
	Conoid	L		Insertion	Fibrocartilaginous
	Trapezoid	L		Insertion	Fibrocartilaginous
Humerus	Supraspinatus	M	Rotator Cuff	Insertion	Fibrocartilaginous
	Subscapularis	M	Rotator Cuff	Insertion	Fibrocartilaginous
	Infraspinatus	M	Rotator Cuff	Insertion	Fibrocartilaginous
	Teres minor	M	Rotator Cuff	Insertion	Fibrocartilaginous
	Pectoralis major	M		Insertion	Fibrous
	Teres major	M		Insertion	Fibrous
	Latissimus dorsi	M		Insertion	Fibrous
	Deltoids	M		Insertion	Fibrous
	Common extensors	M	Various	Origins	Fibrocartilaginous
	Common flexors	M	Various	Origins	Fibrocartilaginous
Radius	Biceps brachii	M		Insertion	Fibrocartilaginous
	Pronator teres	M		Insertion	Fibrous
	Pronator quadratus	M		Insertion	Fibrous*
Ulna	Triceps brachii	M		Insertion	Fibrocartilaginous
	Brachialis	M		Insertion	Fibrocartilaginous
	Supinator	M		Origin	Fibrous
Lower Body					
Femur	Gluteus medius	M		Insertion	Fibrocartilaginous
	Gluteus minimus	M		Insertion	Fibrocartilaginous
	Gluteus maximus	M		Insertion	Fibrous
	Psoas major / Iliacus	M		Insertion	Fibrocartilaginous
	Gastrocnemius	M	Medial Head	Origin	Fibrocartilaginous
Tibia	Patellar Ligament	L		Insertion	Fibrocartilaginous
	Soleus	M		Origin	Fibrous
Calcaneus	Gastrocnemius / Soleus	M		Insertion	Fibrocartilaginous

[1]This list is not mean to be exhaustive as many additional muscles and ligaments can be examined. The muscles and ligaments listed are among the most commonly studied for evidence of activity patterns.
[2]Sources include Benjamin et al. (1986, 2002), Benjamin and McGonagle (2001), Galtés et al. (2006), and Havelková and Villotte (2007).
*Assumed to be a fibrous enthesis as it was not located in the literature consulted.

activities and strains to which entheses are subjected make them prone to overuse injuries in both the young and old (O'Neill and Micheli 1988; DiFiori 1999; Benjamin et al. 2002, 2006). Overuse injuries result from the repetitive loading of muscles, tendons, ligaments, or entheses causing tissue damage. Each of these tissues is resistant to "normal" levels of strain such as those that occur during the performance of typical daily activities. However, when confronted with greater than normal strains, as happens during the performance of certain strenuous or repetitive activities, fatigue damage such as tendonitis and enthesitis can occur (Herring and Nilson 1987). Unlike tendons and ligaments, which are avascular and slow to heal, entheses are often vascular and reactive due to their association with the periosteum (Resnick and Niwayama 1983). Thus, in response to inflammation, stretching, or tearing of the periosteum at the enthesis from increased mechanical loading, bony modeling may be stimulated as the body heals and accommodates to the increased strain (Hoyte and Enlow 1966; Ragsdale et al. 1981; Herring and Nilson 1987; Hawkey and Merbs 1995; Francois et al. 2001; Peterson 2002; Hamrick et al. 2000, 2002).

Clinical, occupational, and sports medicine research indicates that bone responds to mechanically applied loads by altering its structure or shape through periosteal and endosteal modeling and remodeling (Chamay and Tschantz 1972; Bertram and Swartz 1991; Pearson and Lieberman 2004; Ruff et al. 2006; Shaw and Stock 2009a-b). The concept that bone form follows its function was first described in the late 1800s by Roux and later Wolff in what came to be known as Wolff's Law (Roux 1881, 1885; Roesler 1987; Dibbets 1992; Wolff 1988; Lee and Taylor 1999). Further research on the topic during the 1960s resulted in an improved model, the Mechanostat, which states modeling adapts bone architecture and strength to loads so that applied strains will not exceed a given threshold or adapted state (Frost 1997, 2003). Loads applied during the performance of strenuous activity or exercise can rise above the threshold into the overuse window and stimulate modeling. When strains return to the adapted state, modeling ceases (Frost 1997, 2003). Similarly remodeling is initiated if loads fall below the adapted state threshold into the disuse window as occurs with extended bed rest and paralysis (Frost 1997, 2003). Wolff's law and the later Mechanostat are strong models, but they are not always true or applicable in all situations. Though most clinical animal studies support the concept of bone adaptation to applied load, some studies have produced ambiguous and even contradictory results

(Bertram and Swartz 1991; Pearson and Lieberman 2004; Zumwalt 2006). Acknowledging these results, the scientific community commonly accepts a more general interpretation of these concepts, known as *Bone Functional Adaptation*, which proposes that bone adapts to its functional environment by altering its structure or shape (Ruff et al. 2006).

One insight of this research is the realization that adult bone shape is determined predominantly during adolescence through increases in length, shape, and density as the skeleton adapts to its biomechanical and functional environment. Specifically, mechanical loading stimulates the greatest modeling response prior to skeletal maturity, after which modeling continues but at a reduced rate, slowing age-related bone loss while remodeling increases to repair natural damage (Lieberman et al. 2003; Ruff et al. 1994; Pearson and Lieberman 2004; Ruff et al. 2006). Thus external bone morphology continues to change during adulthood, but the most significant influence on adult bone shape occurs during growth and development (Garn 1972, 1980; Pfeiffer 1980; Huiskes 2000; Mays 2001; Marjolein et al. 2002). Therefore, studies whose purpose is the examination or reconstruction of adult activity patterns using entheseal changes are in fact examining the manner in which activity influenced bone development during late adolescence and early adulthood. In other words, entheseal changes may provide a record of individual developmental plasticity and its influence on specific phenotypically expressed traits, to such a degree that observed adult variation provides a relative indicator of differential adolescent involvement in certain activities (Trinkhaus 1993; Jurmain 1999; Pearson and Lieberman 2004).

The study of entheseal changes has a long history dating from the late 1800s with early physicians and anthropologists seeking to interpret distinctive patterns of observed skeletal variation (Kennedy 1989; Capasso et al. 1998). This work produced a series of anecdotal accounts linking skeletal variants to specific occupations often based on limited corroborating evidence. Research on entheses continued sporadically over the decades, but it was not until the 1980s and the work of Merbs (1983), Dutour (1986), and their students that there was a resurgence of interest in the subject. During this period, it was realized that individual entheseal changes could not be directly attributed to specific activities. Instead, it is a broader range of actions that employ the same muscles and ligaments that is under examination (Hawkey 1988; Jurmain 1999). This view is more consistent with functional anatomy, which indicates that in the performance of most actions muscles act in synergistic groups, not individually (Gray 1977; Robb 1998, Stirland 1993, 1998; Jenkins 2002; Peterson 2002; Rhode 2006). Entheseal change research then proceeds by limiting the range of possible activities performed for a given population by critically reviewing appropriate ethnographic or historic sources to identify those repetitive and strenuous activities with the potential to induce entheseal changes. Variation in the performance of these activities and the degrees to which identifiable synergist muscle groups and ligaments are utilized in performing the specified action are expected to stimulate variable entheseal morphology. The quantification and interpretation of this variation throughout the skeleton is the foundation of research whose goal is the examination of activity patterns within and between populations using entheseal change (Stirland 1993; Kennedy 1989, 1998; Capasso et al. 1998). Interest in the field continues with ongoing research and debate over the origins of and appropriate methods to use in studying entheseal changes.

Numerous approaches exist to study entheseal changes ranging from presence/absence and ordinal scale methods to techniques that make use of contour gauges and 3D scanners, but the most common and widely used is the Hawkey method and its variants (Hawkey 1988; Hawkey and Merbs 1995; Robb 1994, 1998; Stirland 1998; Wilczak 1998; 2009; Weiss 2003a-b, 2007; Martiotti et al. 2004, 2007; Villotte 2006; Henderson and Gallant 2007; Molnar 2006, 2010; Lieverse et al. 2009; Cardoso and Henderson 2010; Villotte et al. 2010). The Hawkey method recognizes three types of entheseal changes: robusticity markers, stress lesions, and ossification exostoses,

each of which is scored on an ordinal scale. The benefits of this method are its simple guidelines, easily determined and flexible threshold scores, and low rates of intra- and interobserver error (Hawkey 1988, 1998; Hawkey and Merbs 1995; Robb 1998; Jurmain 1999; Peterson 2002; Weiss 2003a-b).

Robusticity markers are the initial entheseal response to increased strain, expressed as roughened or irregular origins and insertions scored on a scale 0 to 3 (Hawkey and Merbs 1995). Muscle tendons and ligaments exhibiting no entheseal change are scored 0 (absent). A score of 1 (trace) denotes a slightly raised entheseal surface, discernible visually and palpably. A score of 2 (moderate) indicates an enthesis possessing a visually roughened or elevated surface. A score of 3 (severe) identifies an entheseal surface exhibiting a markedly elevated, irregular, or crested surface (**Figures 231 and 232**).

The second type of entheseal change is the stress lesion, which is characterized by grooves, furrows, or pits scored on a scale of 4 to 6 (Hawkey 1988; Hawkey and Merbs 1995). Originally, Hawkey (1988) considered stress lesions to be distinct from robusticity markers but later concluded the two represent a continuum of entheseal change (Hawkey and Merbs 1995). Stress lesions are now considered an extreme form of robusticity markers caused by severe overuse of a muscle tendon or ligament resulting in chronic inflammation leading to bony necrosis at the enthesis (Hoyte and Enlow 1966; Hawkey and Merbs 1995; Galtés et al. 2006). Skeletal lesions with similar properties in other areas of the skeleton are classified as cortical defects. A score of 4 (trace) signifies a shallow groove, furrow, or pit, less than 1 mm in depth. A score of 5 (moderate) indicates a deeper more defined groove, furrow, or pit greater than 1 mm but less than 3 mm in depth. A score of 6 (severe) denotes a very deep groove, furrow, or pit, greater than 3 mm in depth (**Figure 233**).

The third type of entheseal change is the ossification exostosis, enthesopathy, or enthesophtye, which, unlike robusticity markers or stress lesions, results from an abrupt trauma such as a muscle, tendon, or ligament tear or rupture (Hawkey 1988;

Hawkey and Merbs 1995). After such insults, the remaining tissue at the entheses can ossify as the body repairs the injury, forming an irregular bony mound, crest, spur or exostosis coded on a scale from 1 to 3. A score of 1 (trace) denotes a relatively small, rounded, raised, bony exostosis rising less than 2 mm above surrounding cortical bone. A score of 2 (moderate) signifies a more

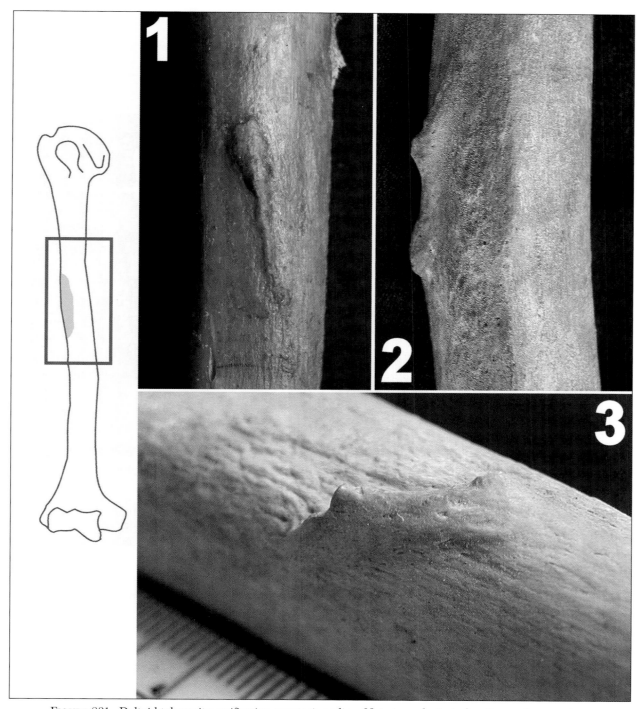

FIGURE 231. Deltoid tuberosity ossification exostosis coding. Note irregularity in location and expression.

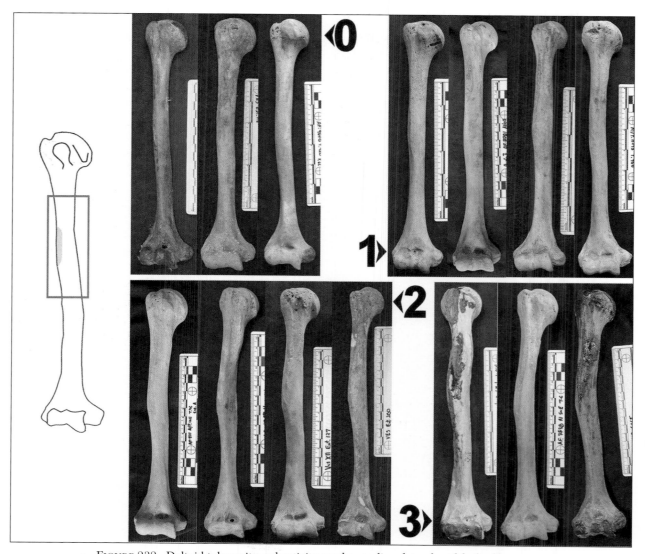

FIGURE 232. Deltoid tuberosity, robusticity marker coding, lateral midshaft of humerus.

distinct bony exostosis that is variable in shape but ranges in size from 2 mm and 5 mm above the surrounding cortical bone. A score of 3 (severe) indicates a strongly pronounced and often sharp bony exostosis, extending greater than 5 mm above the surrounding cortical surface (**Figure 234**). In practice, ossification exostoses are rare when compared with robusticity markers and stress lesions.

Although a variety of methods exist to record entheseal change and numerous projects detailing various aspects of prehistoric behavior and activity patterns, the use of this data in behavioral reconstructions is not free from criticism. The most significant of these is the concern over the lack of a well-defined etiology for entheseal change (Jurmain 1999). Specifically, nearly every enthesis-based study implicitly assumes that the results obtained in bone functional adaptation research, which focuses almost exclusively on bone diaphyseal properties in animals, is directly applicable to studying entheseal changes. Unfortunately, the clinical literature provides only tangential and tenuous support for using this data in activity-based interpretations of entheseal change (Jurmain 1999). Among the few clinical

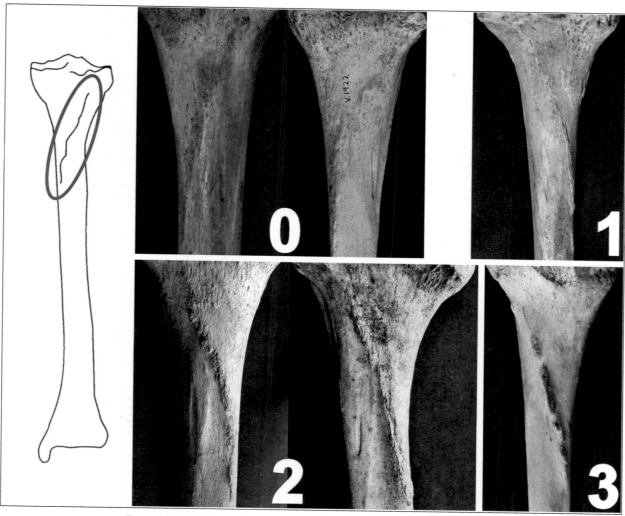

FIGURE 233. Soleus, robusticity marker coding, posterior proximal tibia.

studies examining entheses obtained, results have been negative but suspect (Zumwalt 2005, 2006; Benjamin and McGonagle 2009). Additional concerns exist over the use of animal models and specifically whether they are acceptable human analogs (Pearson and Lieberman 2004). It is obvious, however, that the etiology of entheseal change is multifactoral and includes both genetic and epigenetic components, among which the biomechanical and functional environment is but one component (Jurmain 1999; Henderson 2008, 2009; Villotte et al. 2010).

Lacking a definitive clinical basis, research on entheseal change tends to rely on information from industrial, occupational, sports medicine, ethnographic, ethnohistoric, and historical sources to interpret observed patterns (Kennedy 1989; Jurmain 1999). The use of these sources creates additional complications. Similar to the concern over animal model, there is also is also concern over whether modern athletes are appropriate models for average human development in both prehistoric and modern populations (Jurmain 1999). Parsimony suggests, however, that modern athletes can be considered as extreme, non-pathological examples of physically active individuals (Trinkhaus et al. 1994a-b). Further, questions have been raised over the biased and subjective nature of many ethnographic and historic sources (Boyd 1996; Jurmain 1999). As a

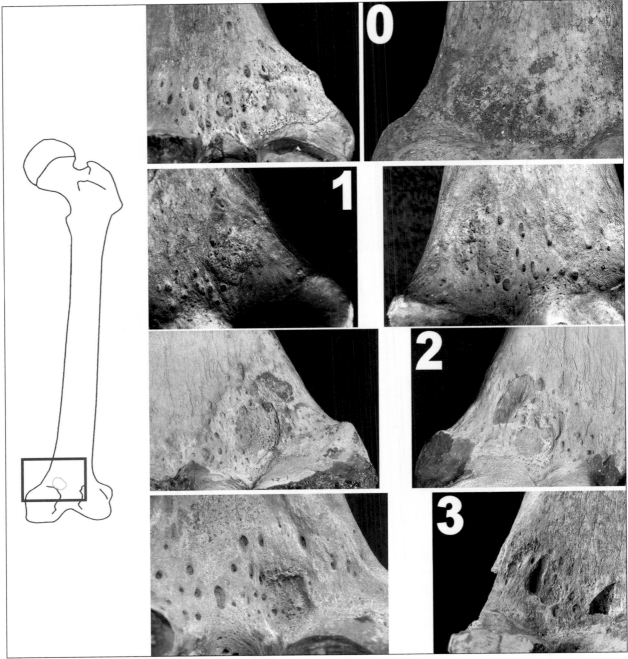

FIGURE 234. Gastrocnemius (medial head) stress lesion coding, posterior distal femur.

result, it is understandable that researchers can often fall prey to the siren call of positing overly simplistic assumption-laden explanations from data that is not always amenable to such fine-grained interpretations (Waldron 1994; Jurmain 1999; Peterson 2002).

Two additional consistently confounding factors observed in entheseal change research are age and body size. Specifically, older and larger individuals tend to exhibit larger, more pronounced entheses when compared to younger and smaller individuals (Hawkey and Merbs 1995; Nagy and

Hawkey 1995; Robb 1998; Steen and Lane 1998; Wilczak 1998; Nagy 1998; Cook and Doherty 2001; Eshed et al. 2004, Molnar 2006, 2010; Rhode 2006; Weiss 2004, 2007; Lieverse et al. 2009). This correlation between age and increasing enthesis severity is interpreted as reflecting the cumulative effects of performing similar activities over the course of a lifetime and as the result of natural age-related degeneration (Nagy 1998; Robb 1998; Wilczak 1998; Ruff et al. 2006). Similarly, the sex differences observed in entheseal data may be associated with body size differences within and across populations. Fortunately, it is possible to correct or adjust entheseal change data to account for some of these confounding factors. These adjustments can range from a simple linear regression on age, to dividing the data by a variable associated with body size like femoral head diameter in addition to others. It is true that age-related degeneration from conditions such as osteoporosis, OA, and more severe conditions such as DISH can significantly alter bone morphology (Ortner and Putschar, 1985; Mann and Murphy, 1990; Jurmain 1999; Mann and Hunt, 2005). Although the influence of such pathological conditions cannot be removed entirely, individuals with obvious pathologies can be excluded from analytical datasets. Such adjustments may remove or obscure integral data relationships, but if interpretable patterns remain, any activity and behavioral-based interpretations established would be much more compelling.

Recently a number of criticisms have been directed at deficiencies in the Hawkey method (Henderson and Gallant 2007; Villotte et al. 2010; Cardoso and Henderson 2010). The most significant is that although the medical literature identifies two types of entheses, the Hawkey method makes no such distinction (Villotte et al. 2010; Jurmain and Villotte 2010; Henderson and Gallant 2007). Fibrous entheses, which occur primarily in the cranium and long bone diaphyses, commonly exhibit irregular origins and insertions (Benjamin et al. 2006). Fibrocartilaginous entheses, which occur near long bone epiphyses, commonly exhibit smooth well-defined origins and insertions (Benjamin et al. 2006). This natural variability in the expression of fibrous entheses is problematic since in the Hawkey method a score of 0 signifies no entheseal development. Yet, if some fibrous entheses are consistently scored as 1 but in reality reflect no entheseal development, unintended error will be introduced into datasets potentially resulting in interpretational inaccuracies. Acknowledging this concern, it may be possible to account for this variability by correcting the Hawkey scores for the variability in fibrous entheses or by analyzing the entheses types separately.

In summary, this review has shown the study of entheseal changes to be an engaging but contentious subject. Although concerns persist over the manner in which entheseal changes are initiated and the appropriate techniques to use in studying them, the analysis of entheseal changes has the potential to provide insights into individual and group level activity patterns. Regardless of the precise etiology of entheseal change, what is indisputable is that many entheses exhibit variable morphologies. This variation is quantifiable and can be examined for patterns within individuals and across population. Further, acknowledging the multifactoral etiology of entheseal change it is possible to adjust the entheseal data to account in part for the confounding influences of age and body size, and degenerative pathological conditions. When these steps are taken, what remains? In some part, the data must include a record of individual bone functional adaptation to its biomechanical environment, which can be tested using activity-based hypotheses. If no significant or interpretable patterns emerge, parsimony suggests either vital information was removed by attempting to correct the data or additional confounding factors remain that obscure relationships. Logically, if this happens, one must conclude that the data is not applicable to research focused on differentiating activity patterns. Yet, if significant and interpretable entheseal patterns emerge from the data, parsimonious indicates that the observed patterns resulted from the performance of certain activities, which have influenced the development of individual skeletal morphology.

Chapter XI

CASE REPORT: DIFFERENTIAL DIAGNOSIS OF SEVERE DIFFUSE SKELETAL PATHOLOGY IN AN ADULT THAI MALE

A LESSON IN INTERPRETATION

Robert W. Mann and Panya Tuamsuk

The authors present an extreme case of diffuse skeletal lesions found in the skeleton of a 58-year-old Thai male (**Figures 235, 236, 237, 238, and 239**). The skeleton exhibits extensive cortical resorption accompanying osteoporosis and marrow hyperplasia resulting in hollow bones with a thin "lacy," "woven," "wicker basket" perforated outer cortex and coarse bars/pillars of bone in the hemopoietic areas. This case allows the researcher a rare opportunity to examine the extensive effects from at least two disease processes. Unfortunately, there are no patient records for this individual; thus the differential diagnosis must be made from the gross pathological evidence in the skeletons (see Note 1 at the end of this chapter). Interpretation from the evidence, thalassemia accompanied by generalized osteoporosis (osteopenia) possibly associated with medications, anemia, Gaucher's disease and hemangioma/lymphangioma, effects of carcinoma and metastatic carcinoma and possibly multiple myeloma are considered. The radiographic findings are not typical of osteomalacia, hyperparathyroidism, multiple myeloma (and/or Waldenstrom's disease) or current-day sarcoid. This case highlights the difficulty of radiographically or grossly distinguishing and diagnosing the possible diseases at work in dry bones, where no soft tissue histology or patient history/records are available.

The skeleton (#698/38) is part of the osteology collection at Khon Kaen University (KKU), Thailand. Khon Kaen University is located some 400 kilometers northeast of Bangkok, in an area known as Isan (Isaan or Isarn). The osteological collection (which the university started collecting in 1972) consists of more than 700 human skeletons donated to the medical school for gross anatomy classes or for the osteological collection. The collection increases by more than 100 individuals each year. After death, the donated body has the soft tissue removed to prepare the skeletons for the osteological collection; the soft tissues receive a Royal Cremation. Most, but not all the donated bodies are accompanied by patient records and histories. The individual reported here consists of a complete skeleton missing only the hyoid and xiphoid bones of a 58-year-old Thai male who died in April 1999. The following information describes his skeletal lesions.

Note: The opinions and differential diagnosis expressed in this chapter are the sole responsibility of the authors and are presented more as points of observation and discussion than a definitive differential diagnosis.

213

DESCRIPTION OF THE SKELETON AND LESIONS

Skull. The skull is mostly unaffected with the exception of the greater wings of the sphenoid and areas around the occipital condyles. The margins of the few lytic lesions are smooth and well defined, and many have coalesced to form larger perforations. The mastoid foramina are symmetrical and, although probably normal variants, are somewhat larger than usual. There is no evidence of bone destruction or porosity ectocranially superior to the nuchal crest. There is a tympanic dehiscence (foramen of Huschke) in the right tympanic plate, several large Pacchionian pits in the endocranium that housed arachnoid granulations, bilateral parietal foramina (a normal feature), and numerous small porosities along the sagittal sinus for veins. The mandible has fewer than five small porosities along the lingual surface of the mandibular canal. There is no evidence of cribra orbitalia or porotic hyperostosis (that sometimes accompanyies iron-deficiency anemia), including "hair-on-end" appearance radiographically. The maxilla and mandible exhibit heavy dental attrition and moderate periodontal disease with a periapical abscess of tooth #9 (maxillary left lateral incisor) and an abscess near the cemento-enamel junction of tooth #30 (maxillary right first molar).

Clavicles. There are localized areas of small porosities in the medial and lateral ends.

Scapulae. Bilateral and symmetrical porosity and lesions without sclerotic margins in all but the thinnest areas, especially along the perimeter of the scapular body. The glenoid cavities are free of porosity.

Sternum. The manubrium and sternum exhibit lesions without sclerotic margins and extensive cortical erosion resulting in coarse strands of bone – all cortical surfaces are affected except the clavicular joints; in essence, the sternum is hollow (Figure 237b).

Humeri. The proximal and distal ends are porous with lesion without sclerotic margins, but the diaphysis and joints are relatively unaffected (Figure 236).

Radii and Ulnae. The proximal and distal ends, with the exception of the joint surfaces, exhibit mild porosity. The right radius and ulna exhibit evidence of prior forearm plate screw fixation and normal healing. There is no radiographic evidence of pseudo-fractures in the ulnae.

Ribs. All ribs exhibit diffuse lesions along their entire length. Most affected are the pleural surfaces, exhibiting a "lacy" appearance of bone resorption. The proximal and distal ends of the ribs are extremely fragile and are nearly hollow. None of the ribs are thickened as a result of remodeling and there is no evidence of fracture. The articular facets are relatively free of involvement (Figure 237a).

Hands and Wrists. All hand and wrist bones exhibit small lesions without sclerotic rims. Most of the destruction involves the proximal and distal ends of the phalanges with the palmar surfaces most affected. Some of the joints exhibit small lytic lesions. Several of the nutrient foramina are enlarged.

Vertebrae. With the exception of the endplates and some articular facets, all of the vertebrae exhibit severe resorption and a "wicker basket" patterning of crisscrossing bars of bone. Most severely affected are the lumbar vertebrae (Figure 235).

Sacrum. The sacrum exhibits the most severe destruction of any bone in the skeleton. The outer cortex of the sacrum exhibits extensive resorption of the cortex and internally is hollow, with an extensive "lacy" appearance by interwoven coarse bars of bone. The dorsal surface exhibits less resorption than the ventral surface, and the promontory and auricular surfaces are unaffected (Figure 238b).

Innominates. The innominates exhibit severe destruction of the pubic bones and iliac plates with large, confluent lesions resulting in interwoven coarse bars of bone with no evidence of an osteoblastic response. The acetabulum is primarily unaffected (Figure 238a).

Femora. The proximal and distal ends of the diaphyses exhibit a few lytic lesions, but the remainder of the diaphyses and articulations are unaffected. Radiography revealed no insufficiency fractures (Looser's zone) of the neck.

FIGURE 235. Lumbar vertebrae showing extensive cortical resorption resulting in a "wicker basket" appearance.

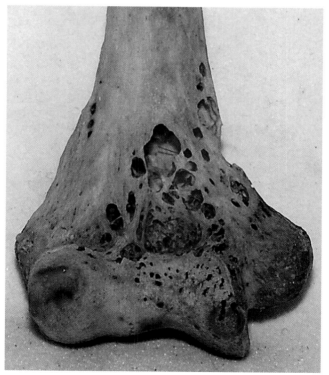

FIGURE 236. Resorptive lesions in the distal right humerus; note the lesions affected the articular surface.

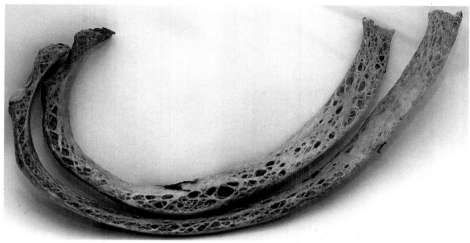

a

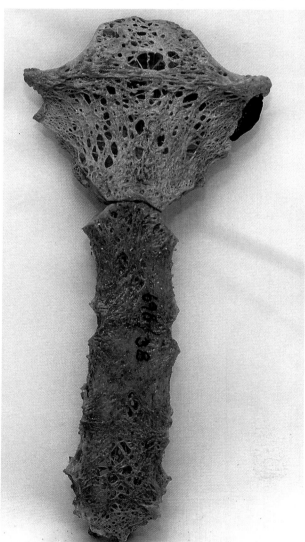

b

FIGURE 237a-b. Cortical resorption of the ribs and sternum. Note the fusiform ("expanded") appearance of the central portion of the top rib.

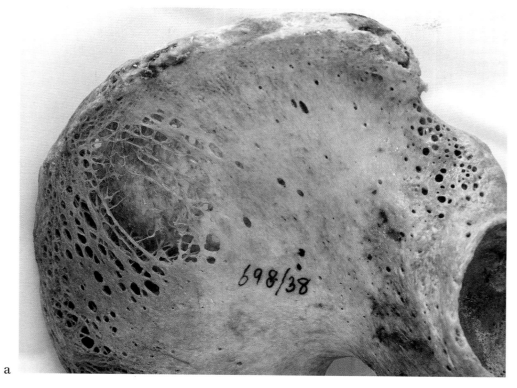

a

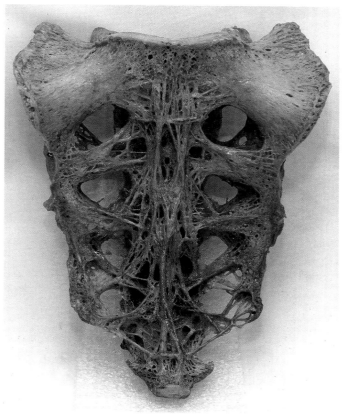

FIGURE 238. Ventral view of the right innominate and sacrum showing massive resorption of the cortex resulting in vertical bars ("buttressing") of bone in the sacrum.

b

Tibiae. The proximal and distal ends of the diaphyses exhibit mild resorptive lesions; the diaphyses and joints are unaffected.

Fibulae. The proximal and distal ends of the diaphyses exhibit mild resorptive lesions; the diaphyses and joints are unaffected. The left fibula has a localized (~3 cm long) lesion of the distal one-third of the diaphysis. Although a predominately resorptive lesion, there is evidence of healing and enlargement of the cortex, suggesting that the lesion may have originated in the endosteum or inner cortex. This is the only lesion in the skeleton that exhibits a discernible expansile, blastic response.

Feet. Severe resorption and lesions without sclerotic rims are found throughout the elements but are primarily seen in the metatarsals and hind foot. The phalanges are least affected and the joints are primarily free of involvement (Figure 239).

Observations

- Absence of weight-bearing fractures
- With exception of one small area on the diaphysis of the left fibula, absence of periosteal reaction (i.e., no periostitis) in the skeleton
- Virtual absence of "punched out" cranial lesions, with evidence of microporosities in the cranial base and sphenoid bone
- No evidence of symmetrical osteoporosis of the skull (parietal bones)
- No evidence of "hair on end" in the skull typical of thalassemia
- Severe nature of the lesions without sclerotic margins and extensive resorption of the cortex, especially the sternum, vertebrae, sacrum and innominates
- Axial skeleton involvement only (appendicular skeleton not involved)
- No Looser's zone fractures

Discussion

Although aggressive lytic skeletal lesions occur in clinical cases and osteological collections, the pattern and distribution of skeletal lesions in this individual are unusual both for their severity and the amount of bony destruction without concomitant fractures. This is one of the most severe case of its kind that the authors have encountered in several thousand modern and ancient skeletons from around the world. A review of the literature found a reference to a 14-year-old girl (Lagia et al. 2007) that exhibited a severe form of resorption of the cortex, especially noteworthy in the axial skeleton, that is consistent with thalassemia. Cortical resorption of this young girl's vertebrae is very similar in appearance to the present case.

While non-expansile lytic lesions without an osteoblastic response often accompany some diseases (and the authors noted a few milder cases of osteoporosis in the Khon Kaen Collection), it is extremely rare to encounter dry-bone specimens affected as severely as the present case. Although a definitive etiology cannot be given for this case, the differential diagnosis would include:

- Thalassemia – A genetic defect resulting in a reduction in the number of red blood cells and hemoglobin. Some individuals have been found to have punched-out lesions as a result of inherited anemia (Hsiao et al. 2005). There are, however, no flask-shaped distal femurs, skull thickening with a "hair-on-end" appearance, or involvement of the facial bones (pneumatisation of the perinasal sinuses) in this specimen that commonly accompanies thalassemia. Another typical skeletal manifestation of thalassemia is expansion of the posterior ribs, which are absent in this individual (but present in skeleton SC130 in the KKU Osteology Collection), (see Figure 240b).
- Gaucher's disease – An autosomal recessive disorder of enzyme deficiency that results in the accumulation of fatty substances in organs and bone, resulting in increased radiolucency of bone and thinning and scalloping of cortical bone.
- Hemangioma/lymphangioma – Although hemangiomas in bone are rare, they are the most common vascular tumor affecting bone and may form on the periosteal, endosteal, or medullary surface of bone.

FIGURE 239. Extensive cortical resorption of the bones of the feet. Note that the lesions primarily affected the mid foot and hind foot, but not the articular surface subchondral bone of the tali and the enlarged nutrient foramen in both second metatarsals (see Plate 140).

- Chronic leukemia, lymphoma, Hodgkins disease, metastatic carcinoma. A case of similar severity, distribution and similar "lacy" morphology of the axial and appendicular skeleton was encountered by D. R. Hunt of a 37-year-old woman who died of cervical cancer, metastasized to the bone (D. R. Hunt, personal communication 2011).
- Multiple myeloma (including Waldenstrom's microglobulinemia) – A malignant tumor of plasma cells causing one or more lytic lesions in bones (Bilsky and Azeem 2008). The small size of the "punched out" lesions ("moth eaten") without sclerotic margins, virtual absence of cranial lesions ("hair on end" is not a sequel of multiple myeloma) and involvement of the axial skeleton are consistent with a variety of forms of multiple myeloma. Nearly one third of patients with multiple myeloma have osteopenia (on radiograph), about 35 per cent present with no radiographic abnormalities and the rest have classical lytic bone lesions (Jeno I. Sebes, personal communication 2011).

While there is no way to definitively identify the disease responsible for the lesions in the present case, the absence of fractures involving the femoral neck (Looser's zone) or pseudofractures in the ulnas argues against osteomalacia or a long-standing congenital process such as X-linked hypophosphatasia. Additionally, there is no evidence of renal osteodystrophy in any of the bones. Given that the forearm fractures healed well after surgery and that there are no other definitive fractures argues against a congenital condition. It is both noteworthy and remarkable that, despite the diffuse "lacy" or "wicker basket" appearance of so many of the weight-bearing bones, this individual didn't experience multiple bony fractures, given the compromised (severely osteoporotic) condition of the bones of the spine, hips, ribs and feet.

The skeleton exhibits two types of lesions: (1) small round or oval lesions lacking sclerotic margins and (2) diffuse resorptive lesions of varying sizes and shapes, many of which have coalesced to form larger lesions (e.g., iliac plates) (See Figure 238a). The relatively consistent size of the lesions is more consistent with multiple myeloma than metastatic carcinoma, since the latter often range in size from small to quite large lytic lesions (Marks and Hamilton 2007).

Review and consultation of this case by several internationally respected pathologists, radiologists, and paleopathologists have left us with no definitive or agreed upon diagnosis for the pathological condition(s) of this individual. The most likely cause of the features and lesions are thought to be due to some sort of pathological disruption of normal bone resorption and remodeling. Disease that would affect the normal function of osteoclastic activity, reformation and mineralization/ossification of the collagen matrix are the most likely culprit. Thus, thalassemia, Gaucher's disease, leukemia, and glandular carcinoma would be the highest for consideration. However, the gross morphological characteristics and radiographic findings have numerous similarities and yet dissimilarities to any one of these diseases, making it difficult to identify one particular disease. As example, the size, distribution and radiographic appearance of the non-sclerotic lesions in this individual are most consistent with thalassemia and /or multiple myeloma, although some of the hallmarks of either of these diseases are not present or consistently distributed in this individual's bones. The diffuse resorptive skeletal lesions, likely occurring secondary to the primary disease, are most typical of an unusual form of osteoporosis, generalized osteopenia, thalassemia, Gaucher's disease, and hemangioma/lymphangioma. The differential diagnosis and etiology may also be a reflection of medications or other malfunction of the body external which have influenced the normal bone function, such as endocrine states, calcium deficiency, alcoholism, chronic liver disease, anemic states, and osteoporosis.

What is to be learned from this case study is a cautionary tale that there very often is no "clearcut" diagnosis of many skeletal pathological conditions. In this particular instance, the severity should have allowed for more concise interpretation. This, as is found, is not the case. It might be because of the severity that numerous other

a
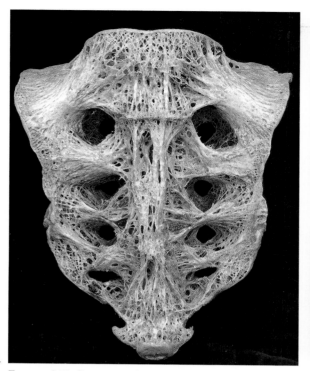
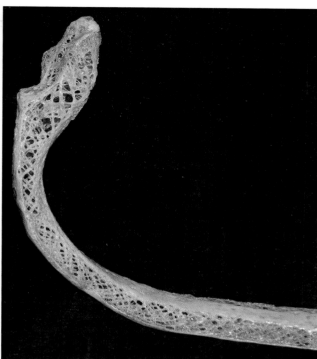
b

FIGURE 240. Sacrum and vertebral end of a left rib in a 33-year-old Thai male diagnosed with thalassemia and anemia. Compare the pattern and appearance of these lesions with those in Figures 237a and 238.

effects could have happened that have now masked the origins of the disease. But in more subtle cases, the evidence for diagnosis will be just as difficult and not as obvious to the researcher.

Note

1. Subsequent examination of the KKU Osteology Collection a few months later revealed another individual with the same distribution and appearance of skeletal lesions. Fortunately, the remains of the second individual, a 33-year-old Thai male, are accompanied by a patient history and diagnosis of "thalassemia with severe anemia" who died of congestive heart failure (with an enlarged liver) – this "spongy bone" condition in Thai is known as "graduk poohk." The only two differences in the two skeletons are expanded ribs near the vertebral border in the 33-year-old male, a condition often accompanying thalassemia, and what appear to be congenitally short humeri. See Figure 240a-b for the pattern and appearance of lesions in the sacrum and rib of the 33-year-old male. Both cases highlight the difficulty of interpreting and diagnosing bone disease in osteological and archaeological specimens.

Chapter XII

HUMAN SKELETON
(VENTRAL AND DORSAL VIEWS)

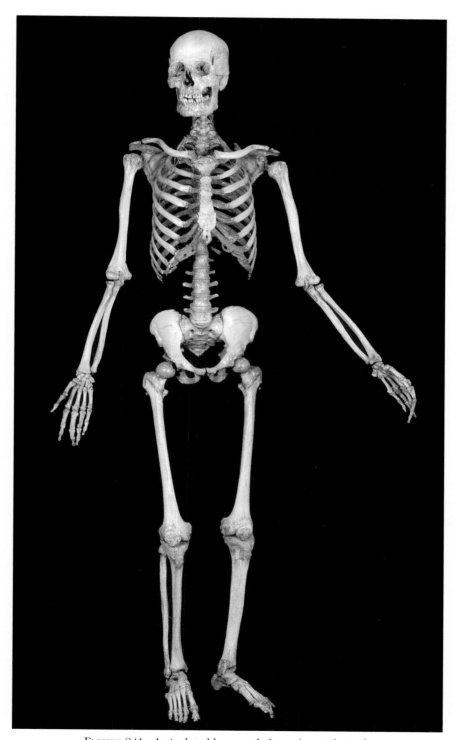

FIGURE 241. Articulated human skeleton (ventral view).

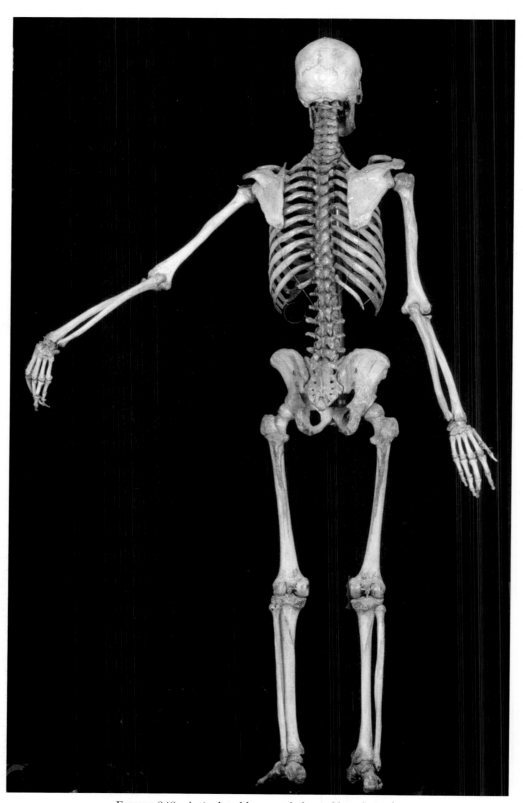

FIGURE 242. Articulated human skeleton (dorsal view).

Chapter XIII

MUSCLE ATTACHMENTS

Skull

1. Temporalis
2. Masseter
3. Temporalis
4. Masseter
5. Buccinator

Clavicle

6. Deltoid
7. Pectoralis major
8. Trapezius

Scapula

9. Inferior belly of omohyoid
10. Conoid ligament
11. Trapezoid ligament
12. Pectoralis minor
13. Coracobrachialis and short head of biceps
14. Long head of triceps
15. Subscapularis
16. Serratus anterior

Humerus

17. Supraspinatus
18. Subscapularis
19. Pectoralis major
20. Latissimus dorsi
21. Teres major
22. Deltoid
23. Brachialis
24. Brachioradialis
25. Extensor carpi radialis longus
26. Pronator teres

27. Common flexor
28. Common extensor

Radius

29. Biceps
30. Supinator
31. Flexor digitorum superficialis, radial head
32. Pronator teres
33. Flexor pollicis longus
34. Pronator quadratus

Ulna

35. Brachialis
36. Flexor digitorum profundus
37. Pronator quadratus

Innominate

38. Erector spinae
39. Iliolumbar ligament
40. Quadratus lumborum
41. Transversus abdominus
42. Internal oblique
43. External oblique
44. Inguinal ligament
45. Sartorius
46. Iliacus
47. Rectus femoris
48. Psoas minor
49. Adductor longus
50. Adductor brevis
51. Gracilis
52. Obturator externus
53. Quadratus femoris

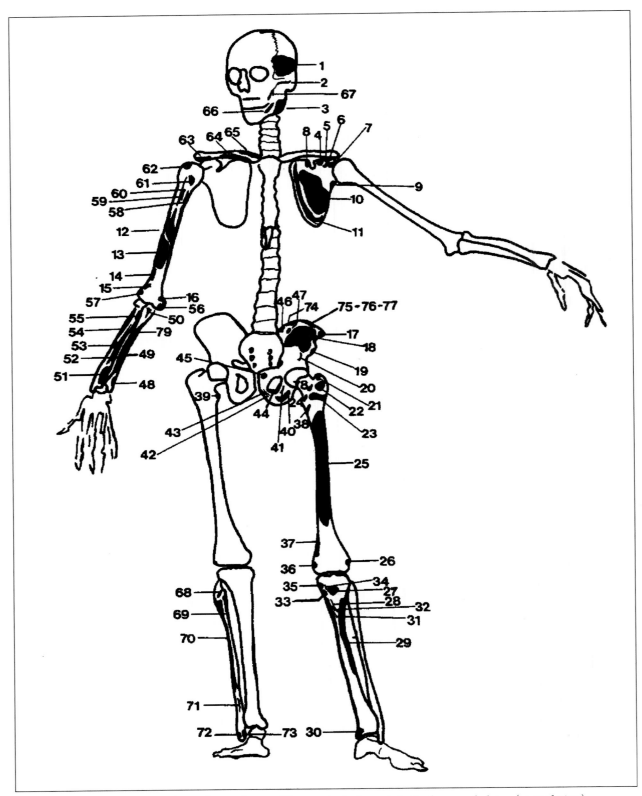

FIGURE 243. Selected muscle, tendon, and ligament attachments in the human skeleton (ventral view).

54. Adductor magnus Femur
55. Piriformis
56. Gluteus minimus
57. Iliofemoral ligament
58. Vastus lateralis
59. Iliofemoral ligament
60. Vastus medialis
61. Vastus intermedius
62. Adductor magnus
63. Fibular collateral ligament
64. Tibial collateral ligament

Tibia

65. Semimembranosus
66. Vastus medialis

67. Patellar ligament
68. Tibial collateral ligament
69. Sartorius
70. Gracilis
71. Semitendinous
72. Tibialis anterior
73. Medial collateral ligament

Fibula

74. Peroneus longus
75. Extensor digitorum longus
76. Peroneus longus
77. Peroneus tertius
78. Calcaneofibular ligament
79. Anterior talofibular ligament

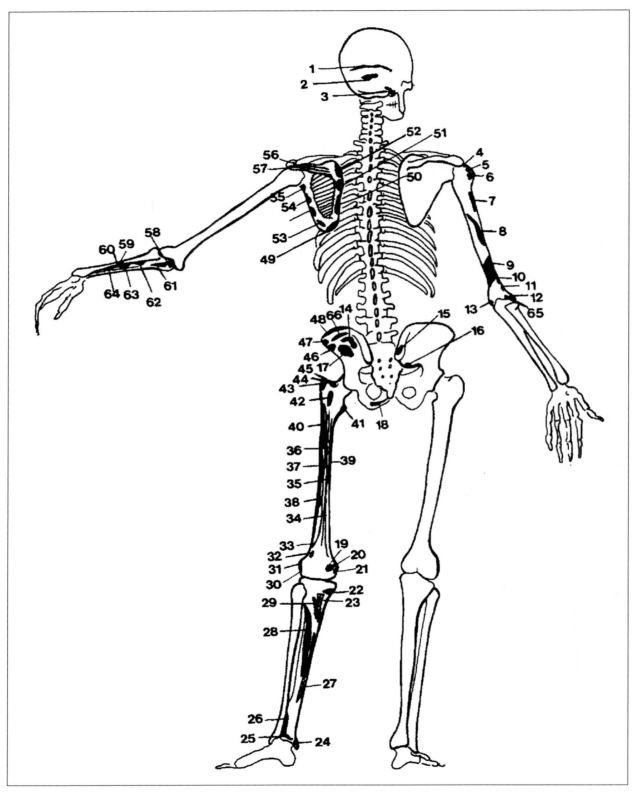

FIGURE 244. Selected muscle, tendon, and ligament attachments in the human skeleton (dorsal view).

Skull

1. Trapezius
2. Semispinalis capitis
3. Sternocleidomastoid

Scapula

4. Levator scapulae
5. Rhomboid minor
6. Rhomboid major
7. Trapezius
8. Deltoid
9. Long head of triceps
10. Teres minor
11. Teres major
12. Latissimus dorsi

Humerus

13. Supraspinatus
14. Infraspinatus
15. Teres minor
16. Lateral head of triceps
17. Medial head of triceps
18. Medial head of triceps
19. Brachioradialis
20. Extensor carpi radialis longus
21. Common extensor
22. Anconeus
23. Common flexor

Radius

24. Extensor pollicis longus
25. Adductor pollicis longus

Ulna

26. Triceps
27. Anconeus
28. Pronator teres
29. Apponeurotic attachment of Flexor digitorum profundus, Flexor carpi ulnaris and Extensor carpi ulnaris
30. Extensor indicis

Innominate

31. Gluteus minimus
32. Tensor fasciae latae
33. External oblique
34. Inguinal ligament
35. Sartorius
36. Gluteus minimus
37. Adductor magnus
38. Gluteus maximus
39. Piriformis

Femur

40. Obturator internus
41. Obturator externus
42. Gluteus medius
43. Quadratus femoris
44. Psoas and iliacus
45. Gluteus maximus
46. Adductor brevis
47. Vastus intermedius
48. Vastus medialis
49. Vastus lateralis
50. Adductor longus
51. Adductor magnus
52. Short head of biceps
53. Plantaris
54. Lateral head of gastrocnemius
55. Fibular collateral ligament
56. Medial head of gastrocnemius
57. Adductor magnus
58. Tibial collateral ligament

Tibia

59. Semimembranous
60. Popliteus
61. Soleus
62. Tibialis posterior
63. Flexor digitorum longus
64. Interosseous ligament
65. Posterior tibiofibular ligament/inferior transverse ligament
66. Medial collateral ligament

Chapter XIV

PLATES

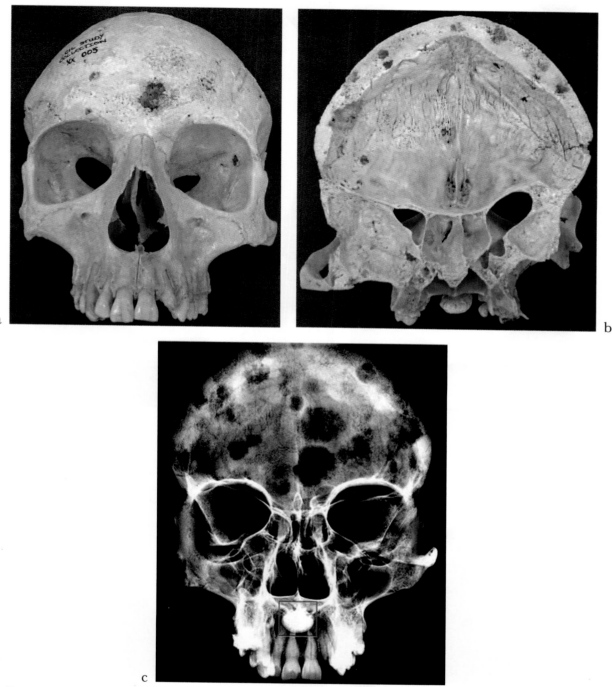

PLATE 1. The varying size and irregular shapes of these lesions are more consistent with metastatic carcinoma than multiple myeloma where the lesions tend to be more similar in size. (b and c) Note torus palatinus (white mass [square]) in a probable adult female (CSC XX005, Matthew P. Rhode). These lytic lesions, most of which can only be seen on x-ray (c), originated in the diploë and spread through the ectocranium (a). Also note the hyperostosis frontalis interna especially noticeable in the upper right photo.

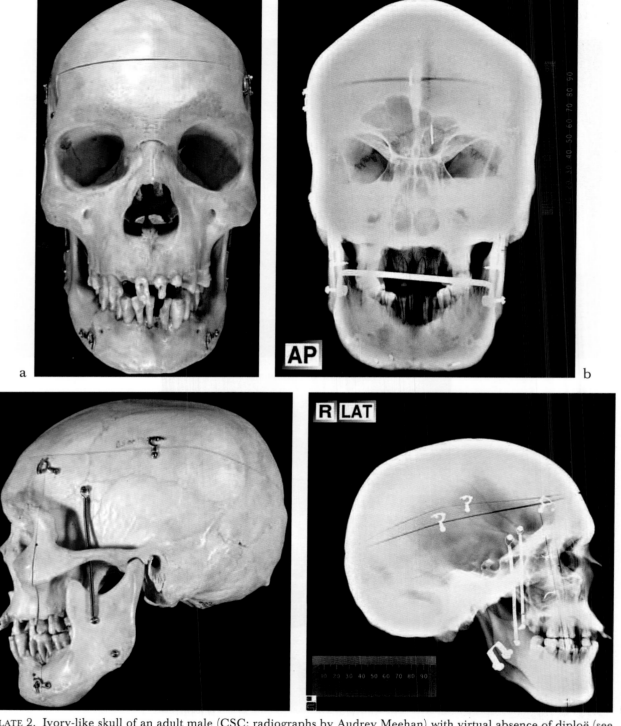

a

b

AP

R LAT

c

d

PLATE 2. Ivory-like skull of an adult male (CSC; radiographs by Audrey Meehan) with virtual absence of diploë (see Plate 5). This skull, which may exhibit fibrous dysplasia, osteopetrosis, or fluorosis, among others, weighed 1482.4 grams, or approximately 3.2 pounds (most dry human skulls weigh between 450 and 800 grams). Note the rounded inferior body of the mandible, narrow and tall "steepled" skull (a) and asymmetrical frontal sinus (b) in this teaching specimen.

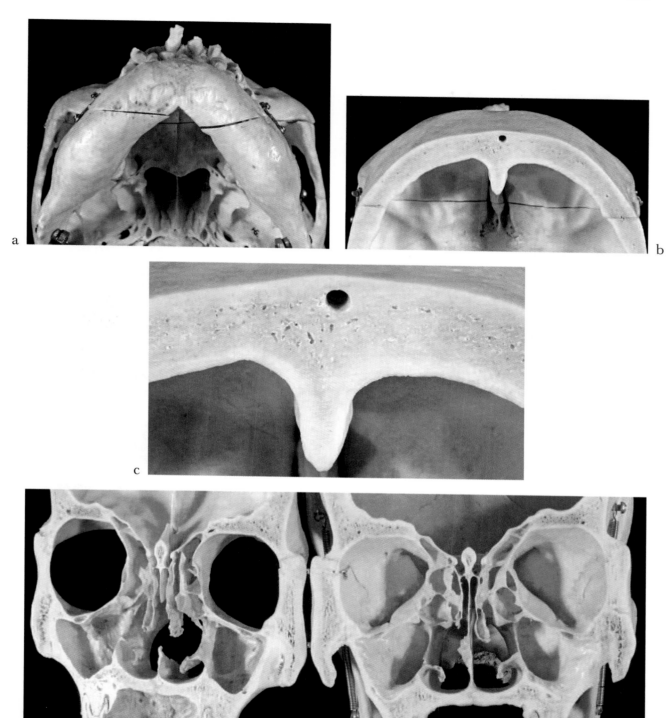

PLATE 3. Extremely robust/hyperostotic and dense mandible (a); virtual absence of diploë (b and c); and normal appearing nasal region, turbinates, ethmoidal and maxillary sinuses in an adult male with an ivory-like skull (d) (this is the same individual as shown in Plate 2).

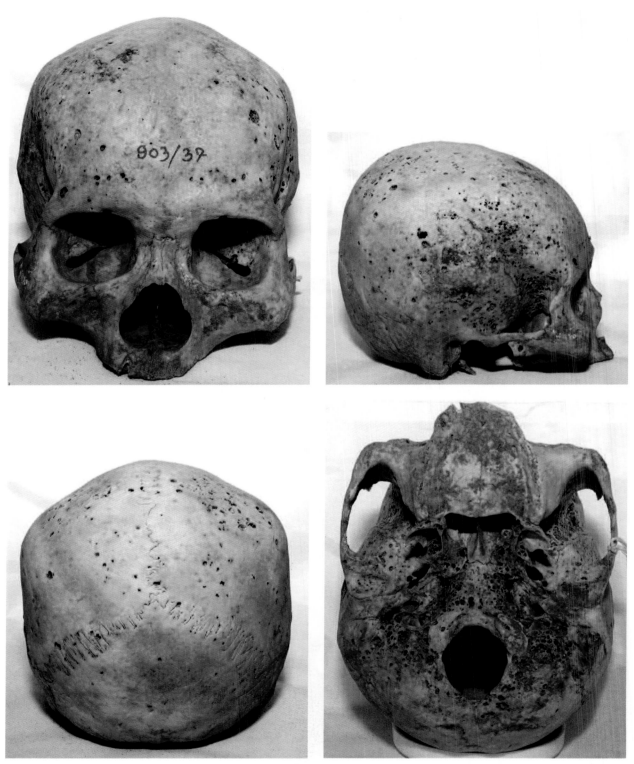

PLATE 4. Extensive lytic lesions in an elderly Thai (KKU).The approximately uniform size of these lesions is more typical of disseminated multiple myeloma than metastatic carcinoma. Note that the lesions have affected the occipital condyles, sutures, and temporomandibular joints (the dark brown areas on the hard palate and cranial base are sand from burial).

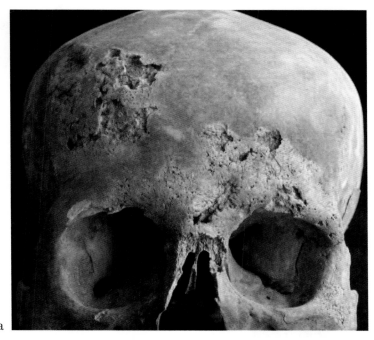

a

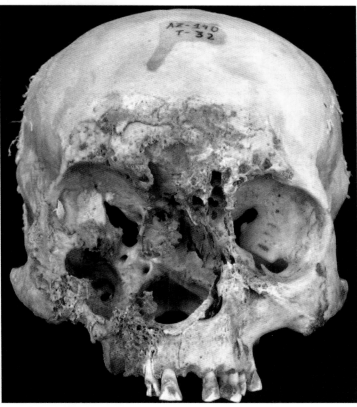

b

PLATE 5. (a) Treponematosis (i.e., syphilis, yaws, bejel, pinta) in an adult male from El Salvador (VES, Matthew P. Rhode); (b) probable mucotaneous leishmaniasis (known locally as Uta) and severely deviated nasal septum in a 45–50-year-old female who died around AD 1000 (Azapa, Matthew P. Rhode). Leishmaniasis is a parasitic infection resulting in significant loss of soft tissue and bone.

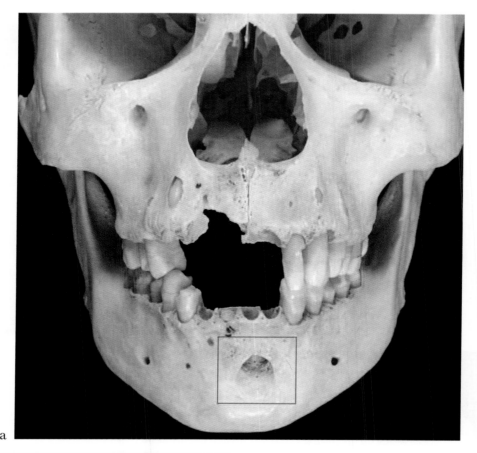

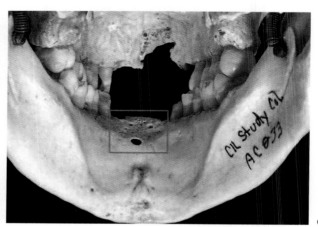

PLATE 6. Large, healed periapical abscess, radicular cyst or granuloma and draining sinus/fistula (rectangle) in the mandibular left central incisor (CSC AC033). The buccal lesion (a and b) has a smooth, non-porous margin and appears to be healed, while the lingual perforation (c) appears to be active and porous (the maxillary alveolar bone in the lower right photo was broken postmortem). Cf. Hillson 2008; Langland, Langlais and Preece 2002.

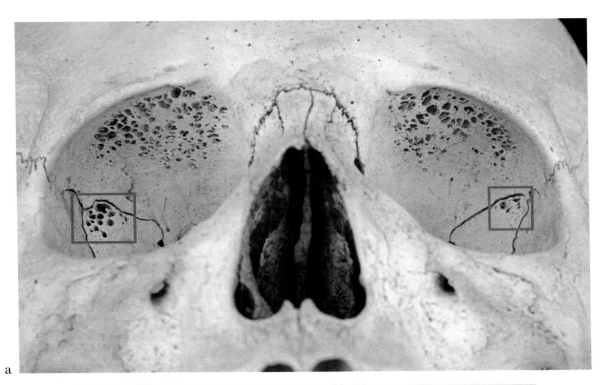

a

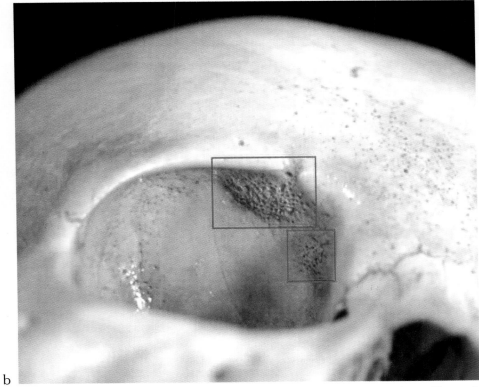

b

PLATE 7. (a) Severe bilateral cribra orbitalia in an adult Peruvian (SI). Note that the sphenoid bones (squares) are also affected. (b) Active bilateral cribra orbitalia in a subadult Thai (KKU). Both orbits exhibit raised mounds of porous bone medially and a few scattered pits elsewhere in the superior orbital plate.

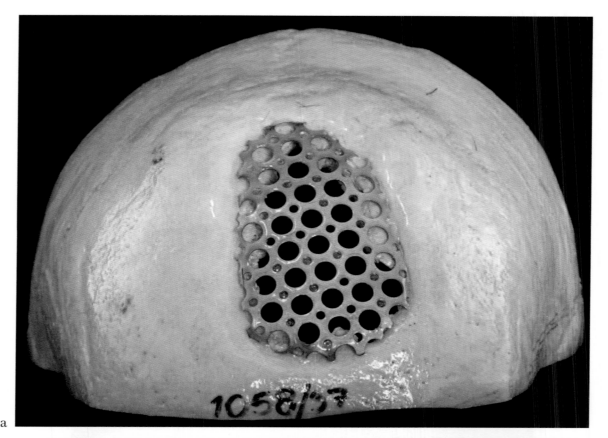

a

b

PLATE 8. (a) Ectocranial and (b) endocranial view of a metal surgical plate (cranioplasty/cranioplastic plate) in the frontal bone of an adult Thai (KKU). Note that the margin of the bony defect is smooth, rounded and healed. See also Verano and Andrushko (2010) for a Prehispanic example of cranioplasty from Peru.

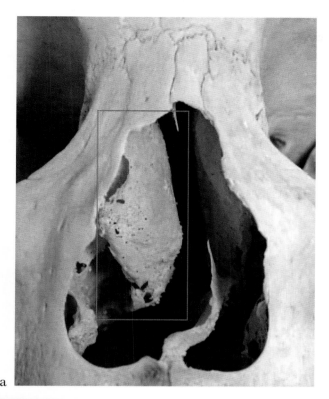

a

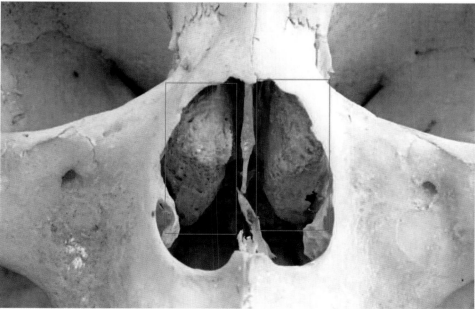

b

PLATE 9. (a) Pneumatized right middle turbinate (concha bullosa) and deviation of the septum to the individual's left (KKU). The large size of this turbinate is consistent with "massive" concha bullosa, a common anatomical variant. (b) Bilateral pneumatized turbinates, which are common findings, may be unilateral or bilateral and often result in deviation of the nasal septum. Bilateral pneumatized middle turbinates (rectangles), also known as concha bullosa, with minimal deviation of the bony nasal septum in an adult Thai (KKU). Cf. Hatipoglu et al. (2005).

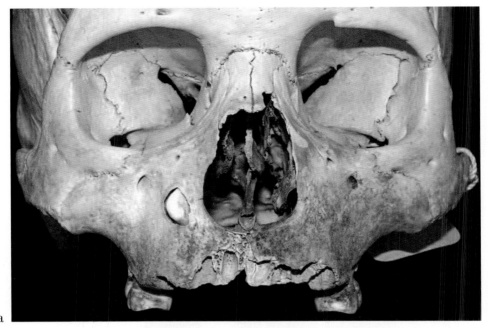

PLATE 10. (a) Ectopic tooth in the right maxilla of an adult Thai male (KKU). (b) Periapical abscesses (periapical lesions of the tip) of the maxillary central and lateral incisors in an adult Japanese male (CUJ, Hugh Tuller). Without histological analysis, it is difficult if not impossible to distinguish a periapical abscess from a periapical granuloma (the most common radiolucent lesion) or radicular cyst. Cf. Hillson (2008).

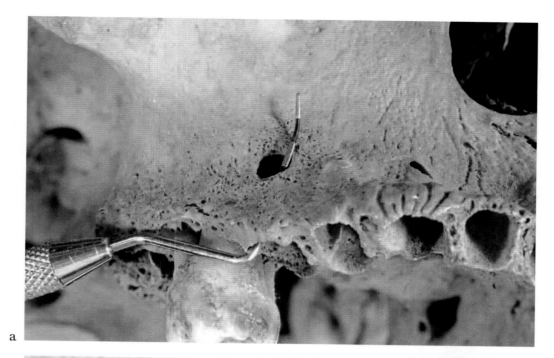

a

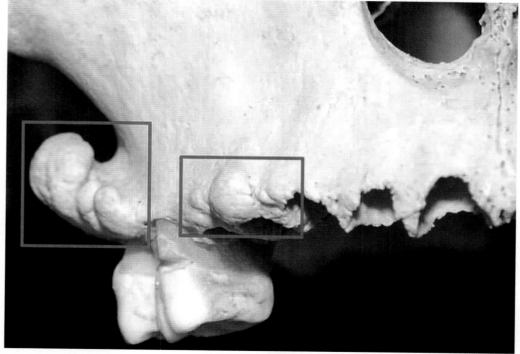

b

PLATE 11. (a) Periapical lesion (probe) in the right first molar of an adult Thai (KKU). (b) Buccal exostoses in an elderly Thai (KKU). These bony growths, which are benign, arise from the cortical plate, may have a genetic component, and are believed to be the result of a variety of causes including nutrition, periodontitis, reduced blood to this area and biomechanical stresses by the teeth. The exostoses in this individual formed along the alveoli of each tooth (nodular) and not as a single bar or platform of bone spanning several teeth (lobular). See also Horning et al. (2000) for a discussion of these lesions, abfraction and buttressing.

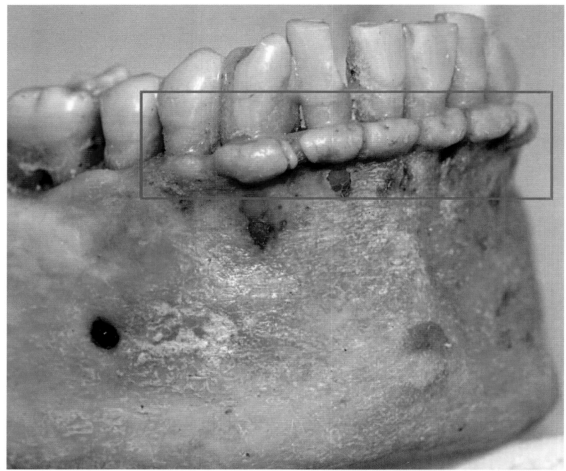

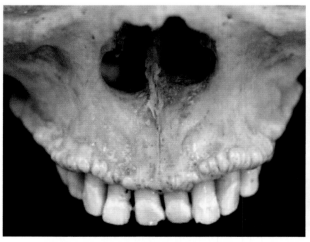

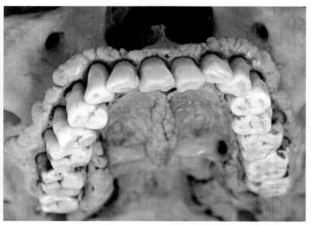

PLATE 12. (a) Nodular buccal exostoses in an adult Thai male (KKU). Note an exostosis along the opening of each alveolus. Exostoses such as these are benign and more commonly form along the lingual than the buccal surface of the mandible. Cf. Horning et al. (2000). (b and c) Rounded nodular buccal and lingual exostoses and a small midline torus palatinus (c) in an adult Thai (KKU). Note that many of the exostoses have formed between the teeth, rather than (a) anterior to them as in the top photo. Cf. Horning et al. (2000).

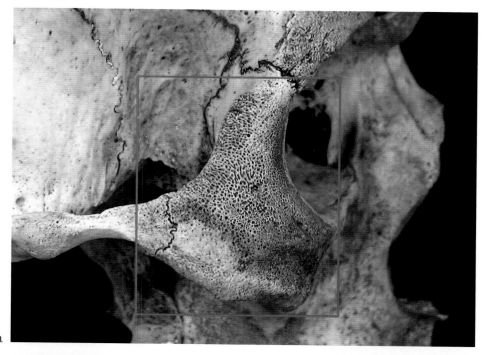

a

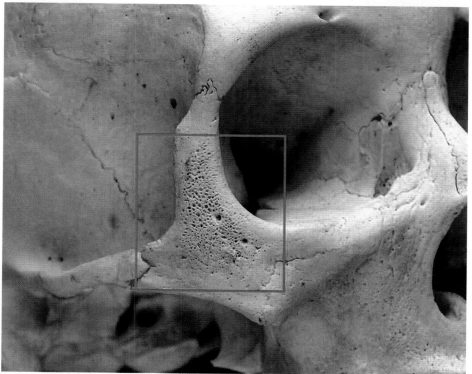

b

PLATE 13. (a) Severe porotic hyperostosis, also known as Tara Grufferty syndrome, of the right malar bone in an adult male – note the affected "spongy" bone is both porous and thickened (Photo courtesy of UTK FAC, WMB UTK, Joe Hefner). See also El Najjar et al. (1975, 1976), Hrdlička (1914), Ortner (2003), Stuart-Macadam (1992), and Walker et al. (2009). (b) Mild to moderate porotic hyperostosis of the malar bones in a Japanese adult (CUJ, Hugh E. Tuller).

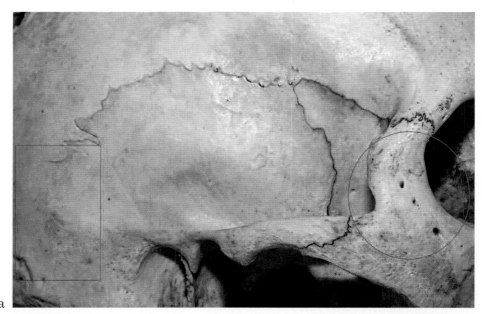

PLATE 14. (a) Partial premature craniosynostosis (rectangle) of the right squamosal suture in an adult Thai male and multiple zygomaticofacial foramina (circle; a normal variant) (KKU). (b) Scaphocephaly, prominent malar tubercle (square; a normal non-metric trait) and frontal grooves in an adult Caucasoid female (Huntington SI). Note that the other sutures are open/patent. Scaphocephaly results from premature synostosis (what some researchers refer to as "craniosynostosis") of the sagittal suture before the brain reaches its full growth size (subadult), resulting in a skull that is long front to back (anterior to posterior).

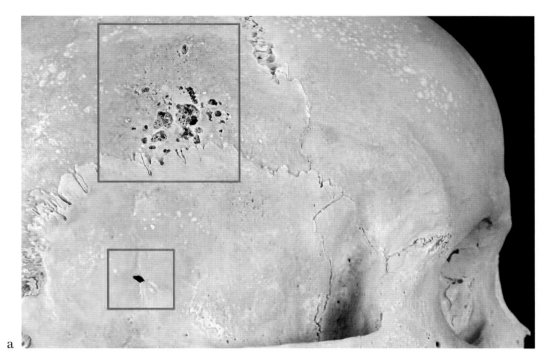

a

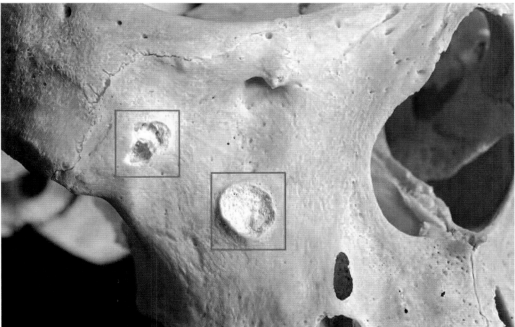

b

PLATE 15. (a) Lytic lesions (squares) of unknown etiology, but possibly reflecting multiple myeloma in the right parietal and temporal bones of a probable 25–40-year-old Asian female (CSC AC036; only the skull is present) (see x-rays in Plate 17). Close examination revealed that the small perforation in the right temporal bone is actually a lytic lesion. The vault lesions originated in the diploë and progressed through the endocranium and ectocranium. (b) Note the circular lytic lesions of unknown etiology in the right maxilla. These two lesions have sharp, raised rims (faintly sclerotic on x-ray) and extreme microporosity encircling the largest defect. The broken canine alveolus occurred postmortem and not as a result of the disease.

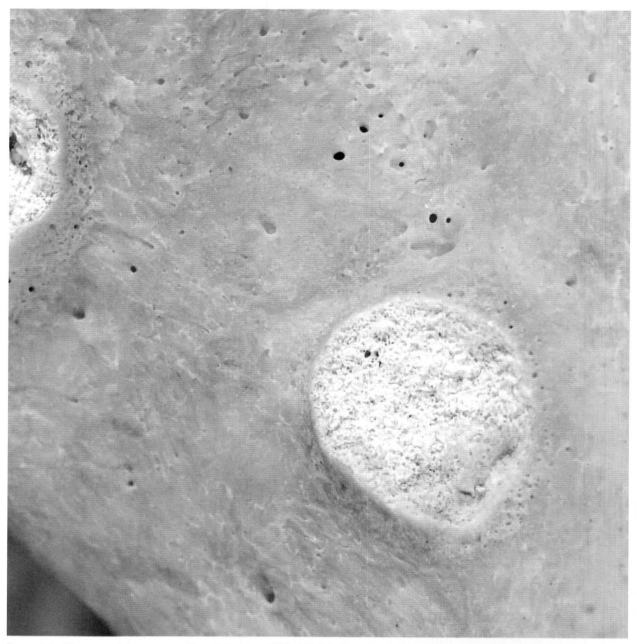

PLATE 16. Close-up of the two maxillary lesions in CSC AC036. Note the raised rim, microporosity and new bone around both lesions.

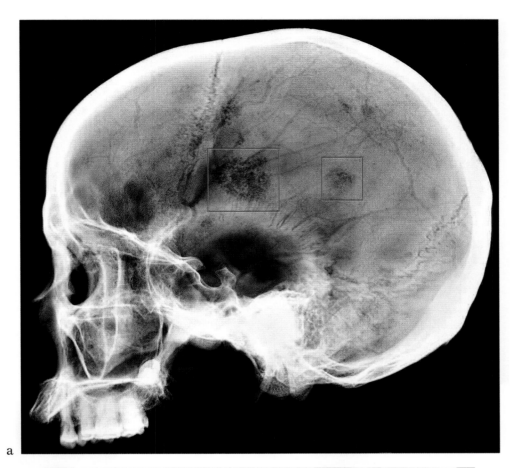

a

b

PLATE 17. (a) Lateral and (b) frontal (AP) views of skull (CSC AC036; radiographs by Audrey Meehan) showing lytic lesions (some of which are indicated by squares) with no evidence of sclerosis (dense bone) possibly reflecting multiple myeloma and a mesiodens (circle) in the incisive fossa (see Plates 15 and 16). The maxillary lesion originated in the diploë and eroded both the anterior/facial and posterior plate. Also note the impacted right third molar. See also Ersin et al. (2004).

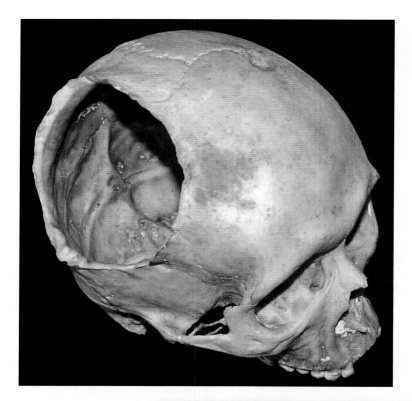

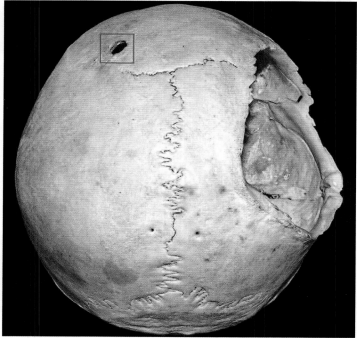

PLATE 18. Possible craniotomy, infection, or tumor in the right parietal and frontal bones in an adult Thai male (KKU). Note the drainage/shunt hole (bottom, square) in the left frontal bone and the raised, outward projecting margin encircling the craniotomy. The flared margin and absence of a burr hole and metal surgical wires suggest the craniotomy was not covered with a bone flap or other cranioplastic material.

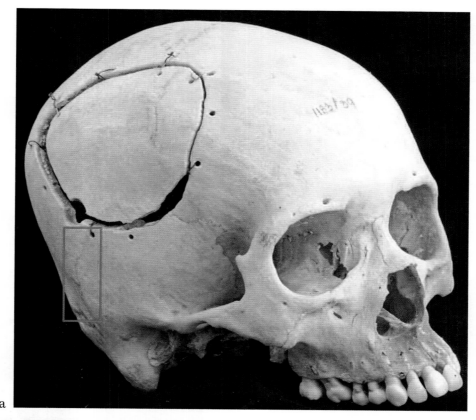

a

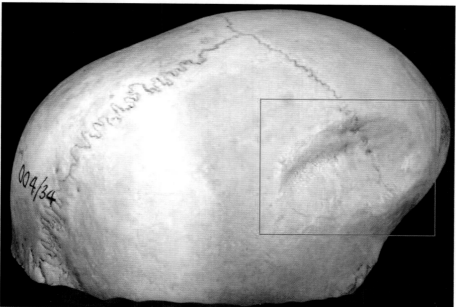

b

PLATE 19. (a) Cranioplasty bone plug and wire sutures securing a portion of the right parietal, frontal and temporal bones in an adult Thai, following cranial trauma (KKU). Note the unhealed perimortem fracture (rectangle). Also note the round burr hole above the rectangle. (b) Large, healed depressed "pond" (reflecting its shape) fracture due to blunt force trauma to the right parietal and frontal bones (KKU). Cf. Lovell (2008).

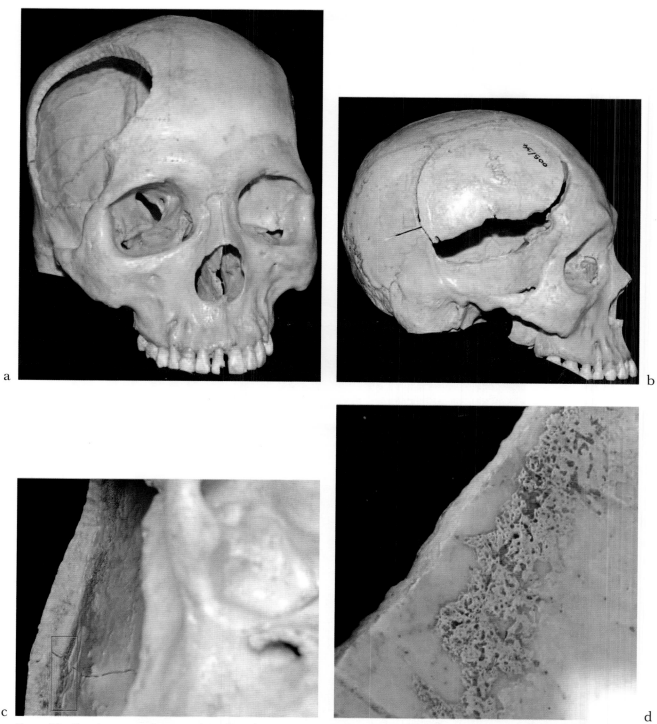

PLATE 20. Craniotomy with cranioplasty (bone) of the right parietal, frontal and temporal bones with remodeling (c and d) in the form of fibrous/woven new bone circling the defect endocranially (c) – craniotomy was performed in this individual following cranial trauma (KKU). Woven bone such as that along the endocranial surface (d) forms as a result of deposited osteoid that mineralizes (Lovell 2008).

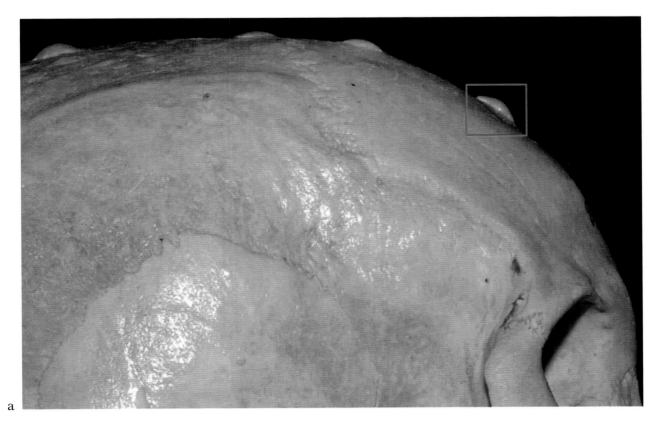

a

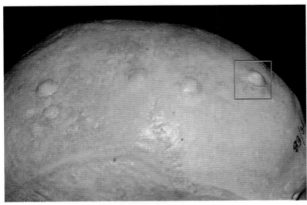

b

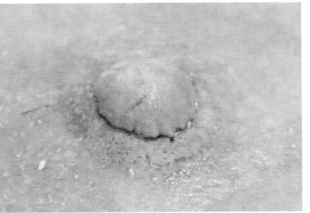

c

PLATE 21. (a) Multiple button osteomas in the frontal and parietal bones of an adult Thai male (KKU). (b and c) Note the dome shape of the lesions (square), raised margins and resemblance to a mushroom where the pedunculated osteoma joins the frontal bone.

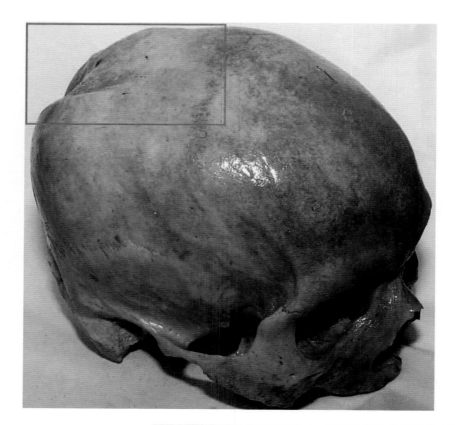

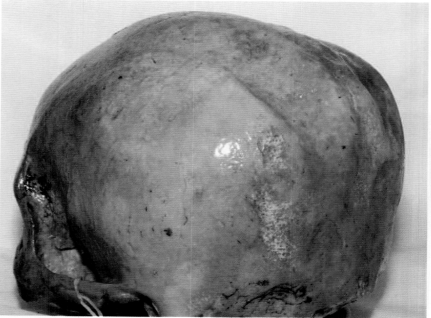

PLATE 22. Biparietal thinning, also known as symmetrical osteoporosis, osteoporosis circumscripta and senile atrophy in an elderly Thai male (KKU). As the name implies, symmetrical osteoporosis almost always affects both parietals symmetrically, causing erosion of the ectocranial plate (outer table) and diploë that may rarely perforate the endocranium (inner table). This condition generally affects middle-aged and older individuals, has been reported in subadults, and is not an anatomical variant (Cederlund et al. 1982).

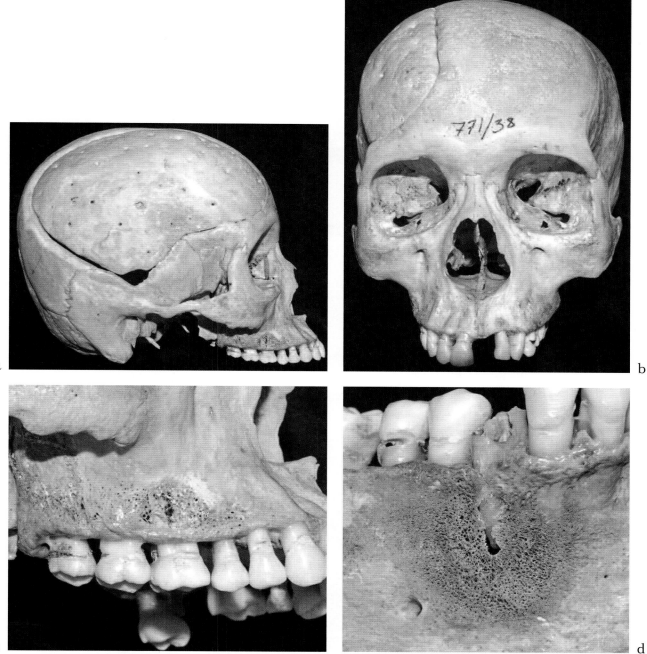

PLATE 23. (a and b) Craniotomy and cranioplasty with evidence of healing of the right parietal and frontal bones and (c and d) active periapical abscesses (periapical lesions) in the right maxilla (right first molar) and mandible (right canine) in an adult Thai male (KKU). Note the porous and dark color of the bone around the root apices and the V-shaped horizontal abrasion grooves along the cementoenamel (cervical) junction of several of the teeth (especially the right second premolar, which has penetrated the pulp. Cervical lesions/grooves are believed to primarily be the result of abfraction (occlusal stress leading to fracturing and erosion of the cervical region) and excessive horizontal tooth brushing, although Bartlett and Shah (2006) found little evidence supporting abfraction as a contributing factor. The small surgical drill holes in the bone flap were done to prevent the accumulation of fluid between the scalp and flap (ectocranium) or the flap and brain (endocranium). Cf. Hillson (2008), Langland et al. (2002), Piotrowski et al. (2001), and Rees (2006).

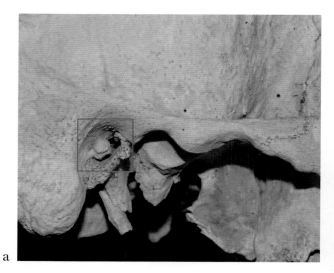

a

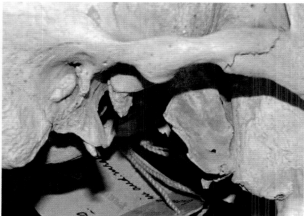

b

c

Plate 24. (a and b) Auditory exostoses, benign tumors, in the right temporal bones of two adult Peruvian males (both of these individuals had bilateral exostoses (SI). Auditory exostoses are often attributed to immersion, such as swimming, surfing, and diving in cold water and is sometimes referred to as surfer's ear in some of the more recent literature. (c) Auditory exostoses (square) that nearly filled the left auditory meatus (KKU).

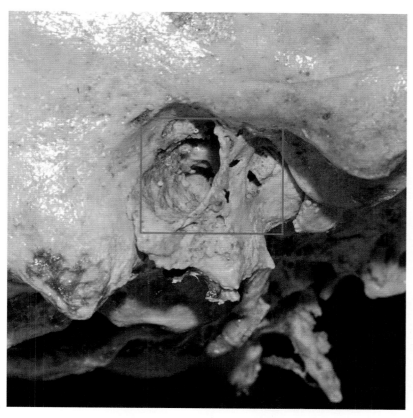

a

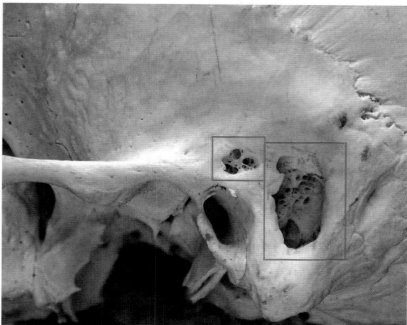

b

PLATE 25. (a) Otitis media resulting in multiple perforations of the tympanic plate and irregular growths in the right external auditory meatus in an adult Thai (KKU). (b) Chronic otitis media of the left mastoid in an adult Japanese (CUJ, Hugh E. Tuller). Note that the margins of the two defects are smooth and rounded, suggesting these are healed lesions.

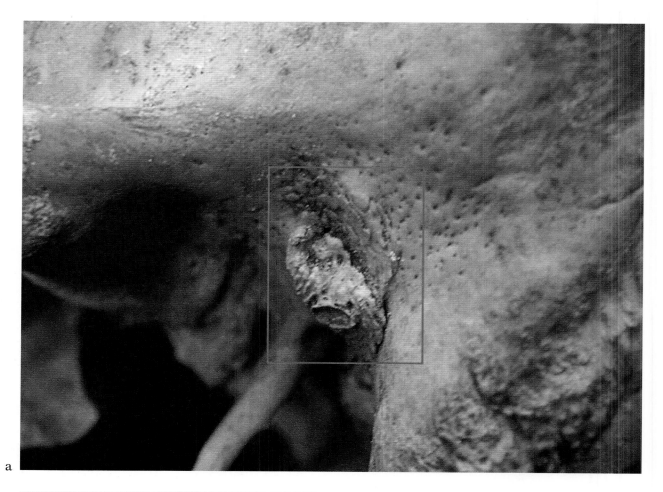

a

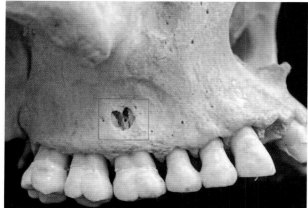

b

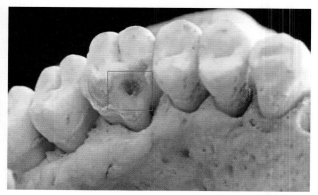

c

PLATE 26. (a) Possible atresia (absence) of the auditory canal in a 30–44-year-old male from Villa El Salvador (VES 173 Matthew P. Rhode). (b) Heart-shaped periapical abscess (periapical lesions) of the maxillary right first molar showing the probable origin (c) of the infection through antemortem fracture of the lingual cusp in an adult Thai (KKU). Cf. Hillson (2008), Langland (2002).

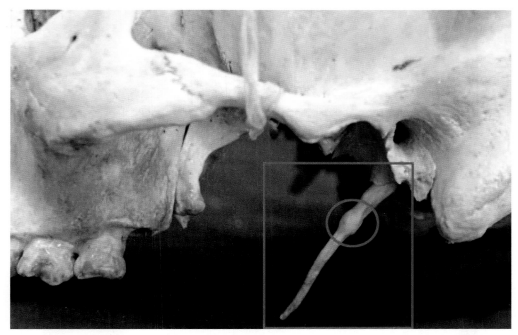

PLATE 27. Elongated styloid process, often referred to as Eagle syndrome, in the left temporal bone of an adult Thai (KKU). Note the segmented joint (circle) where the bony portion of the styloid process ended and the ossified stylohyoid ligament began. The distal end of the stylohyoid ligament attaches to the lesser horns of the hyoid cornu. Similar joints may form along the "stylohyoid chain" as the result of normal ossification or trauma. Cf. Basekim et al. (2005).

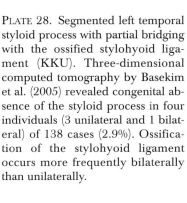

PLATE 28. Segmented left temporal styloid process with partial bridging with the ossified stylohyoid ligament (KKU). Three-dimensional computed tomography by Basekim et al. (2005) revealed congenital absence of the styloid process in four individuals (3 unilateral and 1 bilateral) of 138 cases (2.9%). Ossification of the stylohyoid ligament occurs more frequently bilaterally than unilaterally.

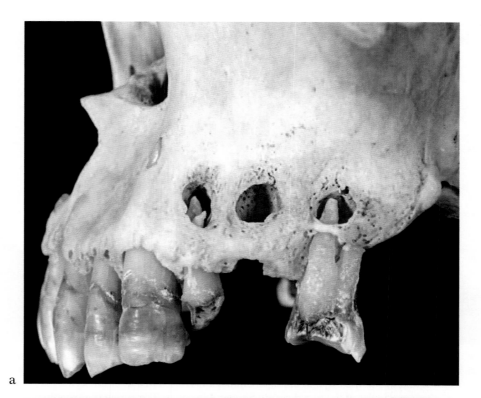

PLATE 29. (a) Periapical abscesses (periapical lesions) with resorption and perforation of the buccal plate housing the maxillary left molars (CSC AC021). (b) Resorption of the lingual alveolus of the left third molar such that the root is no longer anchored in bone. Cf. Hillson (2008).

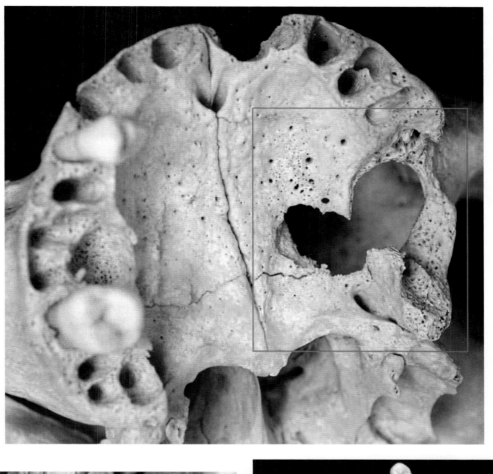

a

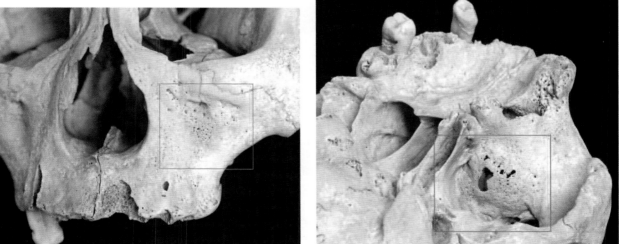

b   c

PLATE 30. (a) Possible periapical abscess (periapical lesion) of the maxillary left molars resulting in perforation and destruction of the hard palate and maxillary sinus, (b) inflammation and periostitis (sinusitis) of the facial surface of the left maxilla, and (c) enlargement and perforation of the maxilla in the region of the posterior dental canals (UHWO 980). Cf. Hillson (2008).

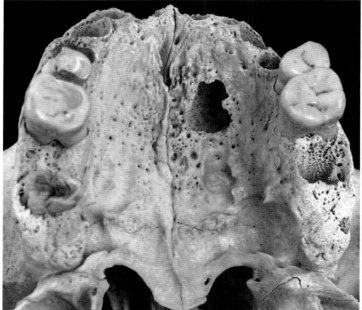

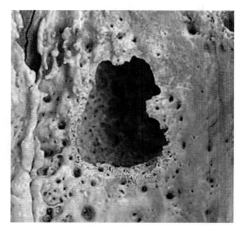

PLATE 31. Periapical lesions (circle) of the left first premolar that perforated the palate in an adult Peruvian male (SI) (Hillson 2008). The jagged nature of the palatal lesion and porosity suggest this lesion was active at the time of death.

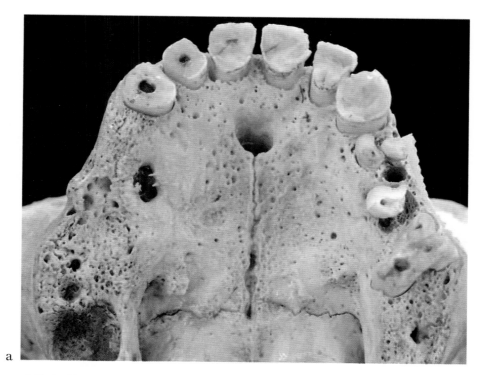

a

b

PLATE 32. (a) Severe alveolar attrition of the teeth in an adult Thai resulting in ante-mortem loss of several teeth, loss of enamel and exposure of the dentine and pulp chambers (KKU). (b) Moderate dental attrition resulting in exposure of the dentine (KKU). Note the small torus palatinus along the midline.

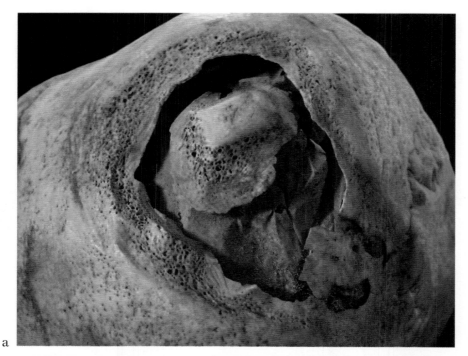

PLATE 33. (a) Healing trepanation (trephination) in the left parietal bone of a middle-aged adult male from Peru (Paracas, Matthew P. Rhode) – note that the dura mater is attached to and holding the bone island in place. (b) Partially healed double trepanation above the right orbit and right parietal bones in an ancient Peruvian male (University of Pennsylvania Museum of Archaeology and Anthropology, 29-144-8).

PLATE 34. Large button osteoma in the left parietal of an elderly adult Thai male (KKU). Note there is a well-defined and somewhat raised margin separating the osteoma from the surrounding cortex.

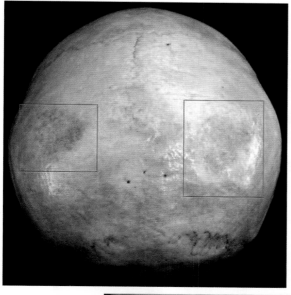

a

PLATE 35. (a) Symmetrical osteoporosis/biparietal thinning/senile osteoporosis in an elderly Thai (KKU). Note that the size and shape of the two resorptive areas is similar, but they differ slightly in position in the parietal bones. Also note that the thinning does not cross the temporal lines laterally, possibly due to biomechanics of the muscles, or blood supply to the muscle region. (b) Backlighting of one of the saucer-shaped depressions reveals that the thinnest portion is in the center of the "crater" – the outer table and diploë have resorbed and all that remains is an eggshell-thin remnant of the inner table. Perforation of the endocranium is extremely rare. Cf. Cederlund et al. (1982).

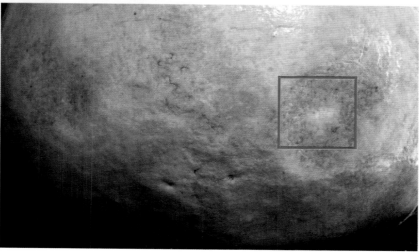

b

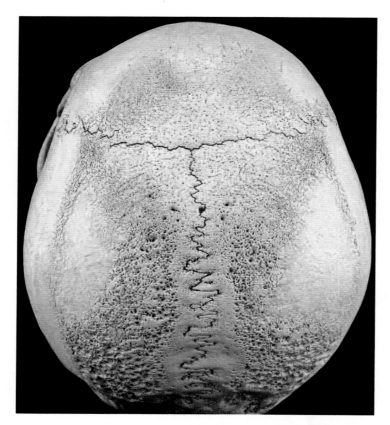

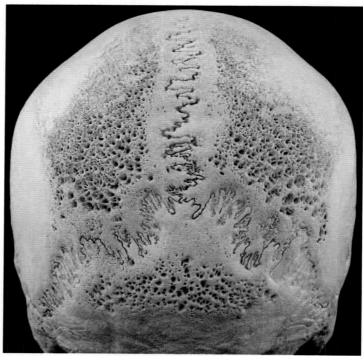

PLATE 36. Severe porotic hyperostosis in an adult Peruvian skull (SI). Note symmetry of the lesions and that they do not affect the sutures or adjacent bone, areas with minimal diploë.

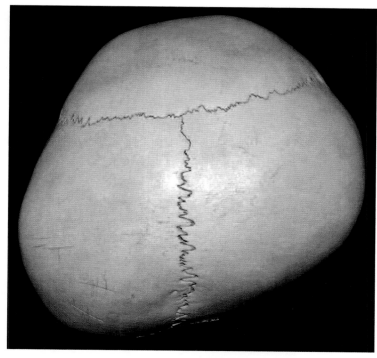

a

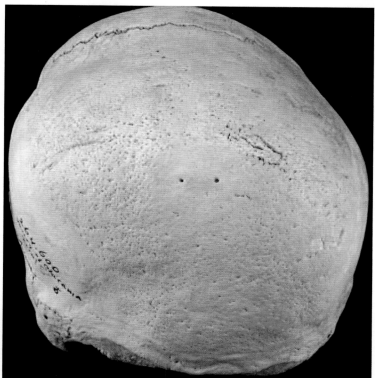

b

PLATE 37. (a) Artificial cranial deformation (the sutures are still patent/open) of right occipital region in a Peruvian subadult (SI). See also Dingwall (1931), Rhode and Arriaza (2006), and Schijman (2005). (b) Cranial asymmetry due to premature synostosis (craniosynostosis, craniostenosis) of the sagittal and lambdoidal sutures, and healed porotic hyperostosis in an ancient Peruvian adult (Peru, SI).

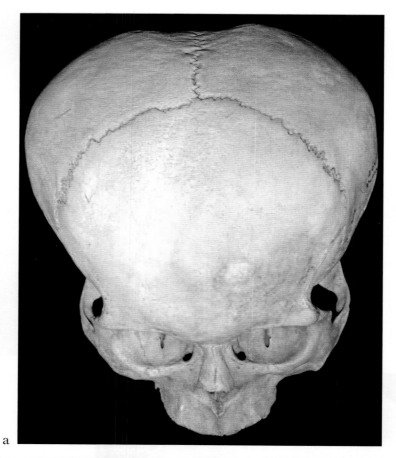

a

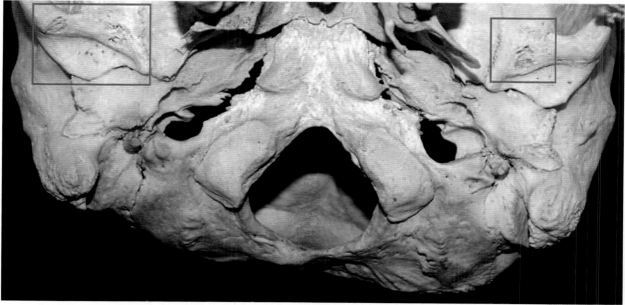

b

PLATE 38. (a) Asymmetrical skull due to cultural modification (binding in the region of the sagittal suture) of the occipital region. (b) Note the asymmetrical occipital bone, degenerative joint disease and osteoarthritic pitting of both temporomoandibular joints and eminences (Peru, SI).

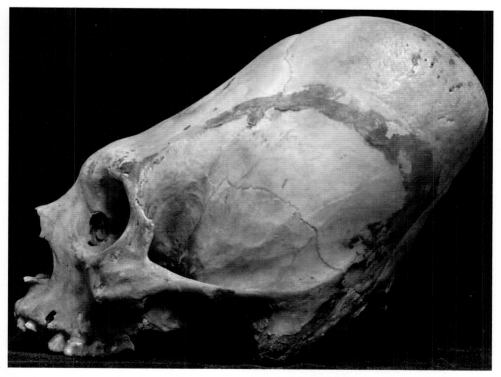

a

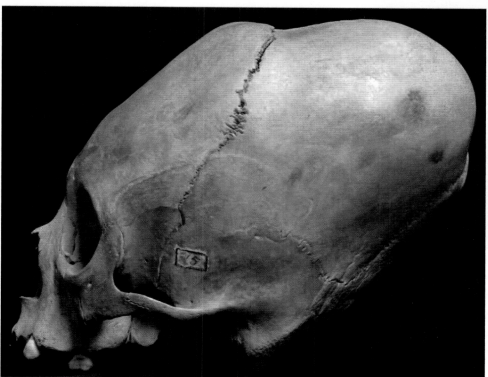

b

PLATE 39. (a) Cultural modification (wrapping), often referred to as artificial cranial deformation, in an adult and (b) a child from Peru (University of Pennsylvania Archaeology and Anthropology Museum, L606 top and 24-142-224 bottom). See also Dingwall (1931), Hoshower et al. (1995), Rhode and Arriaza (2006), and Schijman (2005).

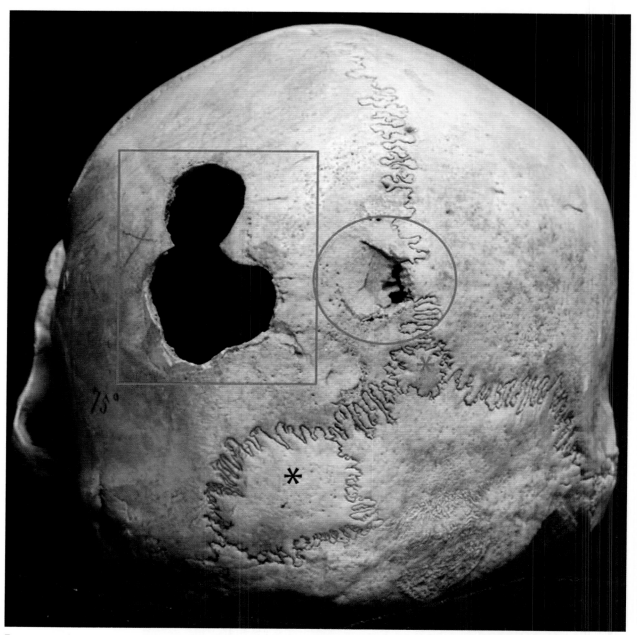

PLATE 40. Three circular blunt force injuries with no evidence of healing in the back of the skull of an adult Peruvian male – two injuries are adjacent (rectangle) to one another and one (circle) is along the sagittal suture (University of Pennsylvania Archaeology and Anthropology Museum, L606 97). Note the small ossicle (red asterisk) at Lambda and large ossicle (black asterisk) along the left half of the lambdoidal suture.

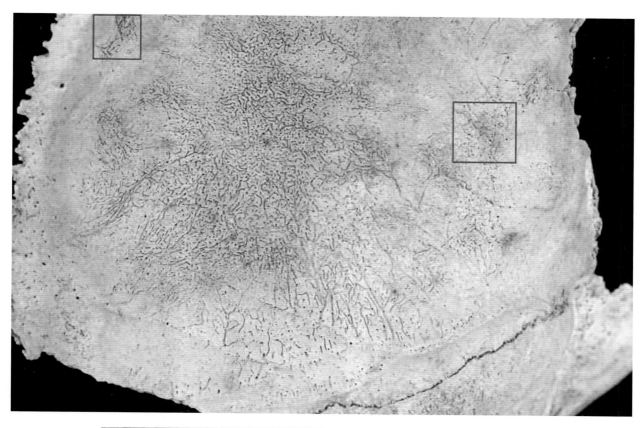

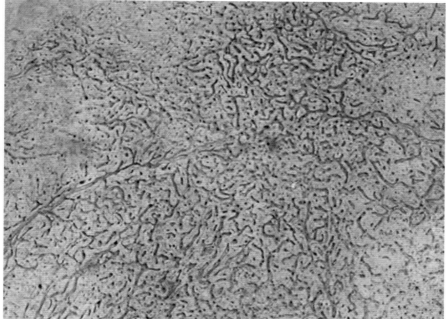

PLATE 41. Serpens endocrania symmetrica (SES) in a subadult right parietal bone, a condition that, according to Hershovitz et al. (2002), may reflect intrathoracic disease or tuberculosis (SI). Note the coral-like appearance of the endocranial surface, that the "canals" are approximately the same width and depth, and that some of these lesions (squares) appear as "islands" situated away from the central focus near the middle of the parietal bone.

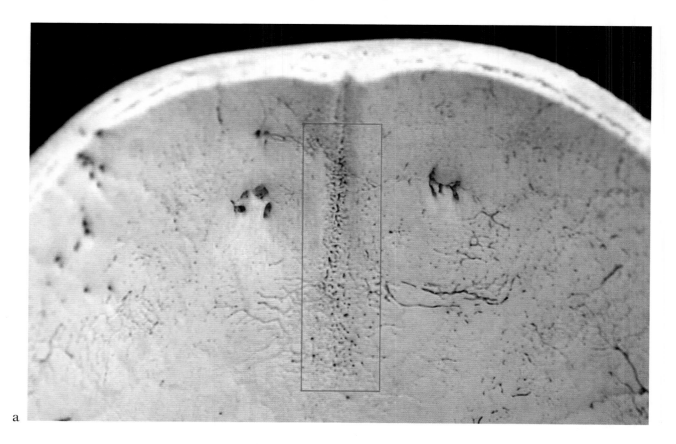

a

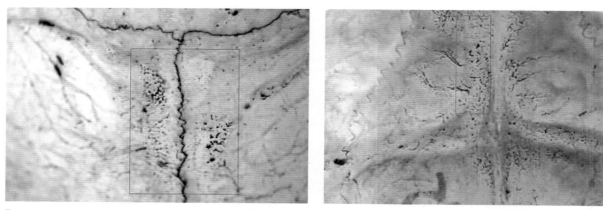

b

c

PLATE 42. Serpens endocrania symmetrica (SES) in a middle-age or older Thai female (KKU). (a) Note that the lesions are along the midline in the frontal bone, (b) sagittal suture and the (c) occipital bone. In this individual the lesions occupied the meningeal artery grooves in the parietal bones and diploic areas in the occipital bone; lesions typically do not affect areas of compact "solid bone" such as the crests in the occipital bone. The term *serpens endocrania symmetrica* is a descriptive term and not the name of a disease or necessarily a differential diagnosis.

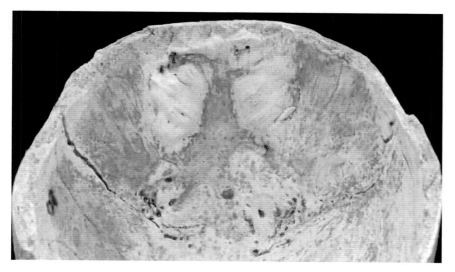

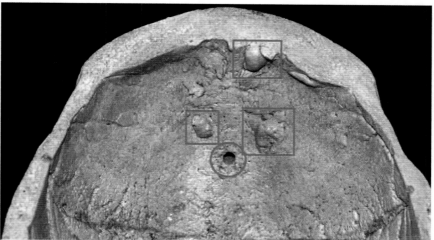

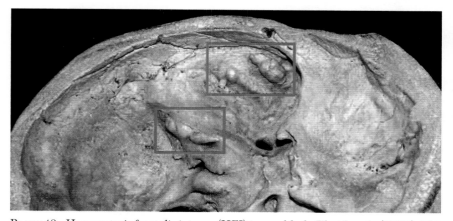

PLATE 43. Hyperostosis frontalis interna (HFI) in an elderly Thai female (KKU). This calotte was much denser and heavier than normal (whitening of the HFI was the result of postmortem cleaning). Hyperostosis frontalis interna (HFI) in an adult Thai female with nodular involvement of the middle cranial fossa, petrous process and sphenoid bone (bottom, rectangles) (KKU). HFI is more often found in older females. The circle in the middle photo indicates a Pacchionian pit, an anatomical variant that houses arachnoid granulations that filter cerebrospinal fluid. Cf. She and Szakacs (2004) (HFI).

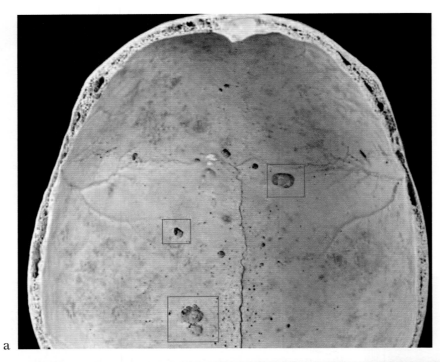

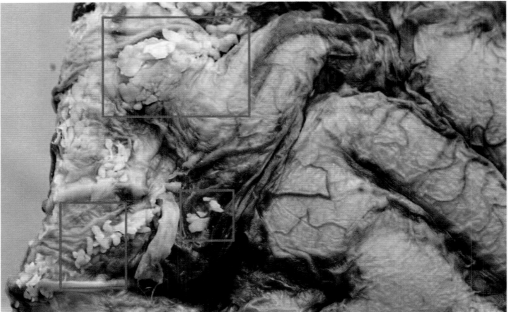

PLATE 44. (a) Pacchionian pits/fovea granulares (squares) in the bone that house (b) arachnoid granulations that filter the cerebrospinal fluid (bone specimen courtesy KKU and soft tissue specimen courtesy FSA). Note that these pits (a) are often located at the end of meningeal artery grooves like a river emptying into a lake, they erode and push through the inner table of the skull, and may appear as a single pit, or as a cluster resembling grapes. Arachnoid granulations (b, from a different individual), so-called because of their resemblance to granules and grains of sand, are herniations of the thin, delicate arachnoid membrane (b, covers the brain), but they probably do not perforate the outer table of the skull in the living. Arachnoid pits and depressions may, however, result in perforation of the outer table as a result of postmortem handling and breakage.

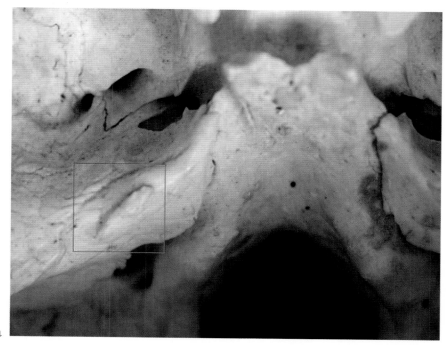

a

b

PLATE 45. (a) Ossification along the superior border/apex of the petrous portion of the left and (b) right temporal bones in two individuals, possibly associated with the superior petrosal or trigeminal nerve (attachment of the cerebellar tentorium) (KKU). This bony projection is superior and medial to the internal auditory/acoustic meatus and varies from a small ridge, to a spiked projection or hook. The authors have noted similar projections in numerous skulls and believe it may be a relatively common feature that may develop with age.

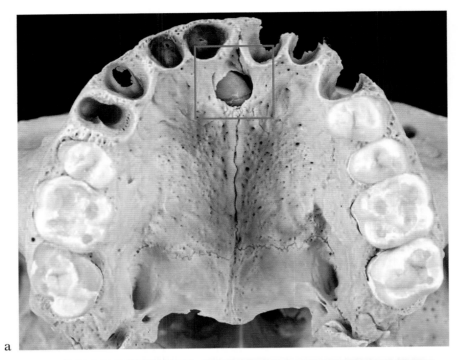

a

b

PLATE 46. (a) Mesiodens (square) in the incisive fossa of an adult probable female (CSC AC-036; only the skull is present). A mesiodens is a supernumerary tooth (odontoma) situated in the midline between the maxillary central incisors, has a frequency of 0.15–1.9% in the general population, occurs more frequently in males, and is the most common type of supernumerary tooth (Buggenhout and Bailleul-Forestier 2008). (b) Mesiodens along the anterior median palatine suture in a subadult Thai (KKU). Note severe resorption and destruction of the right maxilla in the molar region.

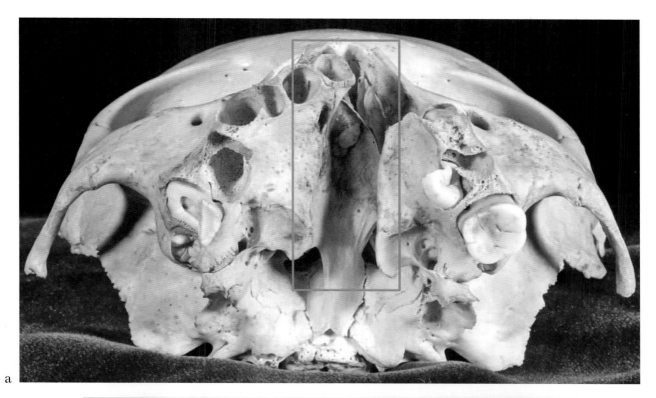

a

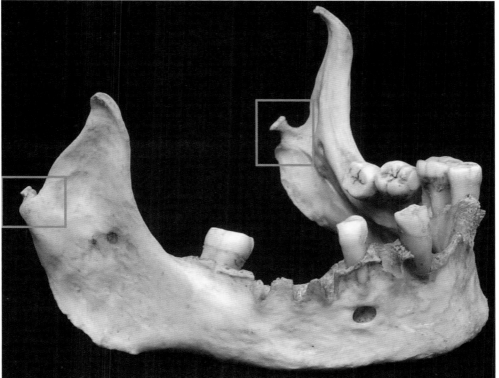

b

PLATE 47. (a) Cleft palate in an 8- to 10-year-old child from Nazca, Peru (SI 293252). (b) Hypoplastic mandibular ramus and condyles and misshapen left coronoid process in an adult (Mütter Museum).

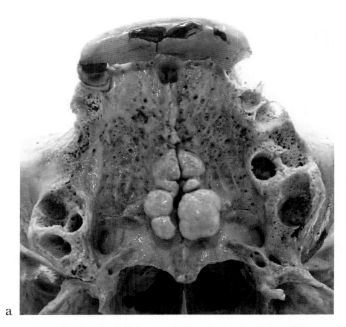

a

b

PLATE 48. (a) Large cauliflower-like torus palatinus and partial denture/dental appliance in an elderly Thai male (KKU). Note that this individual is missing most of his front teeth, hence the dental appliance. The porous nature of some of the tooth sockets, comprised of cribriform plate (lamina dura), indicate remodeling following tooth loss. (b) Small stalk-like enamel pearl (square), also known as an enameloma and an enamel extension (circle) in a maxillary molar of an 18–21-year-old female (VES, Matthew P. Rhode). Enamel pearls, so named because they resemble pearls, are ectopic formations of enamel found along the root bifurcation and are relatively common findings in most groups.

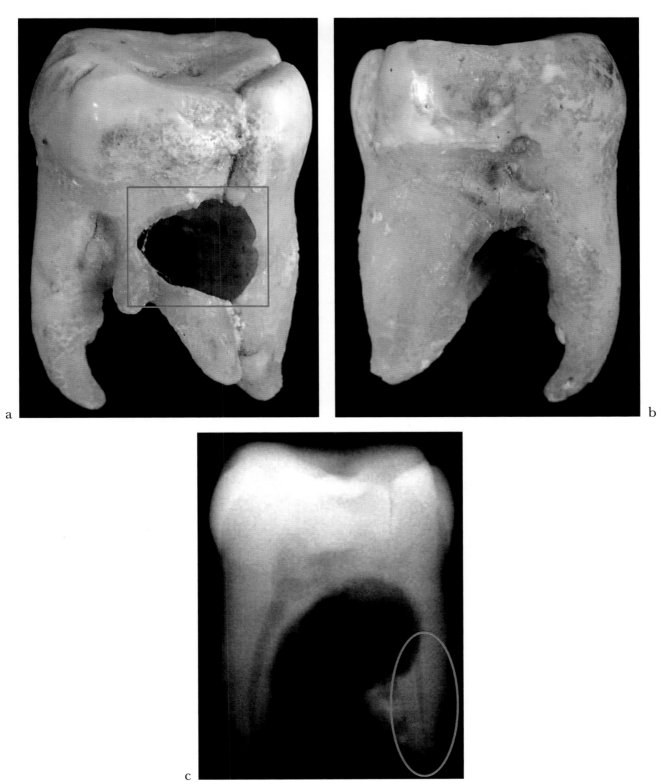

PLATE 49. (a and b) Invasive cervical resorption (a, square) in an upper right third molar. (c, oval) Note resorption of one of the pulp chambers (Thai, not KKU). Cf. Trope (1998) and Heithersay (1998, 2004).

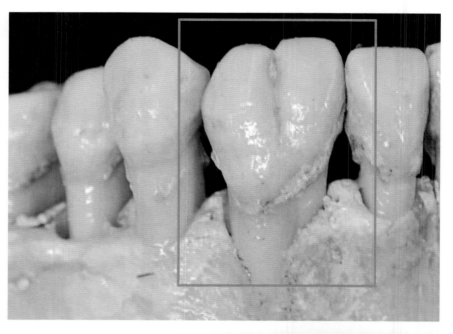

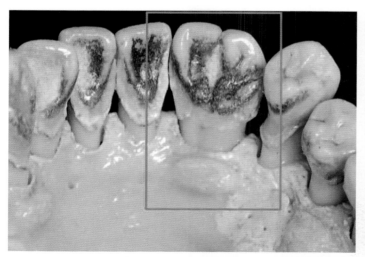

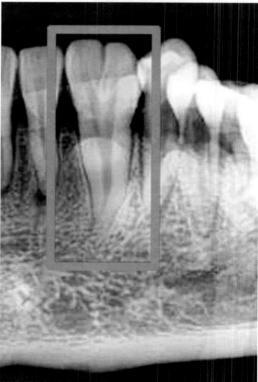

PLATE 50. Geminated (geminate) right lateral incisor and canine in an adult Thai (KKU). Geminated teeth are developmental anomalies arising from one tooth bud, resulting in two crowns that may be totally or partially separated and one root. Two teeth that are fused together is called twinning or twinned teeth – count the number of teeth in the mandible or maxilla with the twinned tooth and if there is not a missing tooth (has the normal number), the condition is gemination; if there is one less than the normal number of teeth, the condition is "fusion." In this example, note in the x-ray (bottom right) that there are two crowns but only one Y-shaped pulp chamber (i.e., one root). Cf. Alves et al. (2010).

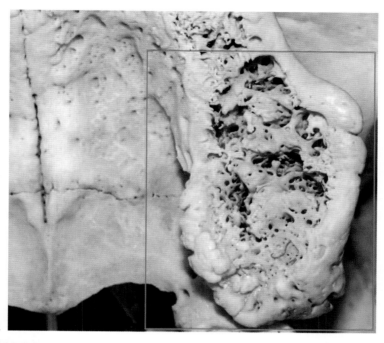

a

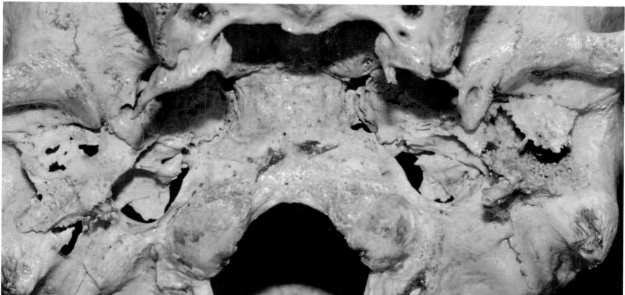

b

PLATE 51. (a) Remodeled maxillary left tooth sockets (alveoli) in an elderly Thai individual. Note that the buccal and lingual plate of the maxilla have extended their margins, resulting in bulging exostoses (KKU). (b) Bilateral otitis media (middle ear infection) with perforation of the tympanic plate in an adult Thai (KKU) – chronic otitis media can be caused by a variety of diseases including tuberculosis and Langerhans cell histiocytosis, among others.

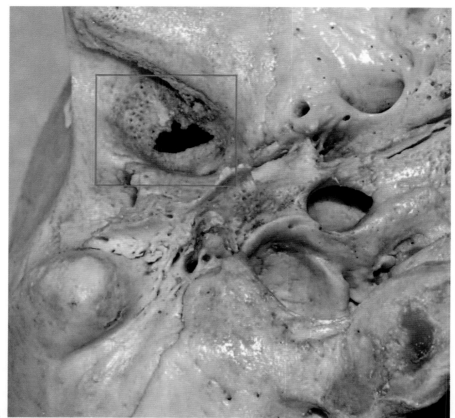

a

b

PLATE 52. (a) Probable infection leading to perforation (rectangle) of the right temporo-mandibular plate (not to be confused with a tympanic dehiscence that affects the more posteriorly positioned tympanic plate of the auditory meatus). Note the dark, loosely adhering reactive bone and porosity surrounding the defect (KKU). (b) Chronic otitis media resulting in destruction of the right tympanic plate, exposing the oval window (rectangle) that accommodates the footplate of the stapes and the round window (circle) that equalizes pressure that enters the membrane-covered oval window (KKU).

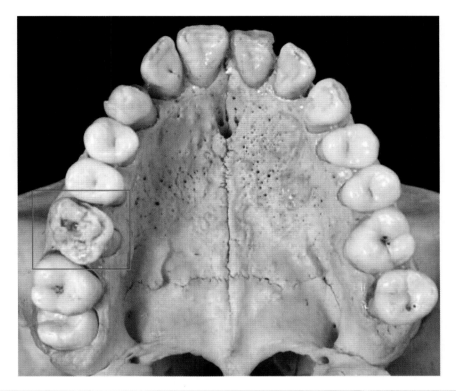

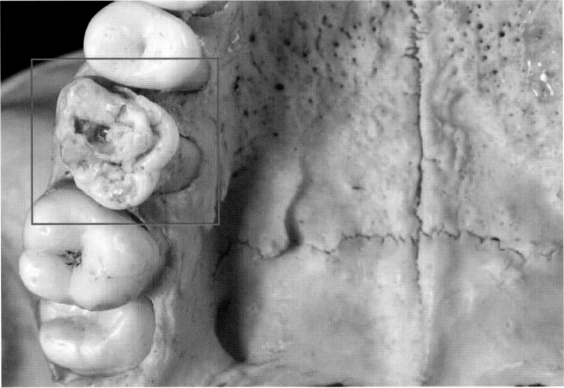

PLATE 53. Amelogenesis imperfecta affecting the maxillary right first molar (squares) and left maxillary central incisor in a young adult Thai (KKU).

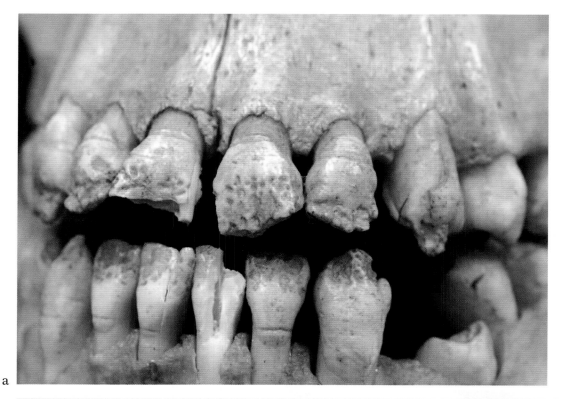

a

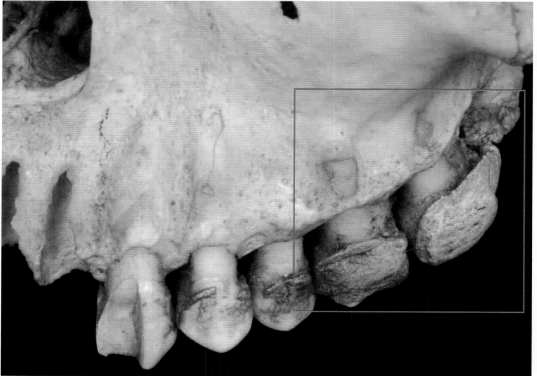

b

PLATE 54. (a) Amelogenesis imperfecta of the teeth in a child (Mütter Museum 1008.57). (b) Calculus (square) in the left maxillary teeth of an adult (Mütter Museum 1008.73).

a

b

PLATE 55. (a) Postmortem breakage of the left mastoid process revealing its normal internal structure of air cells and chambers – note that parts of the bone covering the mastoid process are eggshell thin (CSC). (b) Normal internal architecture of the left temporal styloid process that was broken postmortem (square) (CSC). The styloid process is reinforced internally with trabecular bone, but may be hollow or solid at various points along its length.

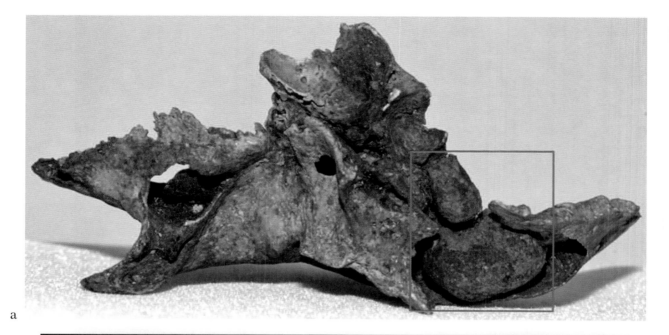

a

b

PLATE 56. (a) Two circular benign tumors (square) in the right temporal bone (mastoid region) in an Asian individual of unknown age and sex. (b) X-ray reveals the two tumors next to one another and a small rectangular enostosis (benign sclerotic bone island). Some of the cortical bone loss was due to postmortem changes, including burial in acidic soils.

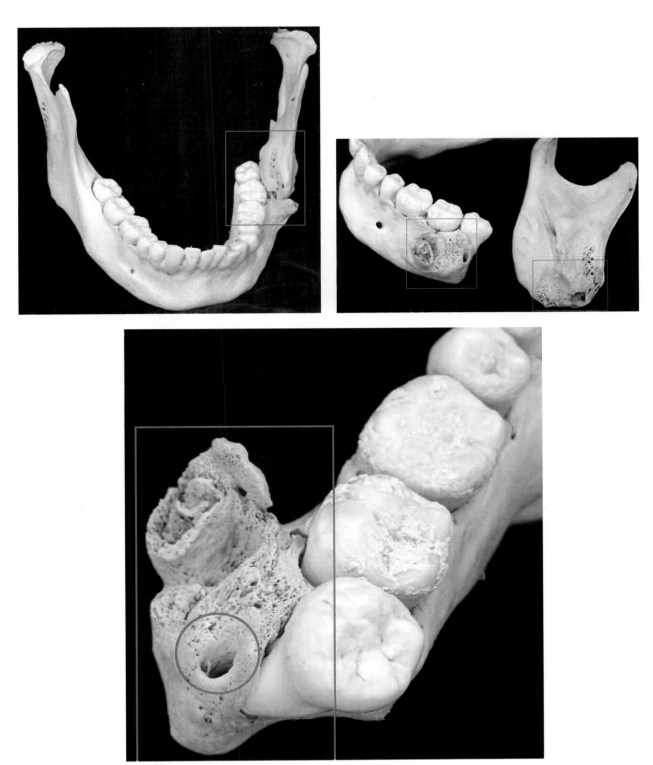

PLATE 57. Healed fracture with displacement (due to muscles) and pseudarthrosis of the left corpus of the mandible in a young adult Thai (KKU). Note that the mandibular canal (bottom, circle) has remodeled and remained in approximate alignment with the ramus. The vast majority of mandible fractures result from direct trauma, primarily vehicular accidents and assault. Cf. Lovell (2008).

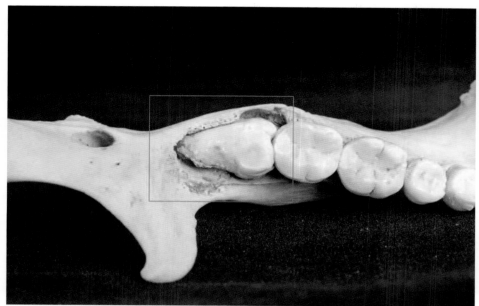

a

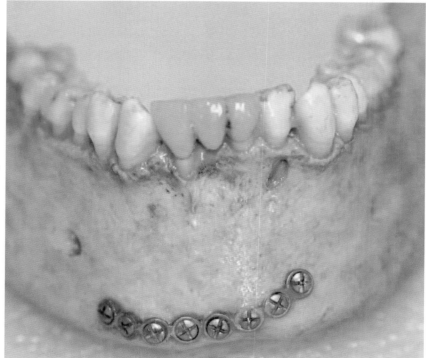

b

PLATE 58. (a) Fully developed, but horizontally impacted mandibular right third molar in an adult Asian male (UHWO). Note how well the cusps of the third molar have formed to fit the contour of the distal surface of the second molar. (b) Surgical reconstruction/straight plate bridging a fully healed fracture of the mandible (KKU). Also note the small buccal exostoses and discolored anterior teeth (crowns) as a result of the trauma. Surgical plate varying in size, shape and configuration may be found elsewhere in the skeleton. Cf. Lovell (2008).

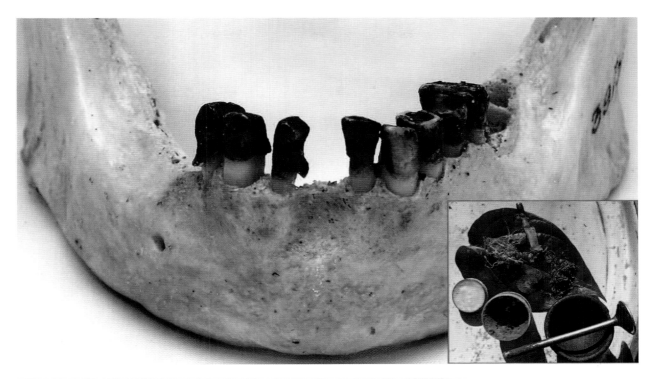

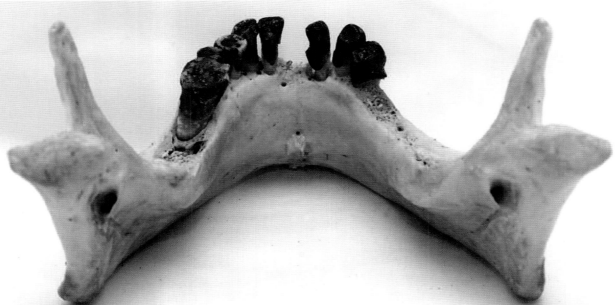

PLATE 59. Black staining of the teeth due to chewing betel nut (Areca catechu species, also known as betel nut palm) in the mandible of an elderly Thai female (KKU). While both Thai men and women chew betel nut, it is predominately an activity of Thai women older than about 50 years who live "up-country." The betel nut "kit" (inset) consists of betel nut, tobacco, a variety of leaves, a paste made of red lime and bark and a mortar and pestle, among other items (the pestle in the above photo is a motorcycle piston valve). White lime is another ingredient of betel nut "kits" in some groups around the world.

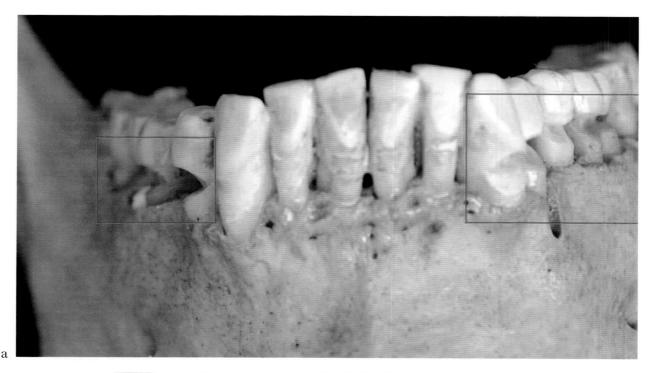

a

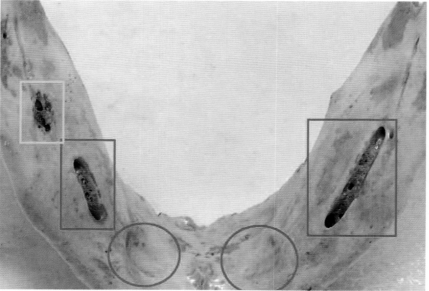

b

PLATE 60. (a) Abfraction (rectangles), sometimes attributed to aggressive toothbrushing, in the teeth of an adult Thai male (KKU 50/2124). Cf. Grippo et al. (2004). (b) Bilateral resorption of the mandibular canals (rectangles) in an adult Thai (KKU) – these lesions are in the region of the submandibular salivary glands, but are not Stafne's defects. Note the active and porous appearance of one of the lesions (yellow rectangle), suggesting that the other elongated lesions/canals formed first. This is the first example of its kind that the authors have seen in more than 10,000 human mandibles from around the world. Although speculative, these lesions may represent aneurysmal-type resorption by the inferior alveolar artery that lies lingual to the vein, or disease that spread from the adjacent submandibular salivary glands. Also note the presence of anterior lingual depressions (circles), sometimes incorrectly referred to as anterior Stafne's defects, that lie adjacent to the sublingual glands. These shallow and mildly sloping depressions are normal anatomical variants with no known clinical significance or symptoms.

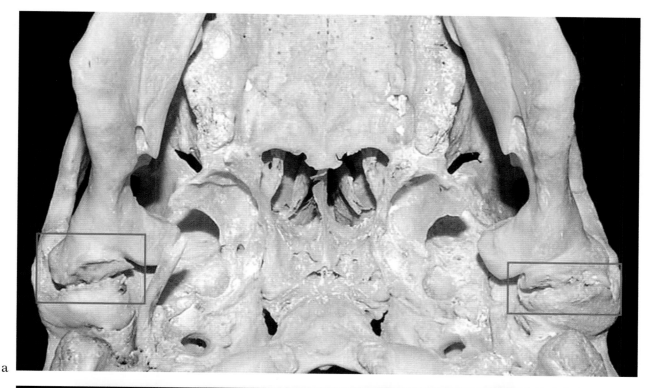

a

b

PLATE 61. (a) Partial bilateral bony ankylosis (rectangles) of the temporomandibular joints (TMJ) in an adult Thai male (KKU). (b) Note the irregular contour of the mandibular condyles. Partial or complete ankylosis of the TMJ may be congenital, but most often results from trauma or infection. Alteration of the mandibular condyles (and concomitant changes to the temporomandibular joints) reflecting osteoarthritis can result in a variety of changes resulting in erosion of one or both condyles, osteophytic lipping encircling the condyle(s), and flattening of one or both condyles.

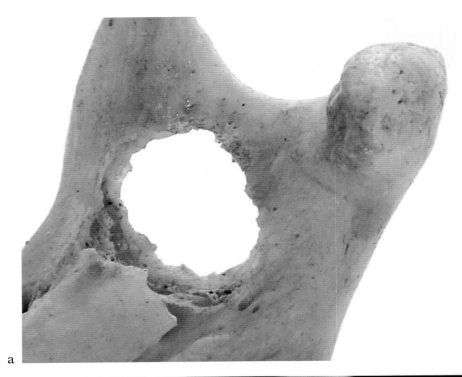

a

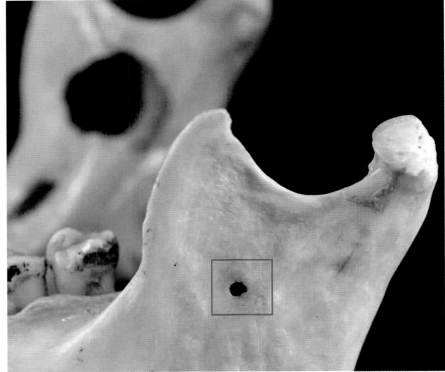

b

PLATE 62. (a) Probable metastatic carcinoma in the right mandibular ramus of an elderly Thai female (KKU). Note the circular and jagged margin of the defect that has a slight lingual bevel and a mild osteoblastic response encircling the defect. (b) This individual has a tiny lytic lesion that has perforated the left ramus.

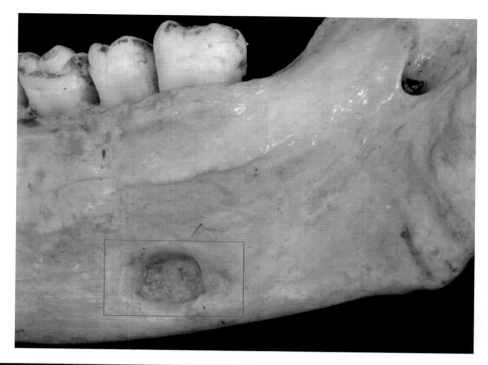

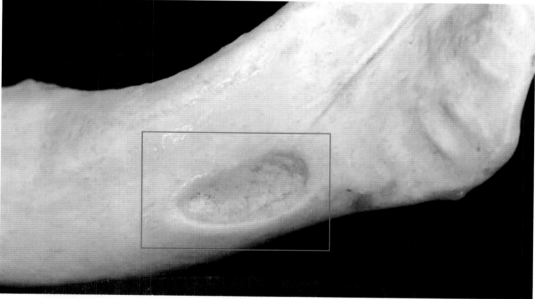

PLATE 63. Two variants of Stafne's defect, also known as static bone cavity and salivary gland inclusion, among others, in the right mandibular corpus (rectangles) (KKU). Posterior Stafne's defects are always located in the region of the submandibular salivary gland and inferior to the molars and mylohyoid ridge. They likely form as a result of contact pressure erosion by the adjacent submandibular salivary gland or other soft tissue, usually have a folded ("sclerotic) rim, occur predominately in middle-aged and older males, and follow the pattern of an X-linked recessive trait (Mann, 2001). The youngest reported individual with a Stafne's defect is an 11-year-old boy from Sweden (Hansson 1980). Similar defects and concavities may be present inferior to the canines and are referred to as anterior Stafne's defects or anterior static bone cavities and may form as a result of contact pressure erosion by the sublingual salivary gland.

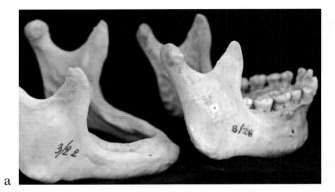

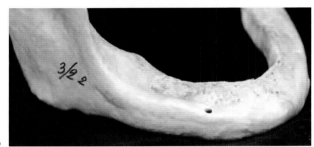

a

b

PLATE 64. (a and b) Edentulous elderly mandible compared with a young adult mandible showing resorption of the bone and "repositioning" of the mental foramen – the foramen doesn't actually "move," but becomes positioned lower on the corpus due to loss of the alveoli (KKU). (c) Ridge-like lingual mandibular torus in adult Thai (KKU). These benign tumors are believed to develop, in part, in response to gum irritation. Torus palatinus (maxillary hard palate) and torus mandibularis are often found in the same individual and typically appear in adults.

c

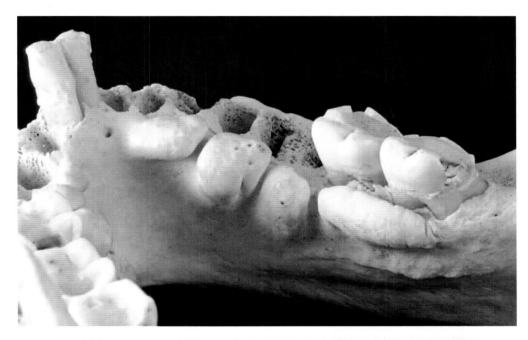

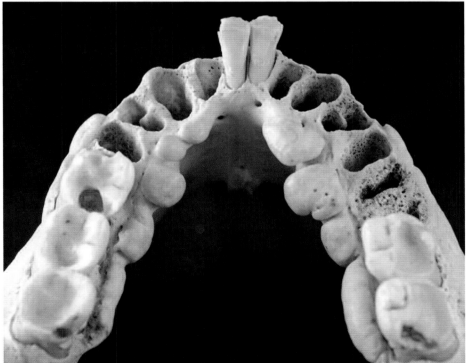

PLATE 65. Mandibular tori (torus mandibularis) adjacent to each tooth socket/alveolus ("one per tooth") in a Thai mandible (KKU). Note the left and right second and third molars each share an elongated bar of bone, rather than single tori for each tooth. The etiology of these benign bony growths, which in this individual formed separately along each tooth and with separate bases, may be in response to irritation, reduced blood flow/ischemia, and changes in biomechanical stresses to the jaws. Tori have a tendency to increase in size with age and may reflect a functional adaptation to mechanical stresses of the jaws and teeth.

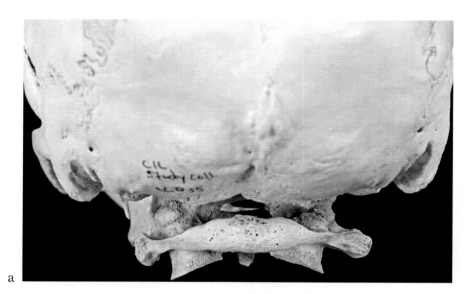

a

b

PLATE 66. (a) Atlanto-occipital fusion of C1 to the occipital condyles in an adult Asian (CSC; only the skull is present). (b) Atlanto-axial synostosis of C1, C2 to the cranial base with angular deformity/twisting in a probable 17–20-year-old Asian female. This condition typically reflects a congenital deformity, a failure of segmentation, although it may also reflect trauma (CSC AC037; only the skull is present).

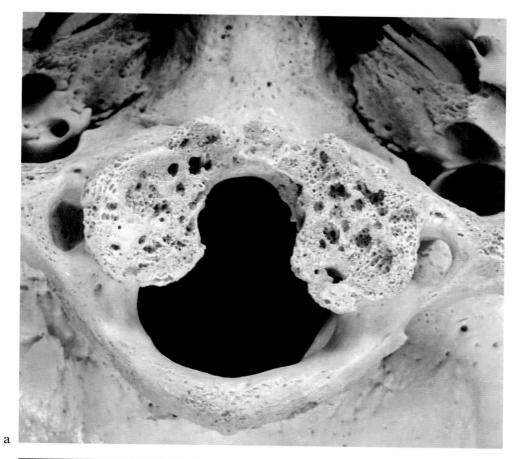

a

b

PLATE 67. (a) Occipito-atlantal synchondrosis/fusion of the atlas and severe degenerative arthritis of the inferior facets (CSC AC032). (b) Bilateral (severe) erosion and alteration of the temporo - mandibular joints extending onto the eminence, bilateral tympanic dehisences (foramen of Huschke), severe resorption of the right maxilla and alveoli, and impaction of a left molar (Peru, SI).

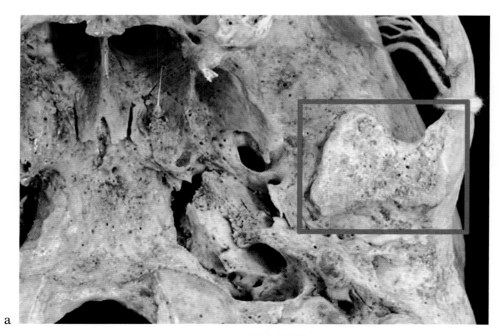

a

b

PLATE 68. (a) Degenerative joint disease of the left temporomandibular joint (TMJ) and articular eminence resulting in alteration and enlargement of the left eminence by the growth of marginal osteophytes (KKU). (b) Circular degenerative joint disease/osteoarthritis resulting in bone loss on the right articular eminence; this is one of the classic locations of degenerative joint disease and osteoarthritis (CSC).

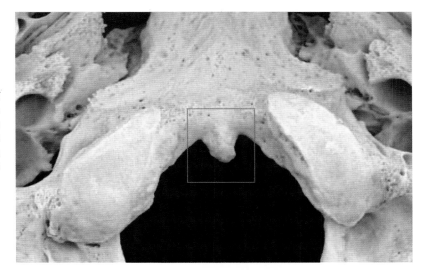

PLATE 69. Ossified apical ligament of the atlas dens (CSC AC023), sometimes referred to as a precondylar tubercle (Barnes 1994). Ossification of this ligament varies greatly in size and shape from a peg or spike, to one that is bifid, double pegs and dish-shaped, among others.

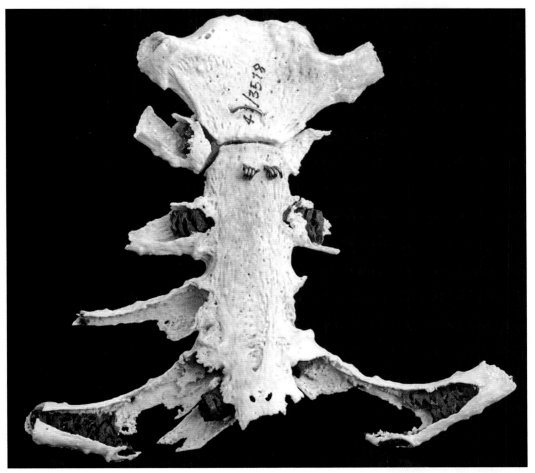

PLATE 70. Normal pattern of marginal (peripheral) ossification/calcification of the costal cartilage in an elderly Thai male (KKU). Note the crab claw pattern of calcification, typical of males ("Swiss cheese" pattern in females), and the dark "cracked" non-ossified hyaline costal cartilage inside the calcifications. Cf. McCormick et al. (1985).

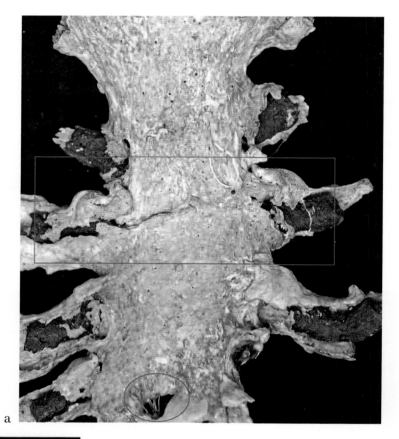

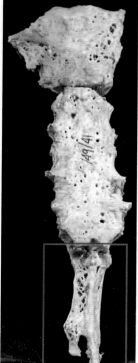

PLATE 71. (a) Ventral view of a healing transverse fracture (rectangle) of the sternum in an elderly Thai male (KKU). Note the small sternal foramen (oval) with ossifying fibrous strands spanning the defect. (b) Ossified xiphoid cartilage in an elderly Thai male (KKU). The scattered porosities/ holes across the sternum reflect osteoporosis. Cf. Yekeler et al. (2006).

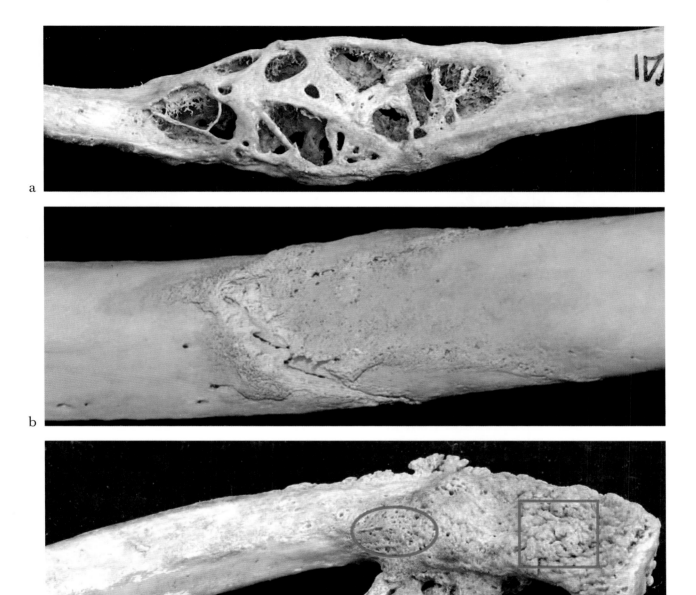

PLATE 72. (a) Enlargement/expansion of a left rib resulting in numerous perforations and remodeling, possibly due to infection (KKU). (b) Healing callus bridging a fractured rib. Note that the new fibrous/woven bone is porous, loosely attached to the underlying cortex, and darker than the surrounding bone (CSC). (c) Exuberant bone growth (possibly reflecting cancer, infection, or traumatic myositis ossificans) of vertebral portion of right 11th rib in an elderly Thai male (KKU). The spiculed bony growths may reflect ossification of the costotransverse ligaments (interosseous ligaments), but the rough surface (square) and porosity (oval) are not typical of the bony response associated with ankylosing spondylitis. This individual exhibited similar growths on the right 12th rib, but not on any of the left ribs.

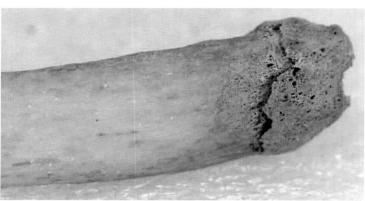

PLATE 73. Fractured ribs with fibrous/woven collars of bone that were healing at the time of the individual's death – this person exhibited several healing rib fractures (KKU). Note the white underlying original cortex in the bottom photo and porous, loosely adhering fibrous bone encircling the site of fracture. (The white chalky substance is adipocere from the postmortem cleaning process.)

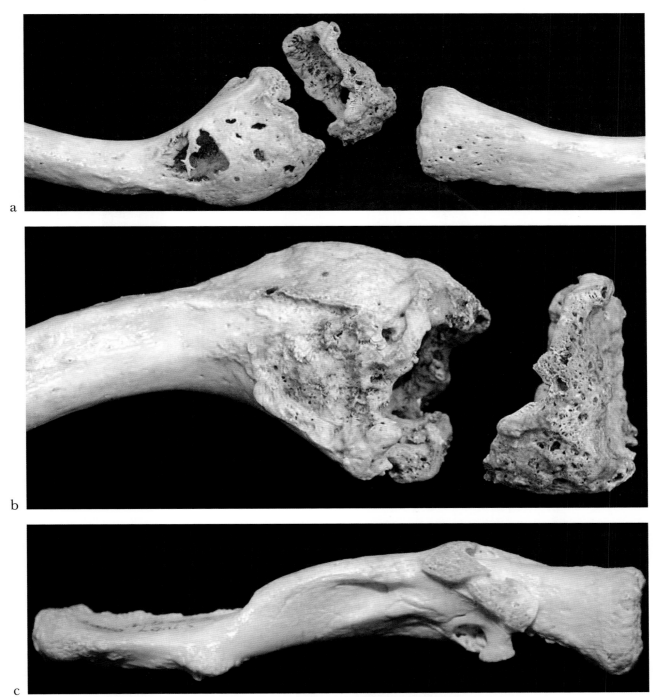

PLATE 74. (a and b) Right clavicle exhibiting a medial pseudarthrosis and "cap" likely reflecting bone infection/osteomyelitis in an elderly adult Thai (KKU). Note the medial portion of the left clavicle exhibits microporosity and mild involvement, but not as severe as the right. (c) Left clavicle with healed fracture and shortened in length (CSC DS007).

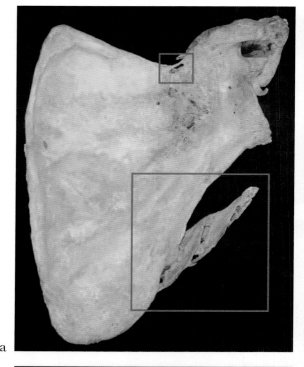

a

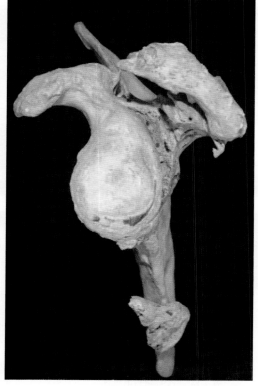

b

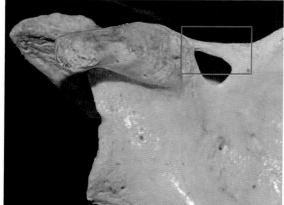

c

PLATE 75. (a and b) Probable fibrodysplasia ossificans progressiva in the left scapula of an elderly Thai male – note ossification of the teres muscle (this male also had ossification of the lower spine and left innominate) (KKU). Note the oval-shaped scapular foramen (small square) formed by ossification of the suprascapular ligament. (c) Ossified suprascapular ligament (bridge) forming a scapular foramen in a right scapula of an adult Thai (KKU). The normal morphology of this area ranges from flat, to a U- or V-shaped notch forming a foramen, often scored as a non-metric trait, when the suprascapular ligament ossifies. (d) Early stage ossification of this ligament in the left scapula of an adult male from China (UHWO).

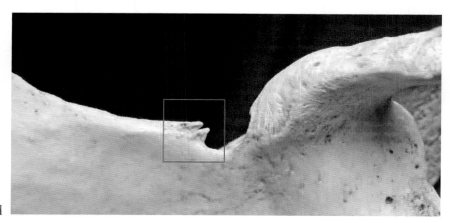

d

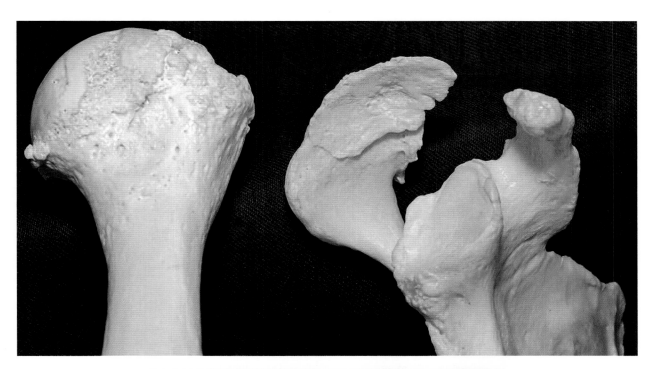

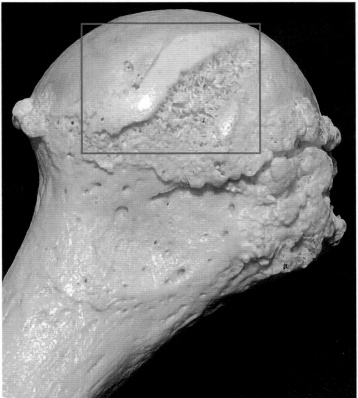

PLATE 76. Probable shoulder impingement resulting in remodeling of the inferior surface of the acromion process assuming the curvature of the humeral head, concomitant arthritic destruction of the glenoid fossa, grooving and erosion of the right humeral head, and eburnation (rectangle) (CSC DS007). Cf. Slatkin (2003).

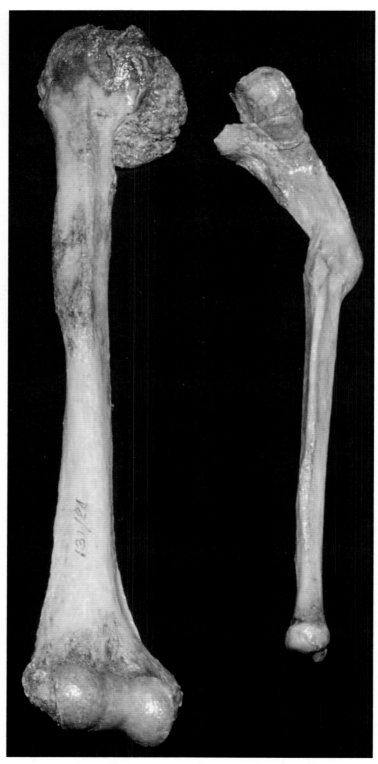

PLATE 77. Healed fracture and angulation of the proximal one-third of the right ulnar diaphysis with severe arthritis and deformation of the humeral head in an adult Thai (KKU).

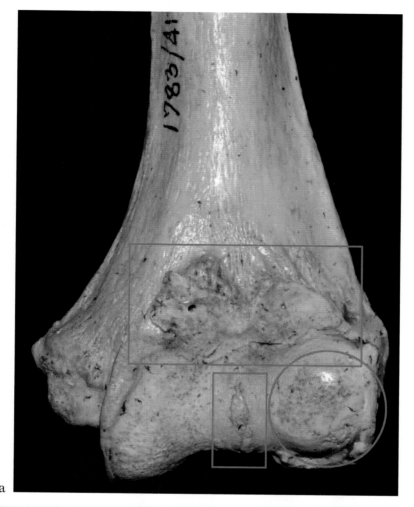

a

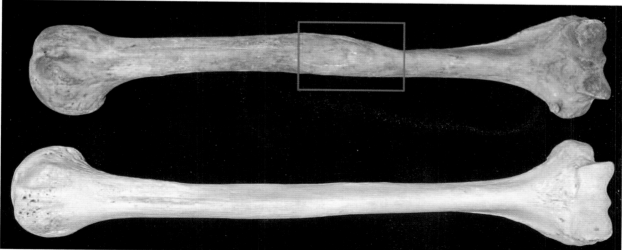

b

PLATE 78. (a) Osteoarthritis in the form of hypertrophic ossification of the coronoid fossa (horizontal rectangle), a ridge of osteophytes between the trochlea and capitulum (vertical rectangle) and eburnation (circle) of the distal left humerus in an elderly Thai male (KKU). (b) Healed fracture of the left humerus (rectangle) resulting in asymmetry and shortening (SI).

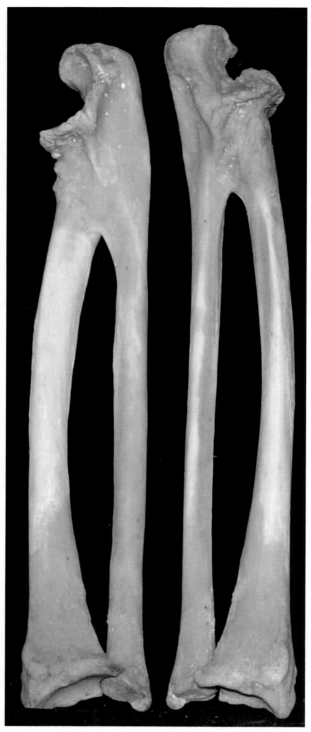

Plate 79. Congenital proximal radioulnar synostosis, a failure of segmentation, resulting in proximal osseous union of both forearms (KKU). Note flattening of the distal joint and narrow, tubular shape of the ulnas. This rare condition may be part of a syndrome, including Crouzon and Apert, has only been reported in a few hundred cases, and may be severely disabling. Cf. Farzan et al. (2002) and Dogra et al. (2003).

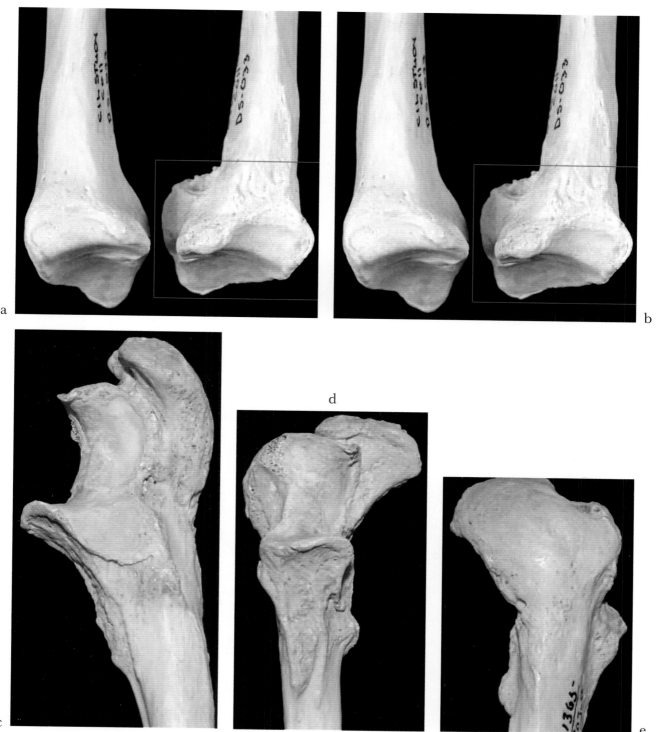

a

b

d

c

e

PLATE 80. (a and b) Healed Colles' fracture (rectangle), also known as distal fracture of the radius, with dorsal displacement of the right wrist (CSC DS033). Dorsal displacement of the radius is a Colles' type fracture, while volar displacement is a Smith's type or reverse Colles' fracture. (c, d and e) Myositis ossificans traumatica of the triceps brachii muscle (left ulna) in a European male (Huntington SI). Mild forms of ossification of this muscle have been referred to as Woodcutter's Lesion, as well as blacksmithing, and baseball playing (Capasso et al. 1998).

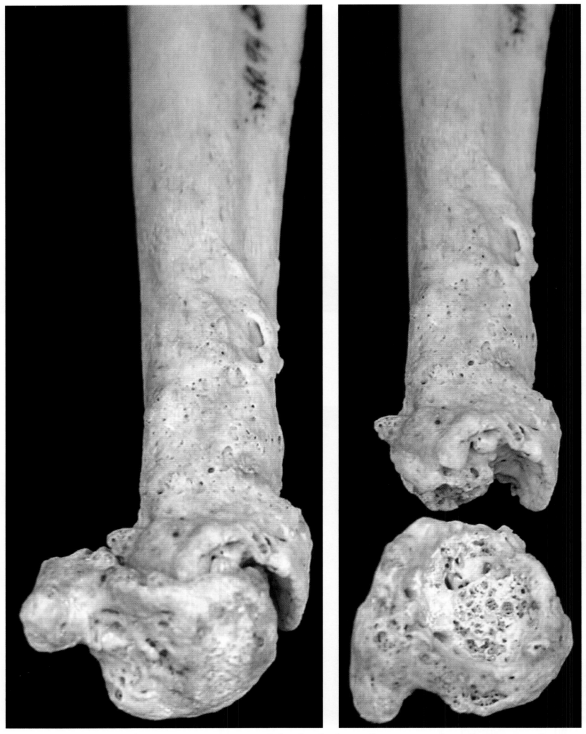

PLATE 81. Epiphyseal plate fracture-separation of the distal ulna with non-union/pseudarthrosis and severe osteoarthritis (KKU). Note the irregular and pitted reactive bone along the diaphysis, proximal to the fracture.

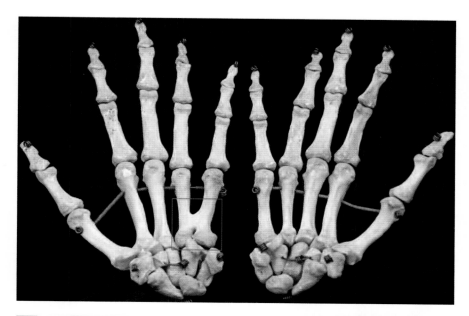

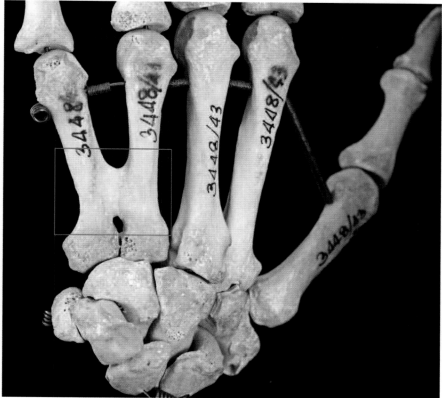

PLATE 82. Palmar and dorsal views of congenital synostosis/syndactyly of the 4th and 5th metacarpals (square) of the left hand in an adult Thai (KKU). Syndactyly may be unilateral or bilateral and may involve fusion of fingers, known as simple syndactyly, or the bones, complex syndactyly.

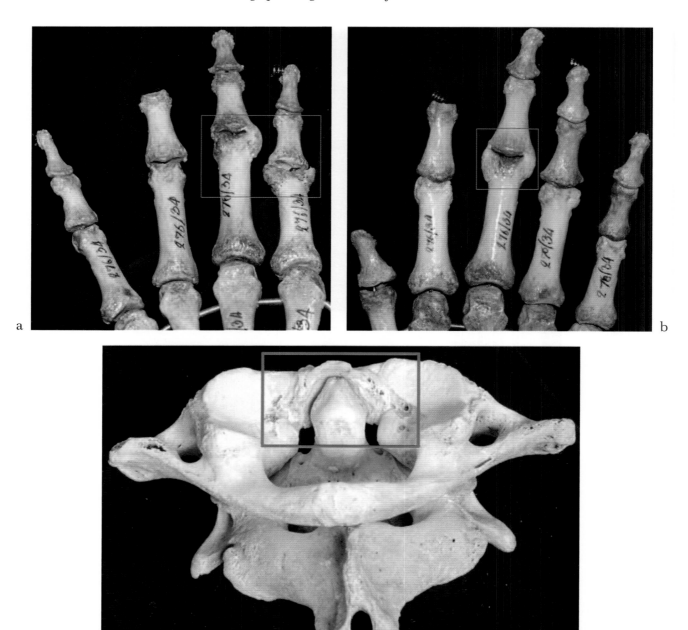

a

b

c

PLATE 83. (a and b) Dorsal view of bilateral calcification (square and rectangle) of the collateral ligaments in the hands of an adult Thai (KKU). The collateral ligaments provide medial-lateral stability for the proximal interphalangeal joints of the fingers and can calcify due to a variety of factors, including trauma. (c) Osteoarthritis and expansion of the articular surface (rectangle) of the atlas in an elderly Thai (KKU). Also note the bifid spinous process in C2, a common normal anatomical variant.

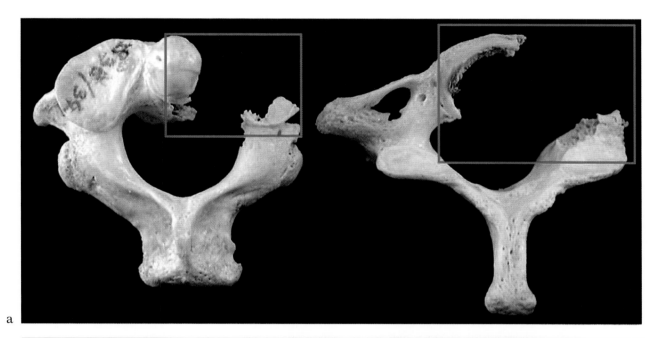

a

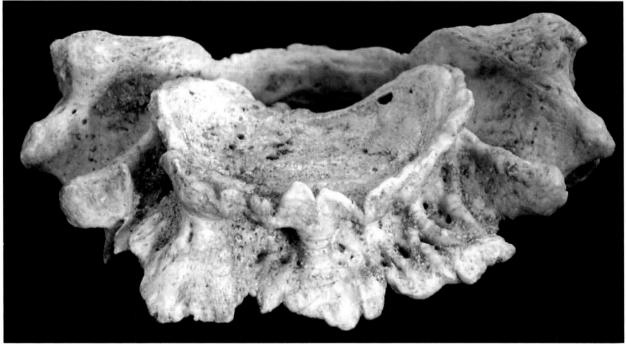

b

PLATE 84. (a) Aggressive lytic lesions of unknown etiology in two cervical vertebrae (KKU). (b) Osteoarthritis of a cervical vertebra in an elderly Thai individual (KKU). Note the vertical pillar-like growths that buttress the middle of the centrum and extend across the intervertebral joint to fuse the vertebrae above and below it. Similar osteophytes may be present in any vertebrae.

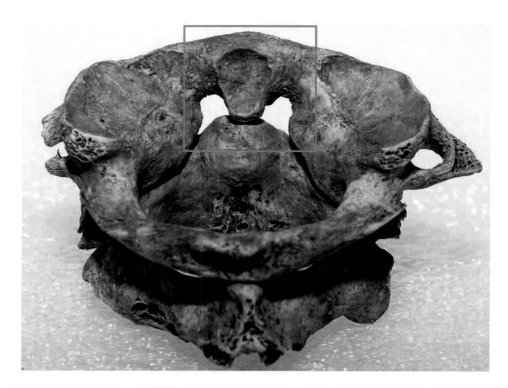

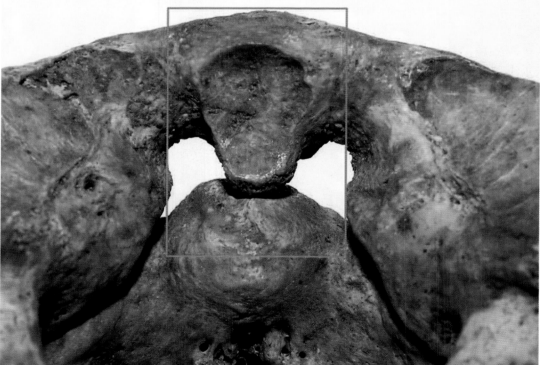

PLATE 85. Os odontoideum (the separate dens/odontoid ossicle was not recovered with the remains) – note well-defined atlas facet (KKU 0202). Research by Sankar et al. (2006) revealed that the os odontoideum has two etiologies as both post-traumatic and congenital – radiographs of 519 patients with abnormal cervical spines revealed 16 (3.1%) with an os odontoideum (rectangle).

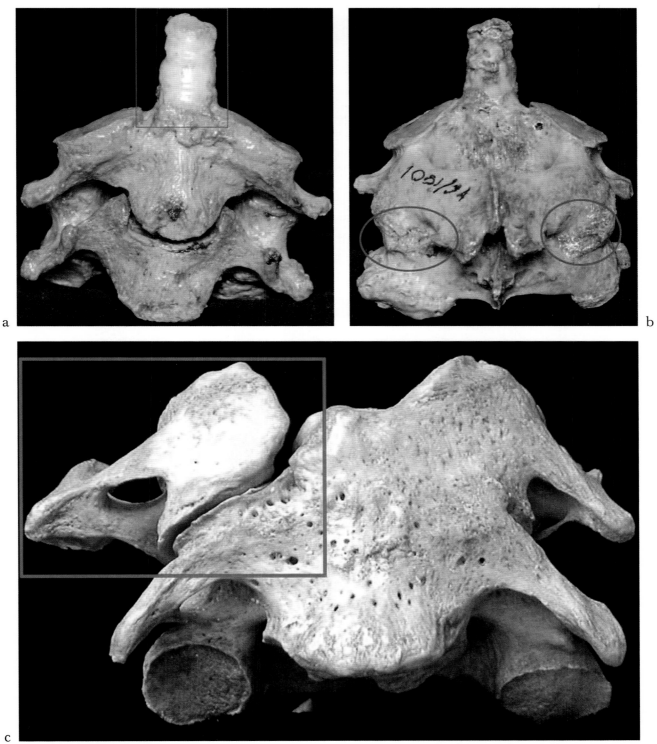

a

b

c

PLATE 86. (a and b) Eburnation of the dens of the second cervical vertebra/axis (rectangle) with vertical elongation due to degenerative joint disease; note fusion of the posterior vertebral facets (circles) in the atlas and axis consistent with DISH, ankylosing spondylitis, and congenital fusion, among others (KKU). (c) Hemivertebra (rectangle) in the cervical spine (Photo courtesy of UTK FAC, WMB UTK, Joseph T. Hefner).

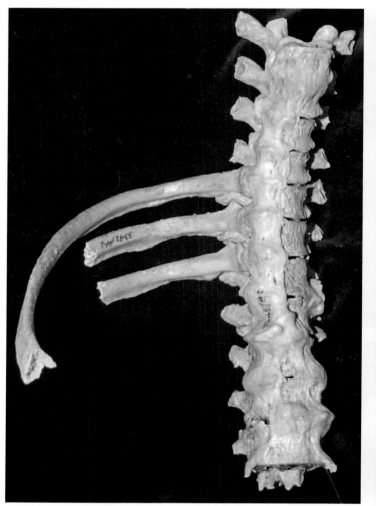

a

b

PLATE 87. (a) Diffuse idiopathic skeletal hyperostosis (DISH, Forestier disease, ankylosing hyperostosis) and ankylosing spondylitis (AS) in an adult Thai male (KKU). Note the flowing calcification of bone (often referred to as osteophytes or syndesmophytes) resembling melted candle wax along the right side of the vertebral centra, preservation of joint space between the centra, and that more than four contiguous vertebrae are affected (consistent with DISH). In DISH, the left side of the spine is free of the ribbon-like growth, likely due to the presence and pulsing of the aorta. Bony ankylosis of the ribs to the vertebrae (costovertebral joints) reflects ankylosing spondylitis (AS). It is not uncommon for an individual to exhibit aspects of both AS and DISH – this individual the sacroiliac joint was fused, a condition consistent with AS and inconsistent with DISH. Compare DISH (a) with spondylosis deformans (b) (Photo courtesy of UTK FAC, WMB UTK, Joseph T. Hefner) where the bony growths (rectangle) are positioned more irregularly, appear to project from the middle of the centra like buttresses, and are not restricted to the right side of the spine. Spondylosis deformans, sometimes referred to as bone spurs or bridging osteophytes, develops in response to degeneration of the intervertebral discs. Both DISH and spondylosis deformans are types of arthritis. Individuals with DISH often exhibit inflammatory responses resulting in enthesophytes (bone spikes) at muscle and tendon attachments such as the innominate, tibia, calcaneus, and patella.

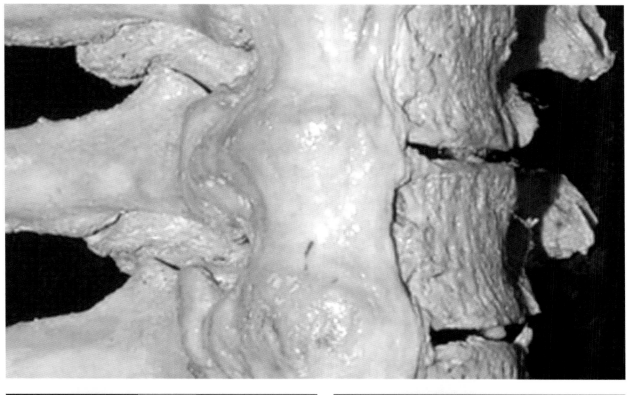

a

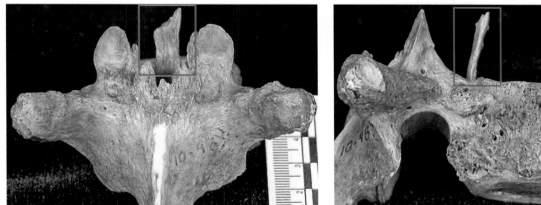

b   c

PLATE 88. (a) Closeup of the flowing osteophytes/syndesmopphtes in the DISH example in Plate 87 (KKU). (b and c) Ossification of the posterior longitudinal ligament (OPLL) (rectangle) in a thoracic vertebra (Photo courtsy of UTK FAC, WMB UTK, Joseph T. Hefner). The posterior longitudinal ligament, which is typically 1–2 mm thick, runs along the posterior portion of the vertebral canal. OPLL, often associated with DISH, is more commonly seen in individuals of Asian descent, with a reported incidence in Japan of 1.7–2.4% and 0.16% in non-Asians (Hida et al. 2008). Also note the spiked projections between the two superior facets indicating ossified ligamentum flavum. Cf. Gaillard (2008).

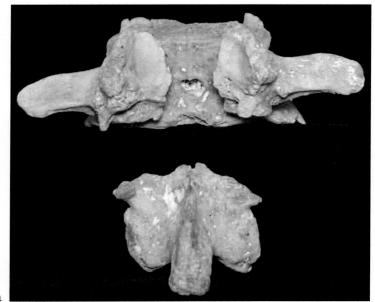

a

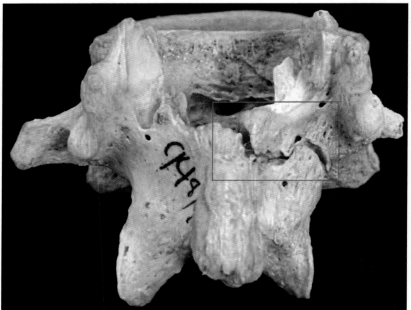

b

PLATE 89. (a) Bilateral spondylolysis (also known as separate neural arch or pars defect) of a lumbar vertebra through the pars interarticularis (KKU). The etiology of spondylolysis is likely due to bending stresses and fracture from mechanical stresses such as flexion and extension of the spine. When articulated, the two pieces typically do not have a "snug" fit, but rather jagged margins that share features with one another and interlock along some aspects of the separation/fracture. This condition becomes spondylolisthesis if the centrum slides forward (which requires that the spondylolysis is bilateral and complete). (b) Unilateral spondylolysis (rectangle) through the pars interarticularis ("between the facets"), separating the superior and inferior facets in a lumbar vertebra (KKU). Bilateral spondylolysis is more commonly encountered than the unilateral form, but either form may result in complete or partial separation of the centrum and arch, likely the result of a fatigue fracture. Note the undulating S-shaped line of separation and saw-toothed appeared of the edges through the pars interarticularis.

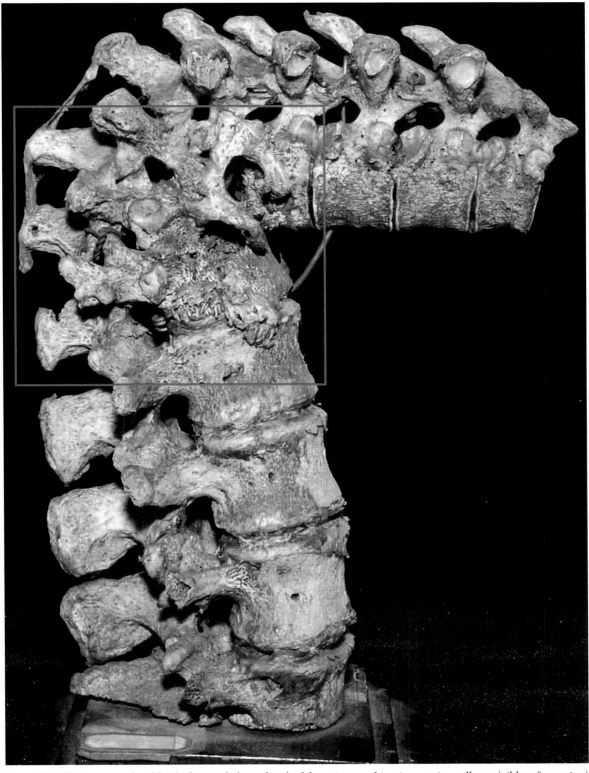

PLATE 90. Classic example of Pott's disease (tuberculosis) of the spine resulting in anterior collapse (gibbus formation) of several thoracic centra/bodies (square) (Mütter Museum 1196.56).

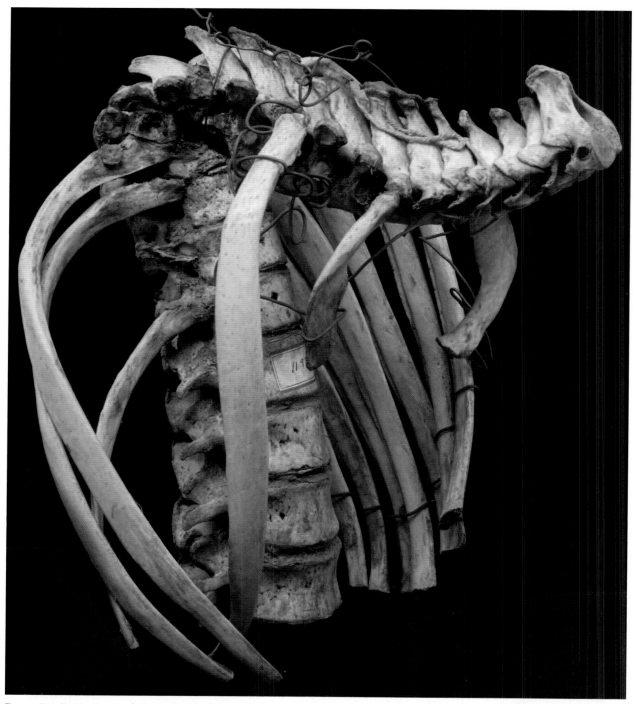

PLATE 91. Pott's disease (tuberculosis) of the mid-thoracic spine resulting in destruction of several and anterior angulation of the upper spine (Mütter Museum 1192.05). Note how the ribs have become angled downward and now lie very close to one another.

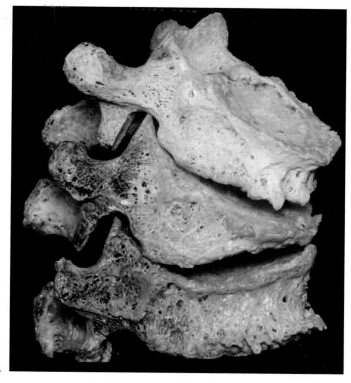

a

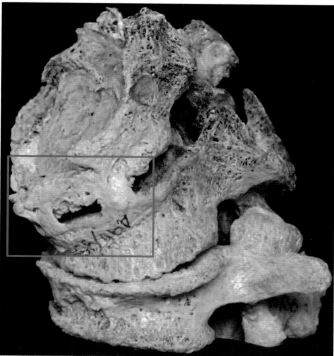

b

PLATE 92. (a) Fracture, collapse and anterior wedging of a lower thoracic vertebra in an elderly Thai (KKU). (b) Anterior wedging, collapse and bridging osteophytes (rectangle) in the thoracolumbar region of the same individual (the two images, however, depict different vertebrae).

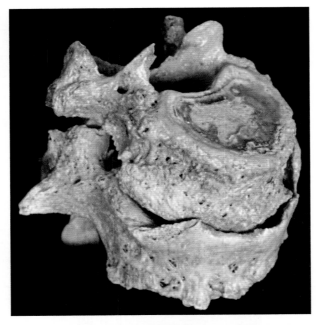

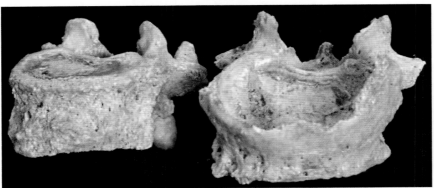

PLATE 93. Endplate collapse of the first lumbar vertebral centrum with resulting large buttressing, compensatory osteophytes "cupping" the vertebra above (KKU).

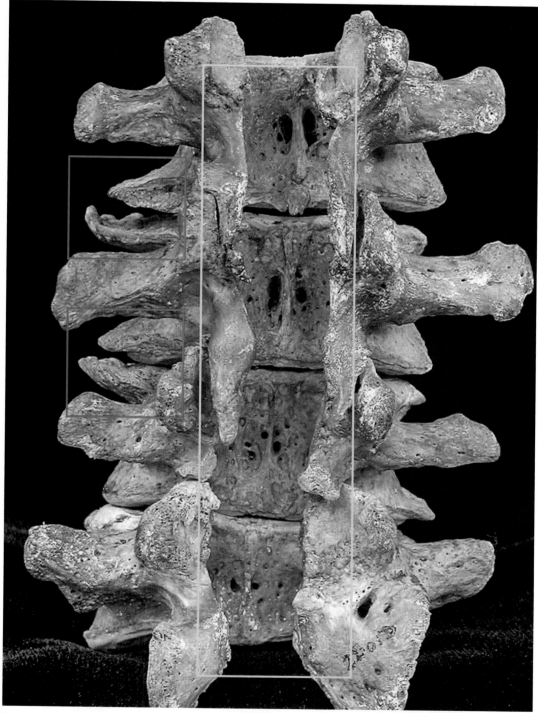

PLATE 94. Absence of the neural arches (yellow rectangle) and compensatory osteophytes (red rectangles) in the lumbar spine (Photo courtesy of UTK FAC, WMB UTK, Joseph T. Hefner). Separation or partial union of the posterior arch (spondylolysis) and cleft spinous process (spina bifida) is a common occurrence, while absence of the posterior arch (variant of spina bifida) is a rare occurrence that may or may not require surgical intervention. Absence of the arches in four vertebrae such as in this example is a rare finding in most skeletal collections, although it may be more commonly encountered clinically and radiologically.

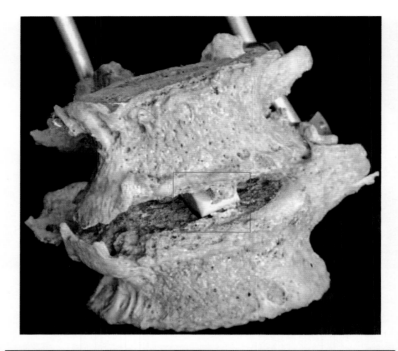

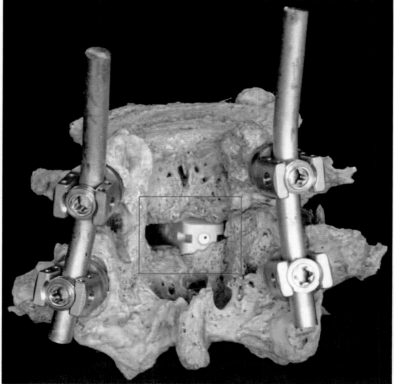

PLATE 95. Discectomy (removal and replacement of the intervertebral disk with an interbody spacer/prosthetic device) (rectangle), laminectomy (surgical removal of one or more posterior arches) and rods and bars fusing the vertebrae by maintaining separation and alignment ("block") of two lumbar vertebrae in an elderly Thai male with DISH (KKU).

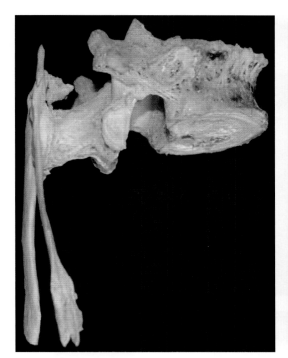
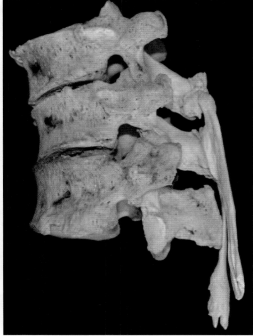
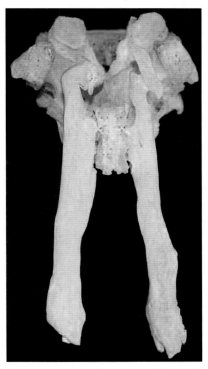

PLATE 96. Fibrodysplasia ossificans progressiva of the spinalis (thoracis) muscles, the most medial erector spinea (deep) muscle group in an elderly Thai male (R. M. Burgos and J. I. Sebes, personal communication) (KKU). Note the "swallowtail" appearance of the ossifications attached to T-12 as they extend downward, passing close to, but not articulating with, L-1 vertebra. The erector spinae muscles help maintain an upright posture, as well as bend and twist the upper body (previously known as myositis ossificans progressiva).

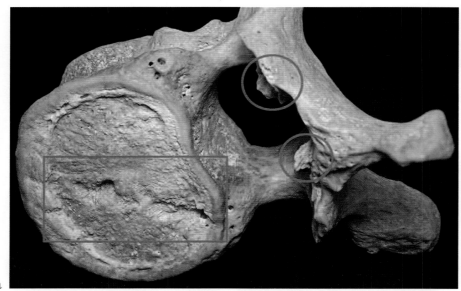

a

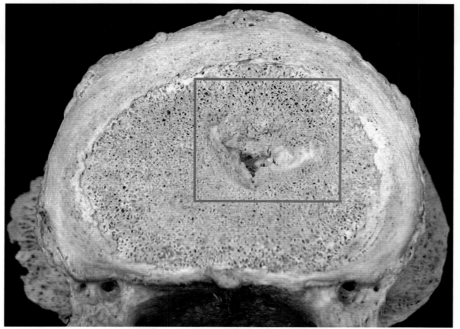

b

PLATE 97. (a) Linear Schmorl's depression (rectangle) and neural canal spurs (circles) in the inferior endplate of a thoracic vertebra (SIH). Neural canal spurs can impinge on the spinal cord and nerves, resulting in pain and limited mobility. (b) Compression fracture of the endplate due to herniation of the nucleus pulposus (KKU). This fracture, which probably occurred as a single traumatic event resulting in a cavernous concavity (fracture) with sloping walls, is not the same as a classic and more commonly encountered Schmorl's depression that typically results in a smooth-walled defect (sclerotic on x-ray) with more vertical walls and that forms more slowly (non-acute) – note the "torn" appearance of the floor of the concavity suggesting fracture, not merely resorption. Schmorl's depressions may, however, be symptomatic or asymptomatic and can form as a result of acute or non-acute trauma. Cf. Schmorl and Junghanns (1971) and Wagner et al. (2000).

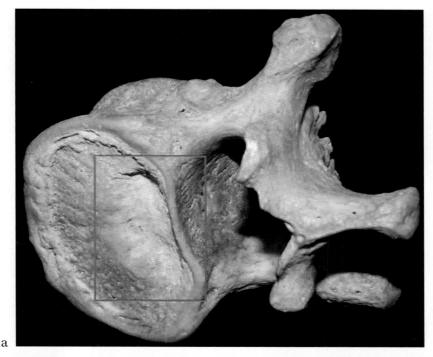

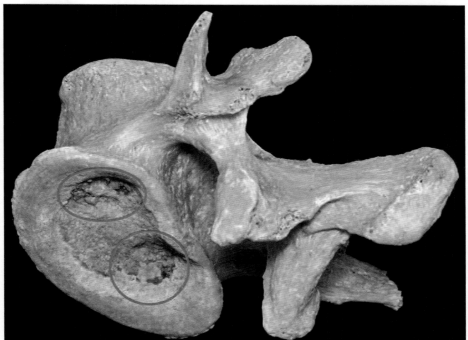

PLATE 98. (a) Posterior Schmorl's depression (rectangle) in the inferior endplate of a lower thoracic vertebra (SIH). (b) Normal bilateral (biconcave) depressions commonly present in the inferior endplate of L3, L4 and/or L5 vertebrae (SIH). Intervertebral discs have been described as biconcave lenses that, in the lumbar region, vary from 10 to 15 mm thick (height), are thicker anteriorly than posteriorly, and reflect the shape of the opposing disc (Postacchini and Rauschning 1999). Normal endplates vary from flat to concave. Cf. Hall et al. (1998).

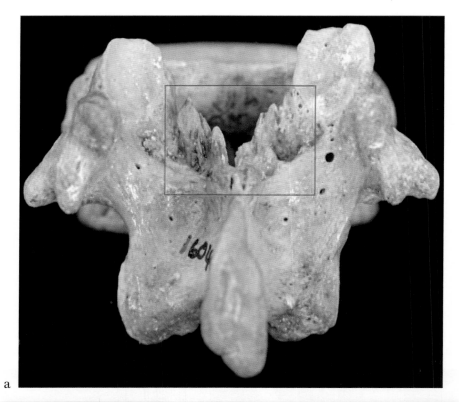

a

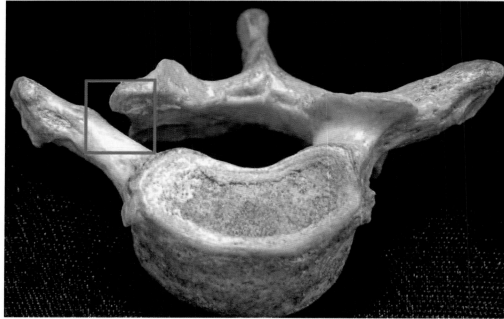

b

PLATE 99. (a) Stalagmite-type ossification of the ligamentum flavum (rectangle), also known as hypertrophied ligamentum flavum and ligamenta flava (in some of the earlier literature) in a thoracic vertebra (KKU). Large ossifications, reportedly a rare phenomenon in young individuals, may result in stenosis/narrowing of the spinal canal and compression of the spinal cord, especially in the thoracic region. (b) Thoracic vertebra with congenital absence of the right pedicle and hypoplasia of the right transverse process (Photo courtesy of UTK FAC, WMB UTK, Joseph T. Hefner).

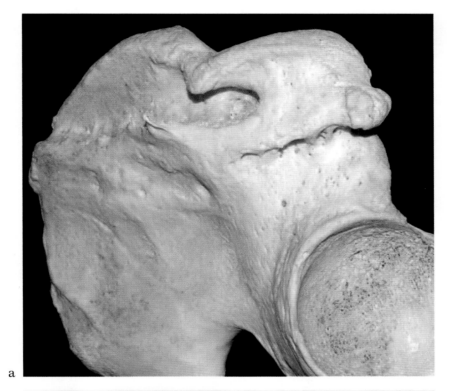

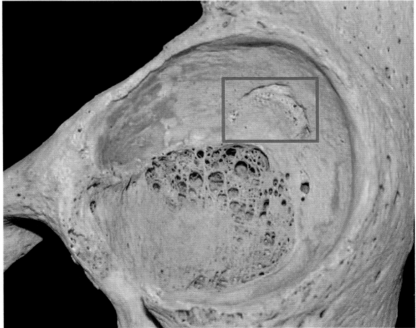

PLATE 100. (a) Healed fracture with anterior displacement of the right ilium in an elderly Thai individual (KKU). Note that the roughened and porous acetabulum has lost its normal C-shaped articular/lunate surface. (b) Porosity and a semi-lunar shaped acetabular crease (rectangle) in the left acetabulum of an elderly Thai (KKU). The porosity in the acetabular pit has been shown to be related to increasing age and is commonly seen in the elderly. Cf. Calce and Rogers (2011).

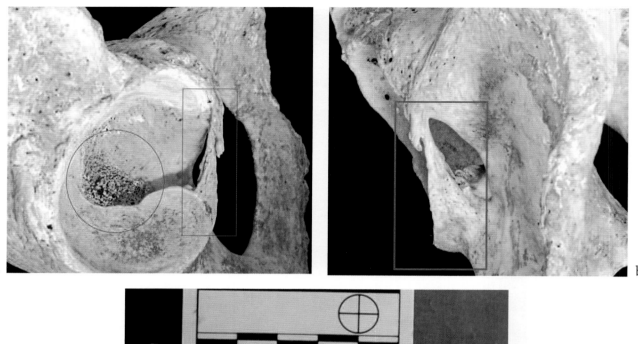

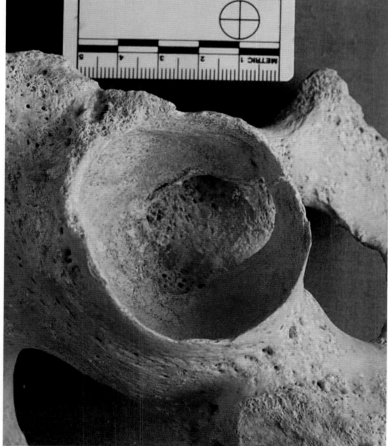

PLATE 101. (a and b) Ossification of the transverse acetabular ligament (rectangles) bridging the acetabular/cotyloid notch and fibrous new bone in the acetabular pit in an elderly Thai male (KKU). (c) "Cupping" and extension/bridging of the normally C-shaped acetabulum as a result of ossification of the transverse acetabular ligament in a Japanese male (CUJ, Hugh E. Tuller).

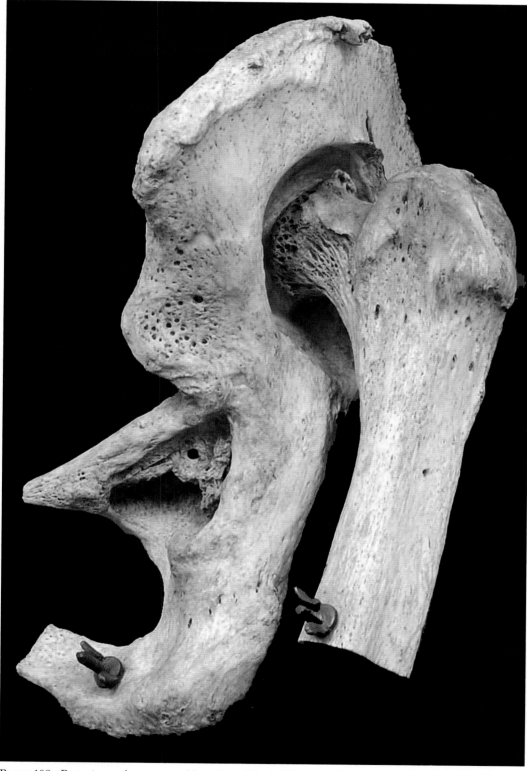

PLATE 102. Posterior and superior subluxation of the left hip, possibly due to trauma to the acetabulum, resulting in remodeling and partial filling in of the acetabulum in an adult (Mütter Museum 1335).

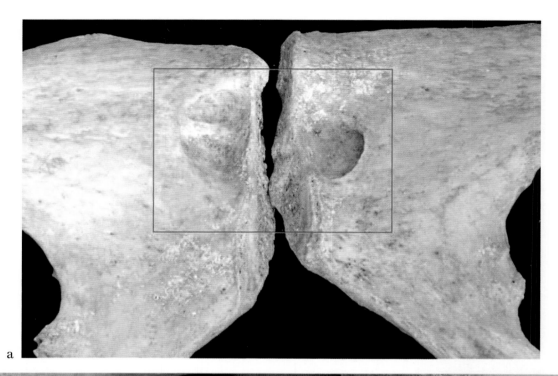

a

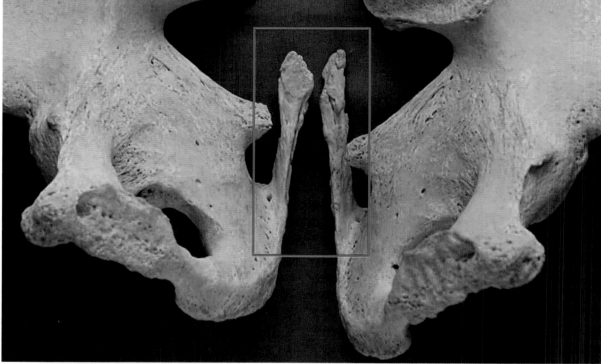

b

PLATE 103. (a) Parturition pits (dorsal pits, birthing pits, parturition scars, parturition scarring) in the dorsal surface of the pubic bones in a Thai female (KKU). The presence of these pits is not necessarily indicative of having given birth, as both nulliparous (documented) and parous females have been shown to have these pits. Cf. Holt (1978). (b) Bilateral heterotopic ossification of the sacrotuberous ligament in an adult Japanese (CUJ, Hugh E. Tuller). This condition often accompanies diffuse idiopathic skeletal hyperostosis (DISH).

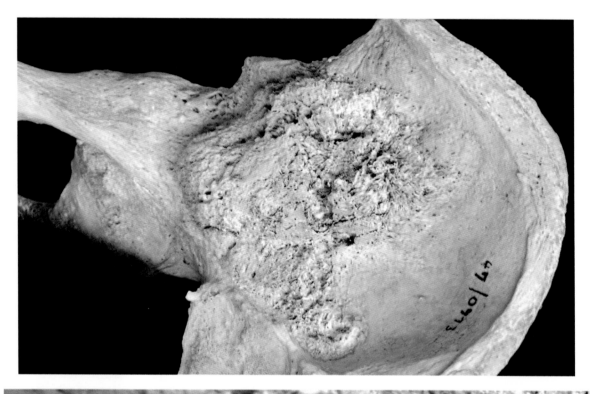

PLATE 104. Starburst pattern of new bone formed in the right supra-acetabular region and iliac fossa in an elderly Thai male (KKU). The age and sex of the individual and radiographic appearance of this tumor are most consistent with metastatic carcinoma of the prostate (Jeno I. Sebes, personal communication).

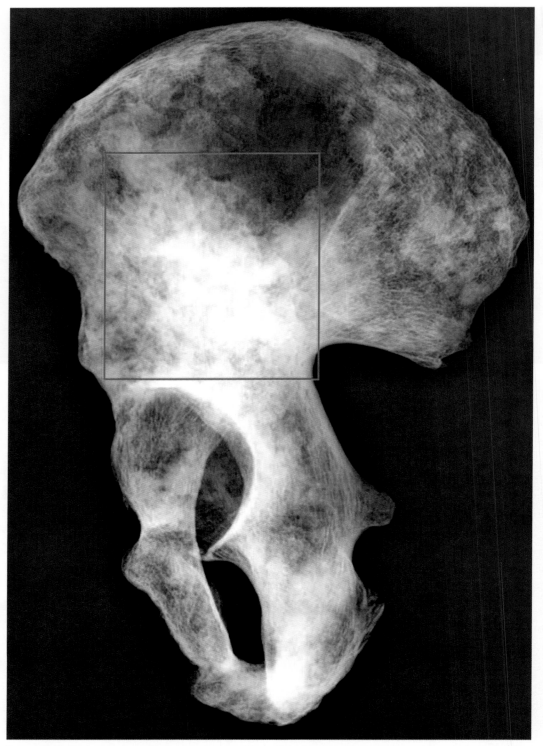

PLATE 105. Radiograph of the innominate in Plate 104, showing the large bony growth (rectangle) in the supra-acetabular region and the iliac fossa, consistent with metastatic carcinoma of the prostate (Jeno I. Sebes, personal communication). Note the numerous small sclerotic lesions (whitish cloud-like areas) in the posterior portion of the bone (x-ray courtesy of Panya Tuamsuk, KKU).

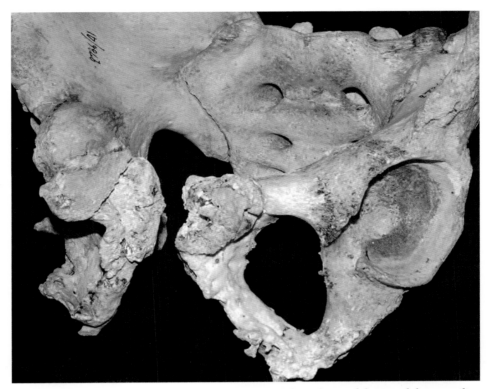

PLATE 106. Unusual osteophytes or enthesophytes, bilateral partial fusion of the sacroiliac joints and massive remodeling and widening of the pubic bones in an elderly Thai male with DISH (KKU). This condition likely reflects more than one disease condition, including ankylosing spondylitis, DISH and possibly exstrophy of the pubic bones.

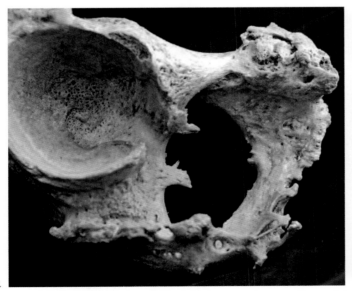
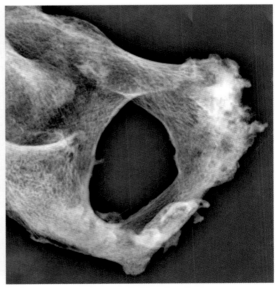

a   b

PLATE 107. (a) Widening and erosion of the symphysis pubis with bilateral and symmetrical "wave-like" bony growths (enthesophytes) and fusion of the sacroiliac joints, possibly due to a combination of congenital or acquired causes including ankylosing spondylitis, DISH and exstrophy, in an elderly Thai male (KKU). (Jeno I. Sebes, personal communication, 2011). (b) Radiograph showing numerous bony growths (enthesophytes) reflecting DISH (KKU).

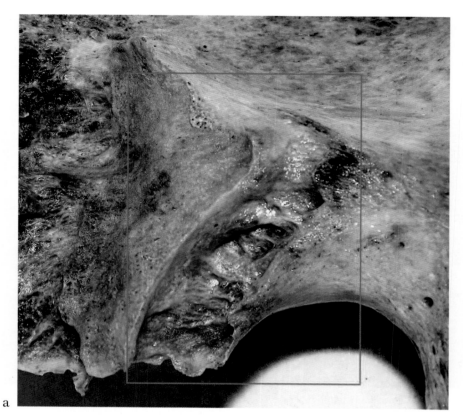

a

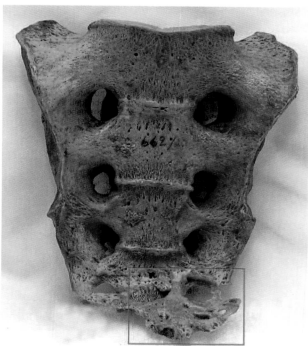

b

PLATE 108. (a) Preauricular sulcus/groove in an adult Thai female (KKU). (b) Healed fracture (rectangle) with left lateral displacement of the lower segments of the sacrum (KKU).

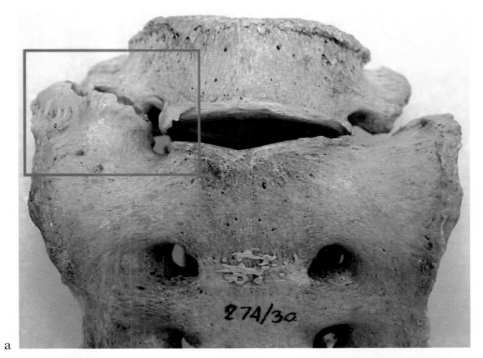

a

b

PLATE 109. (a) Hemisacralization of the sacrum and fifth lumbar vertebra in an adult Thai (KKU). In this individual the right ala of the sacrum has assumed the shape of the morphology of the fifth lumbar vertebra and the fifth lumbar vertebra has assumed the morphology of the sacum. Cf. Barnes (1994). (b) Complete spina bifida of the sacrum and fused left sacroiliac joint (Photo courtesy of UTK FAC, WMB UTK, Joseph T. Hefner).

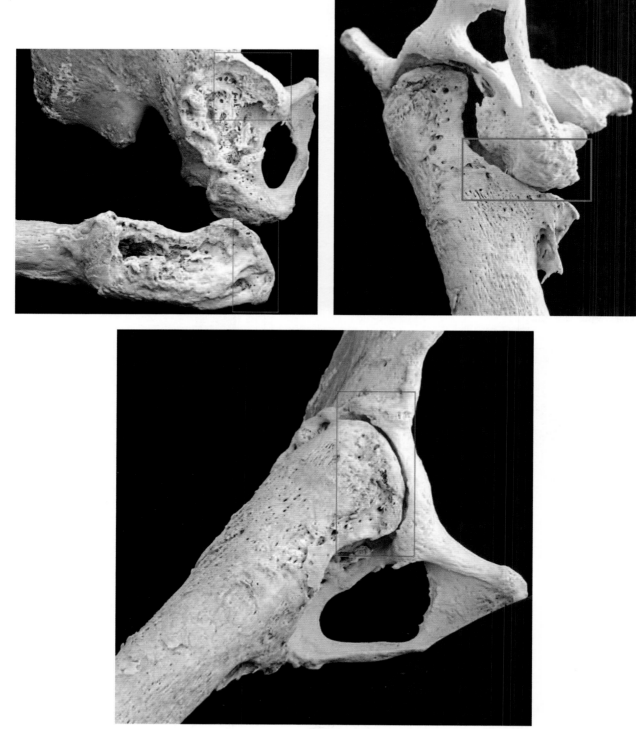

PLATE 110. Destruction and remodeling of the right acetabulum and femoral head in an elderly Thai (KKU). The rectangles indicate areas of contact between the femur and acetabulum and the femur and ischium. Destruction of the hip can result from many factors including infection, trauma, ischemia (decreased blood supply to the hip), rheumatoid arthritis and congenital abnormalities, to name a few.

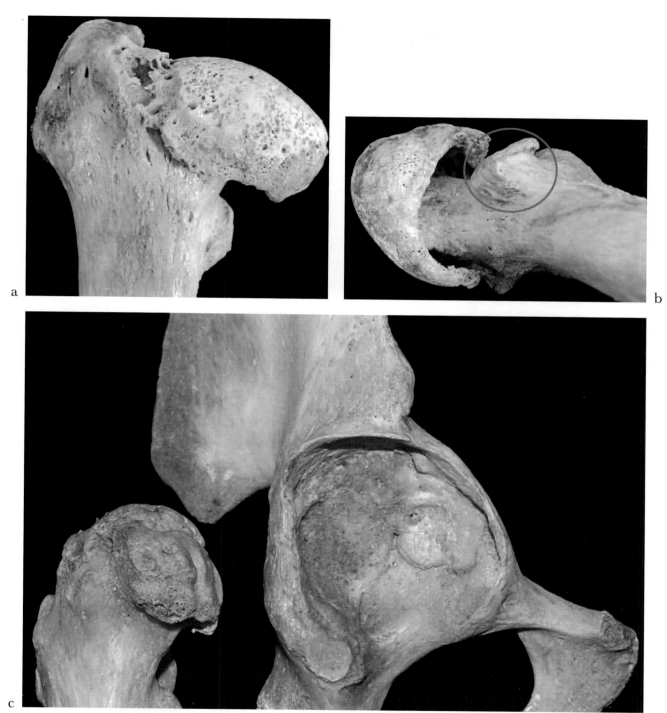

PLATE 111. (a and b) Degenerative joint disease and severe flattening and widening ("mushrooming") of the right femoral head, possibly reflecting Perthes' disease (KKU). Note the flattened articular facet (circle) for the femur and acetabular rim along the posterior surface. (c) Probable Perthes' disease or slipped femoral capital epiphysis in the right hip of a 15-year-old female (SI).

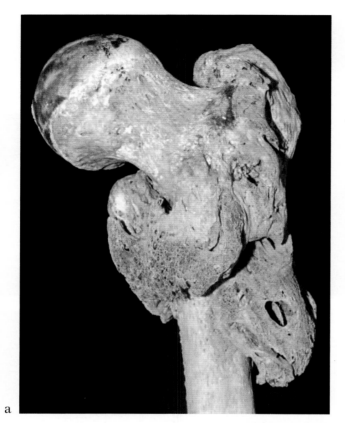

a

b

PLATE 112. (a) Healed fracture of the proximal left femur with shortening of the diaphysis and myositis ossificans traumatica in an adult Thai male (KKU). (b) Osteoarthritis of the left femoral head resulting in an uneven joint surface and irregular margin that extended beyond the normal articular surface of the head reflecting avascular necrosis of the femoral head (Huntington SI).

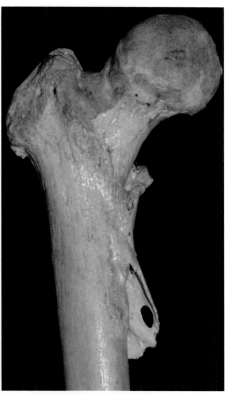

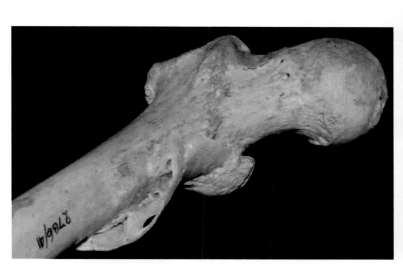

PLATE 113. Myositis ossificans traumatica in the proximal left femur of an adult Thai male (KKU).

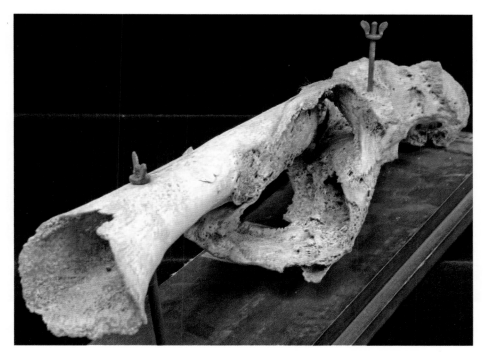

PLATE 114. Fractured and healed right femur with ossification of the muscle (myositis ossificans traumatica) (Mütter Museum 1397.051).

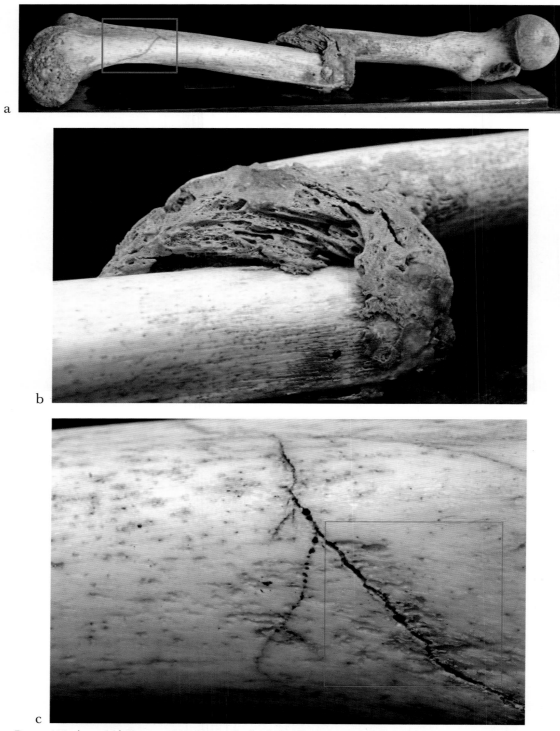

PLATE 115. (a and b) Fractured and partially healed right femur with overriding broken ends and a large callous formation spanning the break (Mütter Museum 1397). Note the perimortem fracture (a) of the distal diaphysis that exhibits early-stage healing (possibly 7–10-day duration) in the form of increased vascularity (surface pitting) and (c) fibrous bone spanning the fracture line. The degree of healing of the lower fracture suggests it may have occurred long after the midshaft fracture that shows a more advanced stage of healing.

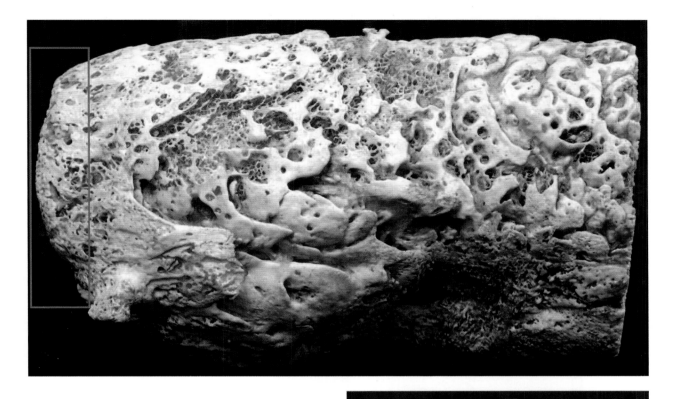

PLATE 116. Double amputation of the femoral shaft (Mütter Museum 1404.31). The rough and irregular surface reflects healing (in response to infection) and increased vascularity. The first cut is indicated by the red square and rectangle. The second cut/amputation is visible in the lower right photo.

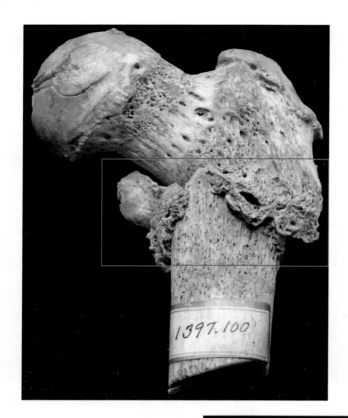

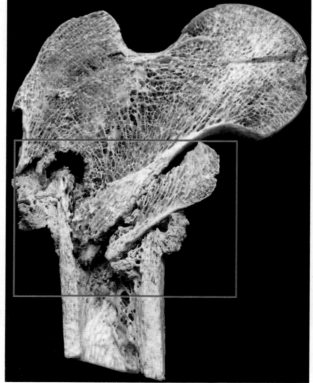

PLATE 117. Healed fracture (rectangles) of the left femoral diaphysis (Mütter Museum 1397.100).

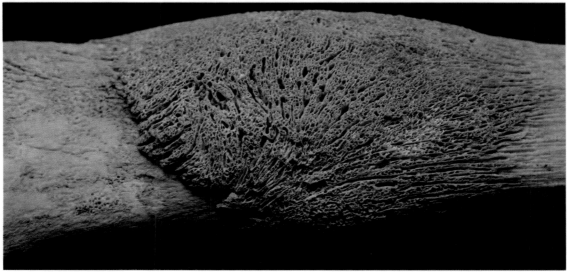

PLATE 118. Syphilis of the proximal portion of the left tibia (Mütter Museum 1449). Note the well-defined margin of the lesion and numerous vascular pits and striations for increased blood supply to this area of the leg.

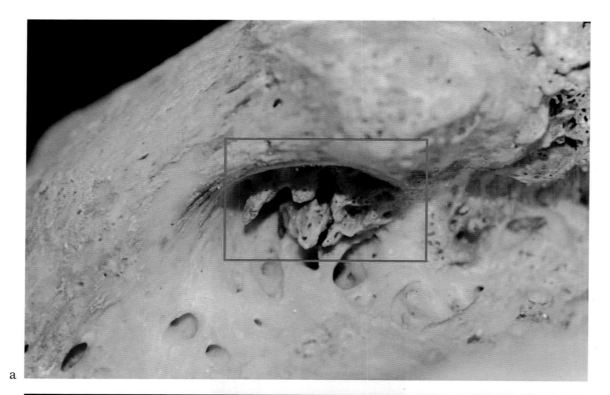

a

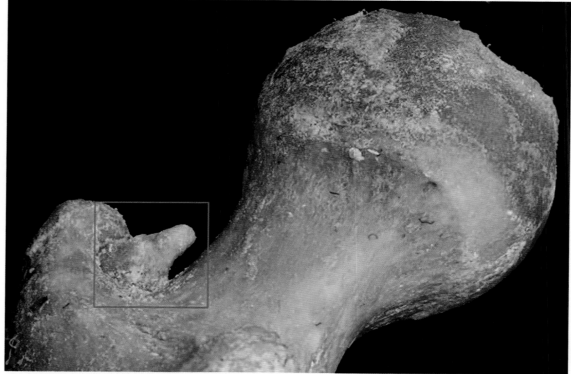

b

PLATE 119. (a) Multiple bony projections/enthesophytes and (b) an unusually large single enthesophyte in the trochanteric fossa of two femurs (KKU). Enthesophytes in the trochanteric fossa typically begin as clusters of small bumps, spikes and projections resembling stalagmites.

PLATE 120. Infection of the right femur with new bone that formed an involucrum (outer sheath or "envelope") and sequestrum (white bone) indicating the original, but now dead, portion of the original femur (Hamann-Todd, Joseph T. Hefner).

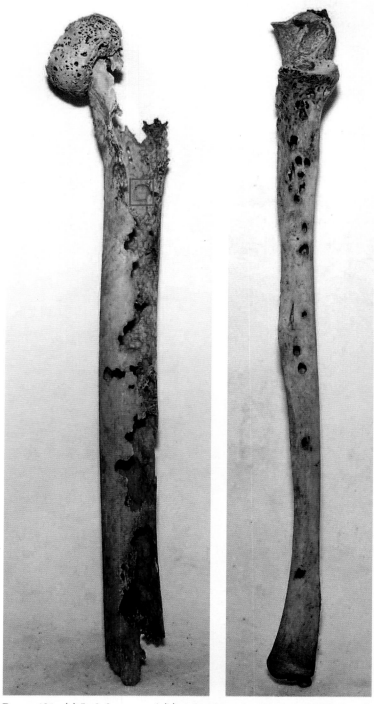

a
b

PLATE 121. (a) Left femur and (b) right ulna of a 57-year-old Thai male exhibiting circular lytic lesions throughout the skeleton as though these areas were removed with an "ice-cream scoop" (square) with destruction of the cortex and femoral head that originated in the medullary cavity and spread outward (KKU). The distribution and appearance of the lesions are consistent with multiple myeloma.

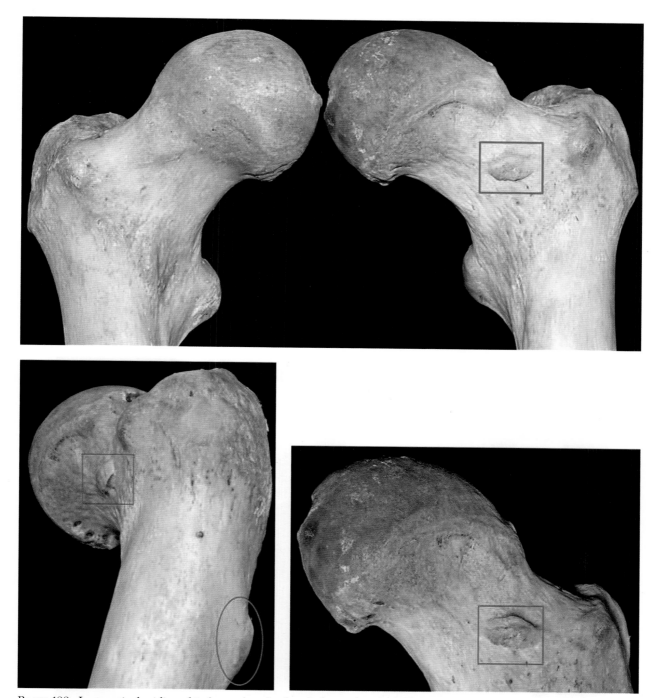

PLATE 122. Juxta-articular idiopathic femoral neck tubercle (square) and small third trochanter (oval) in the left femur of an adult (SI Huntington). This tubercle lacked a central nidus or sclerotic bone. Note that the tubercle is located in an area that has no muscle or tendon attachments, is domed, and that much of its margin is raised. Also note the morphology of the left femoral head with its articular extension (plaque or Poirier's facet) is different than that of the right femur. The authors have only encountered this tubercle in two individuals.

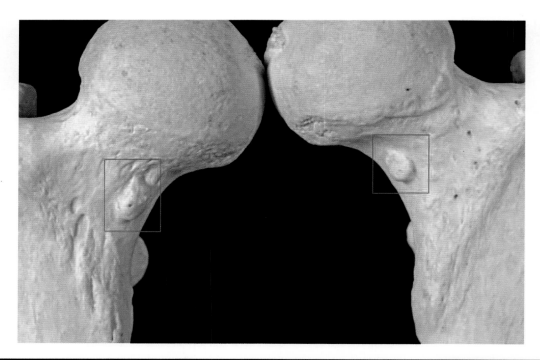

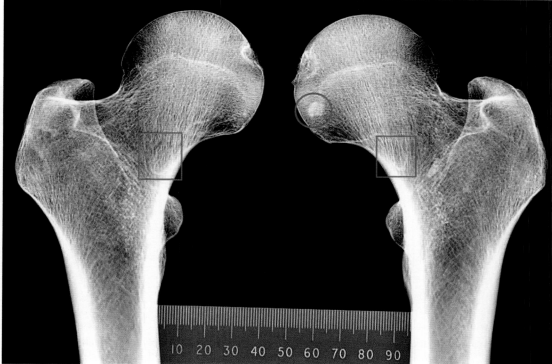

PLATE 123. X-rays of the idiopathic femoral neck tubercles (squares) showing the thin shell of cortical bone in a middle-aged Asian male (CSC DS034). That these tubercles do not communicate with the medulla is inconsistent with osteochondroma. Other possibilities include idiopathic tubercles and femoral-acetabular impingement (less likely). The angles of the left and right necks were within the normal range, neither coxa vara or coxa valga. Note the benign bone island (circle), also known as an enostosis, in the left femoral head – bone islands are comprised of normal compact bone and probably developmental in origin.

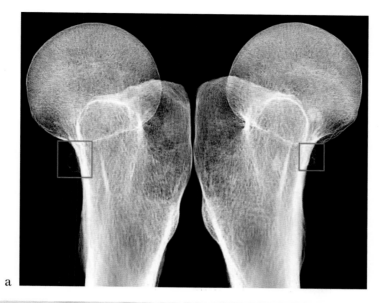

a

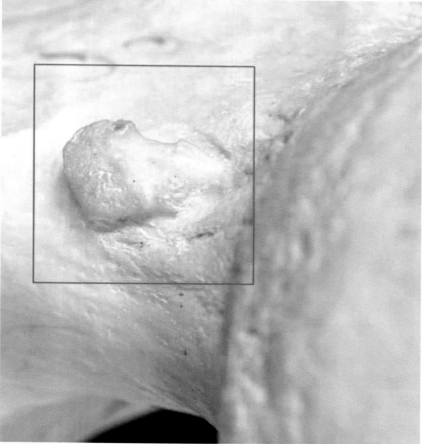

b

PLATE 124. (a) Another radiographic view of the bony tubercles (squares) (CSC DS034). Note that the tubercles have a thin cortical "shell" and a disorganized internal structure that is less dense than the adjacent femoral cortex. (b) Close-up of tubercle on the anterior surface of the right femoral neck (CSC DS034).

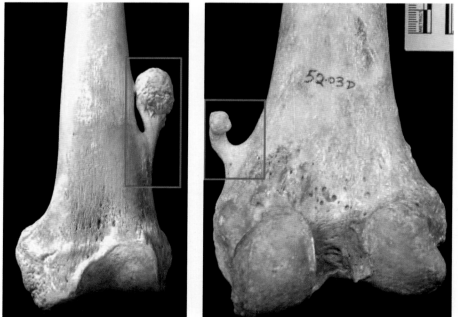

PLATE 125. (a) Advanced and severe osteoarthritis of the right knee (dorsal view) – note the raised marginal and surface osteophytes (Photo courtesy of UTK FAC, WMB UTK, Joseph T. Hefner). The degree of severity in this specimen is not commonly encountered in archaeological specimens, but is a relatively common finding in cadaver and more modern skeletal samples. (b and c) Stalk-like osteochondromas (solitary exostosis, osteocartilaginous exostosis) in the distal femur (Photo courtesy of UTK FAC, WMB UTK, Joseph T. Hefner). Note that these tumorous growths are oriented parallel to the femoral shaft and that it would have been carti-lage capped in the living (osteochondroma should be diagnosed by radiological examination).

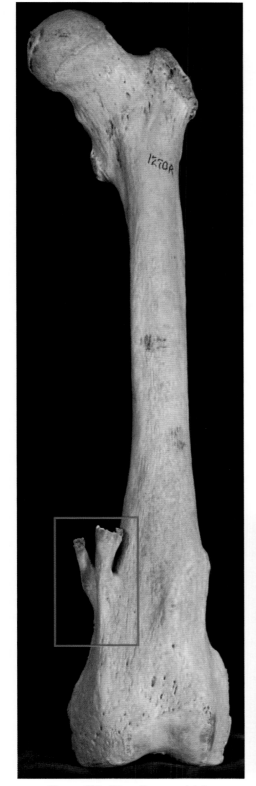

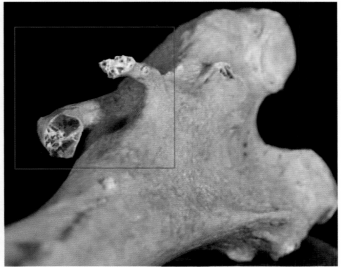

PLATE 126. Hereditary multiple exostoses of the left femur in a 59-year-old white female (SI 1270R).

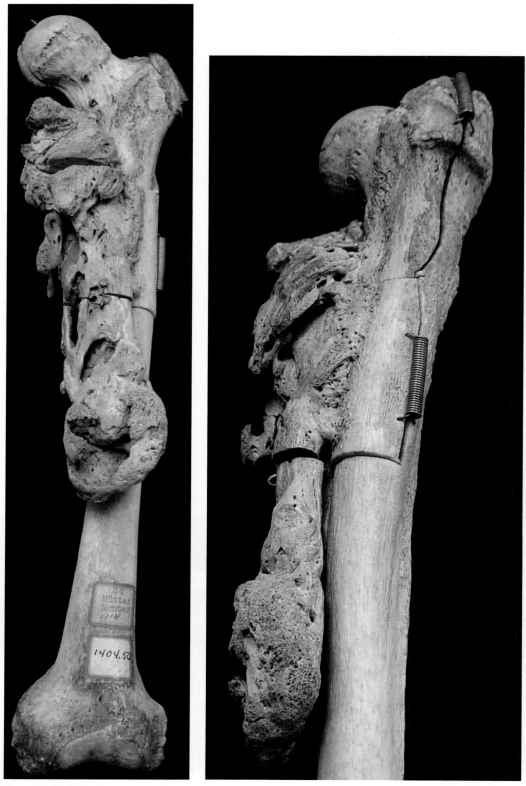

Plate 127. Probable myositis ossificans traumatica of the left femur (Mütter Museum 1404.50) (see Plate 128).

PLATE 128. Cutaway of the anterior surface of the left femur showing the bony growth (myositis ossificans) along the anterior surface and that it is purely periosteal surface – this lesion did not affect the endosteum of the femur (Mütter Museum 1404.50).

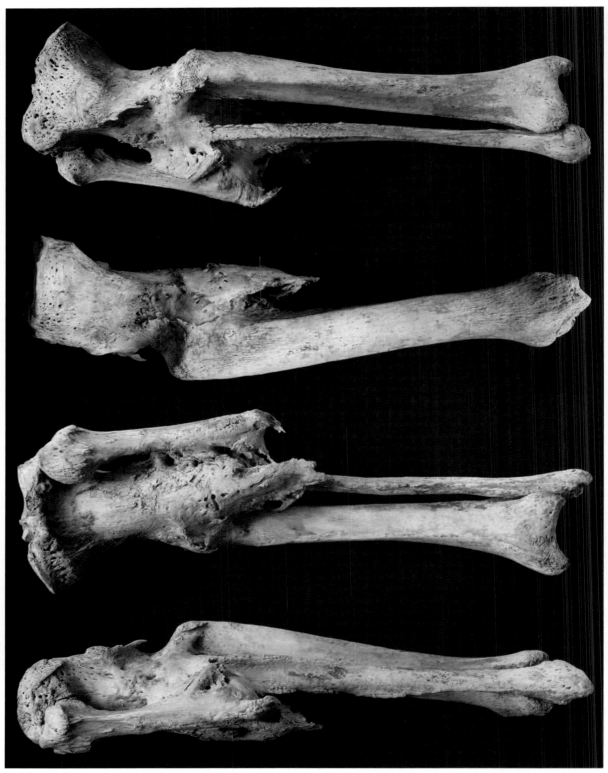

PLATE 129. Four views of healed fractures of the right tibia and fibula from a 33–43-year-old Peruvian male (Ancon I T960, Matthew E. Rhode).

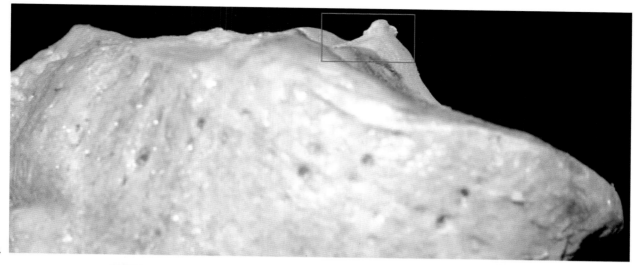

PLATE 130. (a) Spiking/spurring of the tibial spine, also known as intercondylar eminence, and (b) ossification of the posterior cruciate ligament in the tibia (KKU). Spiking of the tibial spine is a common accompaniment to osteoarthritis, and heterotopic ossification of the posterior cruciate ligament often occurs as a result of trauma to the knee and ankylosing spondylitis, among others.

PLATE 131. (a) Healed periostitis in a left tibia (KKU). Note the pitting, striated appearance of some areas, well-defined margin separating the periostitis from the normal cortex, and a darkening of the periosteal bone giving it a woody, bark-like appearance; (b) actual tree bark). (c and d) Common form of arthritic lipping of the margin and porosity of the articular surface in a left patella and flowing calcification/ossification of the quadriceps tendon on the anterior surface (KKU). This is a common condition in most populations, especially in elderly individuals and those with DISH.

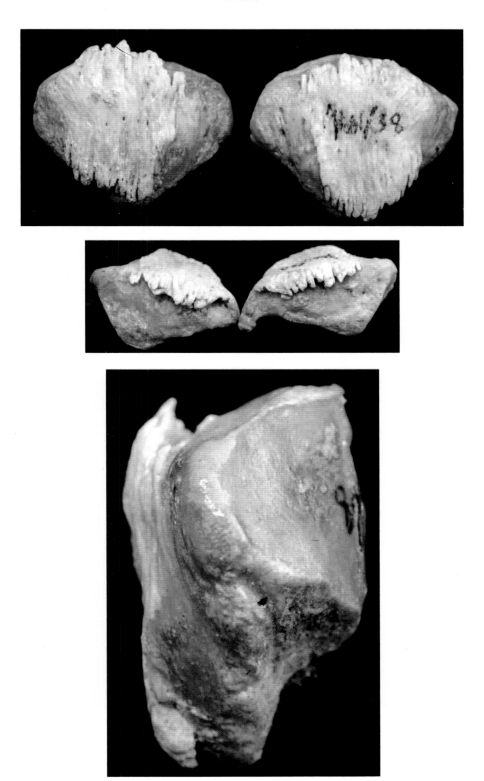

PLATE 132. Flowing enthesophytes reflecting ossification of the quadriceps tendon that is typical of DISH, in an elderly Thai male (KKU). (The white bone areas are artifacts of the cleaning process.)

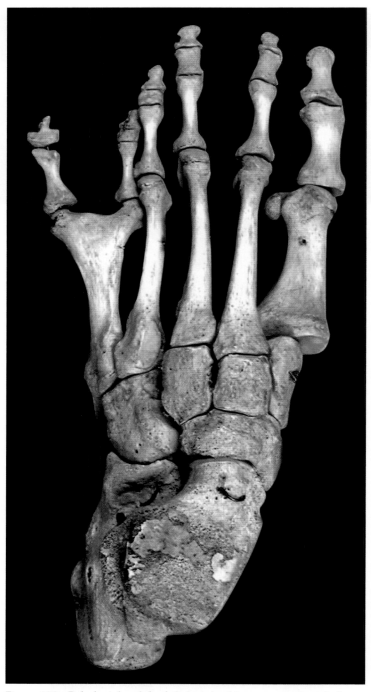

PLATE 133. Polydactyly of the left foot, in this case a postaxial abnormality of the lateral foot resulting in a T-shaped fifth metatarsal – note the six toes (Cf. Venn-Watson, 1976) (Civil War specimen, Polydactyl Foot NMHM 1996.0017, AMM Med Sec 0724; photo by Joseph T. Hefner). Polydactyly is duplication of a digit or metatarsal and is the most common congenital anomaly of the foot, occurring in 1of 500 to 1 of 1000 live births, although higher in some ethnic groups. Cf. Castilla et al. (1973), Shiota and Matsunago (1978), and Huang et al. (2000).

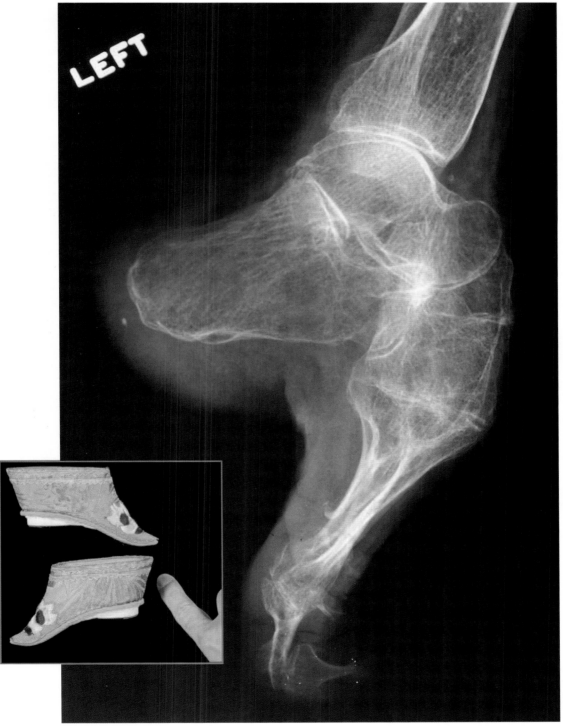

PLATE 134. Radiograph of foot binding in an elderly Chinese female (unbound foot in this x-ray) reflecting changes over a period of many decades. Note the soft tissue folds, thin ("eggshell") bone with loss of trabecular structure, horizontal position of the distal phalanx and extension of the foot. The shoes, which were made for an adult, measure 10 cm long and 3 cm wide. Inset shows an example pair of shoes worn by Chinese women with bound feet.

PLATE 135. Example of Chinese foot binding (unbound in the photo) (Mütter Museum 8429). Note the extremely high arch and position of the toes both in front of and wrapped under (square indicates the little toe) the right foot. (Not the same specimen as Plate 134.)

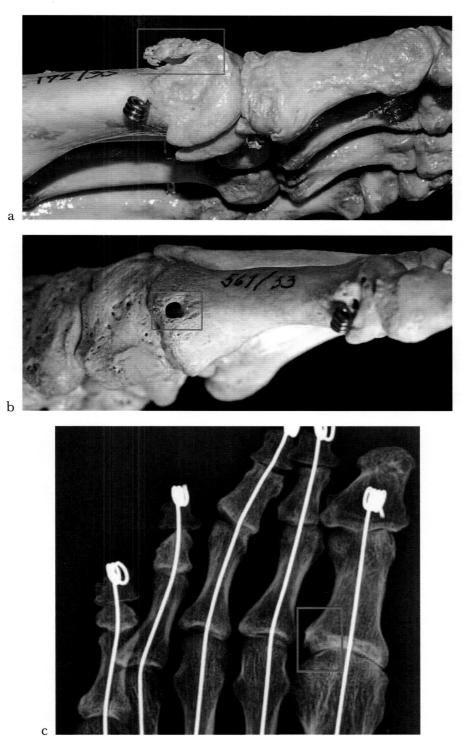

PLATE 136. (a) Large osteophytes (rectangle) reflecting osteoarthritis in the left first metatarso-phalangeal joint in an elderly Thai (KKU). (b and c) Periarticular lytic lesion (rectangle), possibly reflecting Rheumatoid Arthritis of the left first tarsometatarsal joint in an adult Thai male (KKU). This lesion lacks the characteristic overhanging margin of gouty arthritis (Jeno I. Sebes, personal communication).

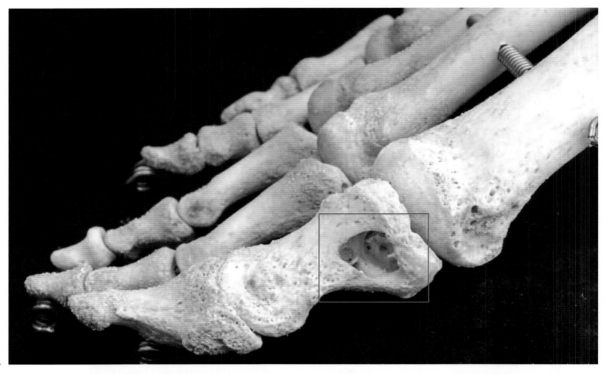

a

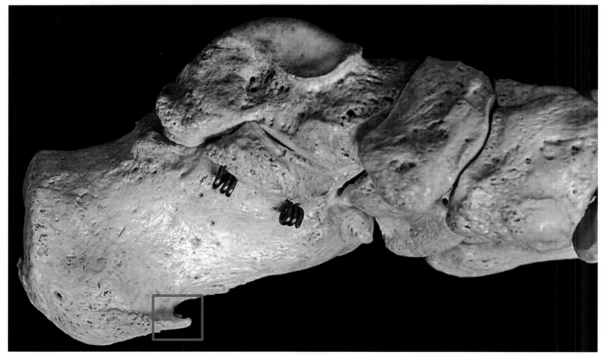

b

PLATE 137. (a) Large lytic lesion in the metatarso-phalangeal joint of the right first phalanx of unknown etiology (possibly gout). The lesion perforated the articular surface (base) of the phalanx (KKU). (b) Inferior calcaneal spur (square) at the attachment of the plantar fascia ligament, which is a common finding in middle-aged and elderly individuals, especially those with ankylosing spondylitis (AS) and DISH, but also commonly occurs in athletes, especially runners (KKU).

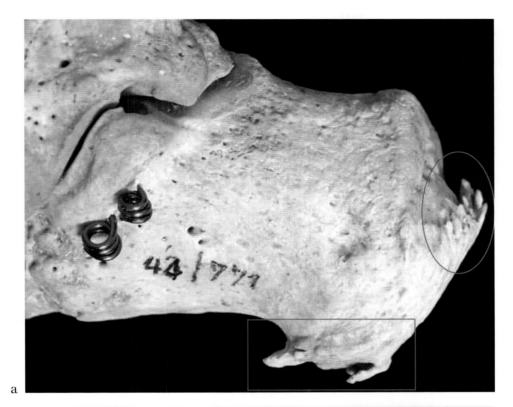

a

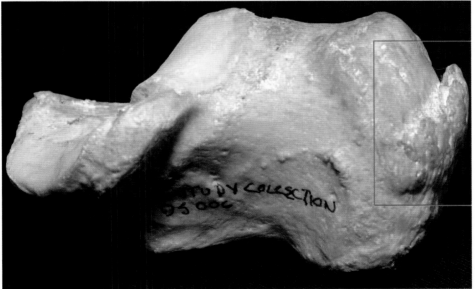

b

PLATE 138. (a) Posterior (oval) and inferior (rectangle) enthesophytes in an elderly Thai male (KKU). These spiked or tooth-like lesions are common findings in the elderly and individuals with ankylosing spondylitis (AS), diffuse idiopathic skeletal hyperostosis (DISH), mechanical pulls and of the plantar fascia or Achilles tendon, and degenerative arthritis. Synonyms for these lesions include infracalcaneal spur, traction spur, heel spur, calcaneal spur, plantar fasciitis, calcaneal exostosis, and calcaneal traction osteophytes, among others. (b) Large posterior calcaneal spur (rectangle) at the attachment of the Achilles' tendon, typical of elderly individuals or those with AS or DISH (CSC).

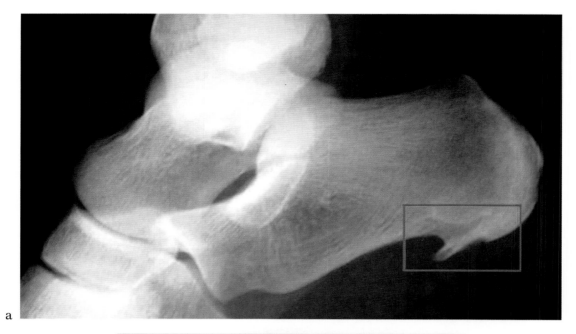

a

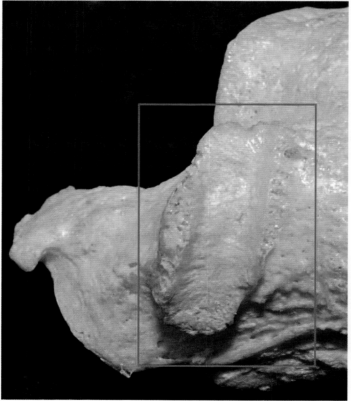

b

PLATE 139. (a) Radiograph of an inferior heel spur (rectangle) (SI). (b) Hypertrophied trochlear process, also known as the peroneal tubercle, groove or peroneal trochlea, that accommodates and guides the peroneal tendons (rectangle) in a left calcaneus (CSC DS 007). Most calcanei exhibit little or no evidence of the peroneal tubercle, but this groove or tubercle may become prominent in some individuals. Note the nearly vertical groove running through the defect and raised margins (lateral view). Cf. Wang et al. (2005).

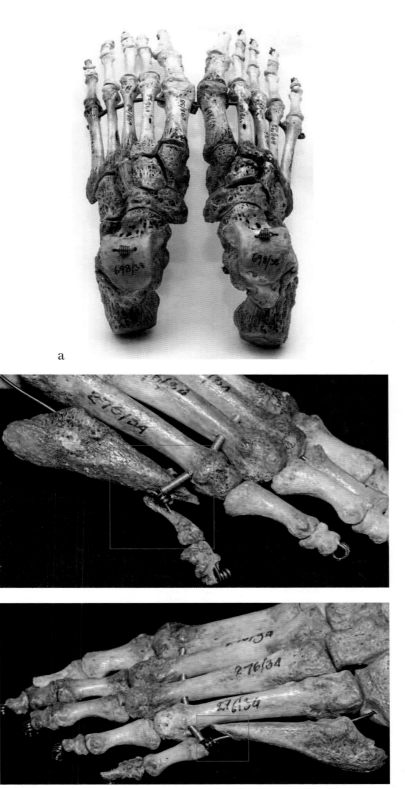

PLATE 140. (a) Feet of 58-year-old Thai male with thalassemia and osteoporosis (KKU 698/38). (b and c) Bilateral resorption and penciling of the distal portions of the 5th metatarsals in an adult Thai, due to leprosy (KKU). Cf. Ortner (2008).

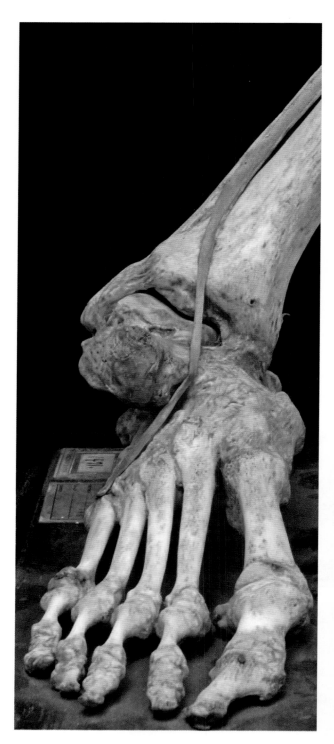

PLATE 141 *(left)*. Complete dislocation of the right talus (Mütter Museum 1471). Note the position of the peroneus brevis tendon to the talar head.

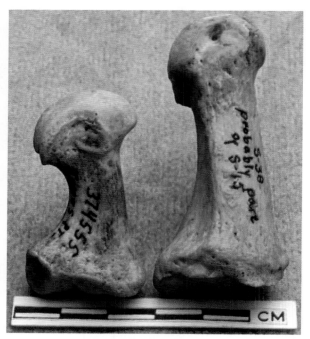

PLATE 142 *(above)*. Hypoplastic and shortened left first metatarsal compared to the same individual's normally-sized right first metatarsal (SI).

BIBLIOGRAPHY

Abraham, E., Verinder, D. G. R., and Sharrard, W. J. W.: The treatment of flexion contracture of the knee in myelomeningocele. *Journal of Bone and Joint Surgery 59-B*(4):433–438, 1997.

Adams, J. D., and Leonard, D. R.: A developmental anomaly of the patella frequently diagnosed as fracture. *Surgery, Gynecology, Obstetrics 41*:601–604, 1925.

Adams, I. D.: Osteoarthrosis and sport. *Journal of the Royal Society of Medicine 72*(3):185–187, 1977.

Adams, J. L.: The supracondyloid variation in the human embryo. *Anatomical Record 5*(9):315, 1934.

Adelaar, R. S. (Ed.): *Complex foot and ankle trauma.* Philadelphia, Lippincott-Raven, 1999.

Adirim, T. A., and Cheng, T. L.: Overview of injuries in the young athlete. *Sports Medicine 33*(1):75, 2003.

Aguiar, L. B. V., Neves, F. S., Bastos, L. C., Crusoé-Rebello, I., Ambrosano, G. M. B., and Campos, P. S. F.: Multiple stafne bone defects: A rare entity. *International Scholarly Research Network*, Article ID 792145, 2011.

Ait-Ameur, A., Wakim, A., Dubousset, J., Kalifa, G., and Adamsbaum, C.: The AP diameter of the pelvis: A new criterion for continence in the exstrophy complex? *Pediatric Radiology 31*(9):640–645, 2001.

Akrawi, F.: Is bejel syphilis? *British Journal of Venereal Disease 26*:115–123, 1949.

Albright, F., Smith, P. A., and Richardson, A. M.: Postmenopausal osteoporosis: Its clinical features. *Journal of American Medical Association 116*:2465, 1941.

Alexander, C. J.: Osteoarthritis: a review of old myths and current concepts. *Skeletal Radiology 19*:327–333, 1990.

Allbrook, D. B.: The East African vertebral column. A study in racial variability. *American Journal of Physical Anthropology 13*:489–513, 1955.

Allen, H.: *A System of Human Anatomy, Sec. II, Bones and Joints.* Philadelphia, 1882.

Allison, M. J., and Gerszten, E.: *Paleopathology in South American mummies.* Richmond, Virginia Commonwealth University, 1982.

Allison, M. J., Gerszten, E., Munizaga, J., and Santoro, C.: La practica de la formacion craneana entre los pueblos andinos precolumbinos. *Chungara 8*:238–260, 1981.

Altchek, M.: Congenital clubfeet (letters to the editor). *Clinical Orthopaedics and Related Research 130*:303–305, 1978.

Alves, N., Nascimento, C. M. O., and Patriarca, J. J.: Dental germination in the inferior canine in both dentition. Case Report. *International Journal of Morphology 28*(3): 873–874, 2010.

Anand, V. T., Latif, M. A., and Smith, W. P.: Defects of the external auditory canal: A new reconstruction technique. *Journal of Laryngology Otology 114*(4):279–282, 2000.

Andersen, J. G.: *Studies on the Medieval diagnosis of leprosy in Denmark. An osteoarchaeological, historical and clinical study.* Copenhagen, Costers Bogtrykkeri, 1969.

Andersen, J. G.: *The medieval diagnosis of leprosy.* In Ortner D. J., and Aufderheide, A. C., (Eds.): *Human palaeopathology: Current syntheses and future options.* Washington, DC: Smithsonian Institution Press, pp. 205–208, 1991.

Anderson, J.E.: The people of Fairty. *National Museum of Canada Bulletin 193*:128–129, 1963.

_____: The osteological diagnosis of leprosy. Proceedings of the Paleopathology Association 4th European Meeting, Middleberg, Antwerpen, 1982.

Anderson, T.: Calcaneus secundarius: An osteo-archaeological note. *American Journal of Physical Anthropology 77*:529–531, 1988.

_____: A medieval example of a sagittal cleft or "butterfly" vertebra. *International Journal of Osteoarchaeology 13*(6): 352–357, 2003.

Angel, J. L.: Osteoporosis: Thalassemia? *American Journal of Physical Anthropology 22*:369–373, 1964.

_____: Porotic hyperostosis, anemias, malarias and marshes in prehistoric eastern Mediterranean. *Science 153*:760–763, 1966.

_____: Colonial to modern skeletal change in the USA. *American Journal of Physical Anthropology 45*:723–735, 1976.

_____: History and development of paleopathology. *American Journal of Physical Anthropology 56*:509–515, 1981.

Angel, J. L., Kelley, J. O., Parrington, M., and Pinter, S.: Life stresses of the free black community as represented by First African Baptist Church, Philadelphia. *American Journal of Physical Anthropology 74*:213–229, 1987.

Antoniades, D. Z., Belazi, M., and Papanayiotou, P.: Concurrence of torus palatinus with palatal and buccal exostoses: Case report and review of the literature. *Oral Surgery Oral Medicine Oral Pathology 85*(5):552–557, 1998.

Apley, A. G., and Solomon, L.: *Concise system of orthopaedics and fractures.* London, Butterworth and Company, 1988.

Arey, L. B.: *Developmental anatomy: A textbook and laboratory manual of embryology* (6th ed). Philadelphia, W. B. Saunders, 1966.

Arnay-de-la-Rosa, M., Gonzalez-Reimers, E., Castilla-Garcia, A. and Santolaria-Fernandez, F.: Radiopaque transverse lines (Harris lines) in the prehispanic population of El Hierro (Canary Island). *Anthropologischer Anzei ger 52*(1):53–57, 1994.

Arnay-de-la-Rosa, M., Velasco-Vazquez, J., Gonzalez-Reimers, E., and Santolaria-Fernandez F.: Auricular exostoses among the prehistoric population of different islands of the Canary archipelago. *Annals Otology Rhinology Laryngology 110*(11):1080–1083, 2001.

Arnott, R., Finger, S. and Smith, C. U. M. (Eds.). *Trepanation: History, discovery, theory.* Lisse, Swets & Zeitlinger, 2005.

Aronica-Pollak, P. A., Stefan, V. H., and Mclemore, J.: Coronal cleft vertebra initially suspected as an abusive fracture in an infant. *Journal of Forensic Sciences 48*(4): 836–838, 2003.

Arriaza, B. T., Salo, W., Aufderheide, A. C., and Holcomb, T. A.: Pre-Columbian tuberculosis in Northern Chile: Molecular and skeletal evidence. *American Journal Physical Anthropology 98*:37–45, 1995.

Ashley, G. T.: The relationship between the pattern of ossification and the definitive shape of the mesosternum in man. *Journal of Anatomy 90*:87–105, 1956.

Attah, C. A., and Cerruti, M. M.: Aspergillus osteomyelitis of the sternum after cardiac surgery. *New York State Journal of Medicine 79*:1420–1421, 1979.

Aufderheide, A. C., and Rodriguez-Martin, C.: *The Cambridge encyclopaedia of human paleopathology.* Cambridge University Press, 1998.

Aydogan, M., Karatoprak, O., Mirzanli, C., Ozturk, C., Tezer, M., and Hamzaoglu, A.: Severe erosion of lumbar vertebral body because of a chronic ruptured abdominal aortic aneurysm. *The Spine Journal 8*(2):394– 396, 2008.

Baker, B. J., and Armelagos, G. J.: The origin and antiquity of syphilis. *Current Anthropology 29*(5):703–737, 1988.

Balen, P. F., and Helms, C. A.: Bony ankylosis following thermal and electrical injury. *Skeletal Radiology 30*:393–397, 2001.

Banadda, B. M., Gona, 0., Vaz, R., and Ndlovu, D. M.: Calcaneal spurs in a black African population. *Foot and Ankle 13*(6):352–354, 1992.

Banks, A. S., Downey, M. S., Martin, D. E., and Miller, S. J.: *McGlamry's comprehensive textbook of foot and ankle surgery* (Vol. 2, 3rd ed). Philadelphia, Lippincott, Williams & Wilkins, 2001.

Baraliakos, X., Listing, J., Rudwaleit, M., Haibel, H., Brandt, J., Sieper, J., and Braun, J.: Progression of radiographic damage in patients with ankylosing spondylitis: Defining the central role of syndesmophytes. *Annals of the Rheumatic Diseases 66*:910–915 doi:10.1136/ard.2006.066415, 2007.

Barbian, L., and Sledzik, P. S.: Healing following cranial trauma. *Journal of Forensic Sciences 53*(2):263–268, 2008.

Barnard, L. B., and McCoy, S. M.: The supracondyloid process of the humerus. *Journal of Bone and Joint Surgery 28*:845, 1946.

Barnes, E.: *Developmental defects of the axial skeleton in paleopathology.* University of Oklahoma Press, Niwot, 1994.

Barnett, C. H.: Squatting facets on the European talus. *Journal of Anatomy 88*(4):509–513, 1954.

Barrie, H. J.: Osteochondritis dissecans 1887–1987: A centennial look at Konig's memorable phrase. *Journal of Bone and Joint Surgery 69-B*(5):693–695, 1987.

Barry, H. C.: *Paget's disease of bone.* London, E&S Livingstone Ltd., 1969.

Bartlett, D. W., and Shah, P.: A critical review of non-carious cervical (wear) lesions and the role of abfraction, erosion, and abrasion. *Journal of Dental Research 85*(4):306–312, 2006.

Basekim, C. C., Mutlu, H., Güngör, A., Şilit, E., Pekkafali, Z., Kutlay, M., Çolak, A., Özturk, E., and Kizilkaya, E.: Evaluation of styloid process by three-dimensional computed tomography. *European Radiology 15*(1):134–139, 2005.

Bass, W. M.: *Human osteology: A laboratory and field manual* (4th ed.). Columbia, Missouri Archaeological Society Special Publication No. 2, 1995.

Bassiouni, M.: Incidence of calcaneal spurs in osteoarthrosis and rheumatoid arthritis, and in control patients. *Annals of the Rheumatic Diseases 24*:490–493, 1965.

Battistone, M. J., Manaster, B. J., Reda, D. J., and Clegg, D. O.: The prevalence of sacroiliitis in psoriatic arthritis: New perspectives from a large, multicenter cohort. *Skeletal Radiology 28*:196–201, 1999.

Baykara, I., Hakan, Y., Timur, G., and Erksin, G.: Squatting facet: A case study of Dilkaya and Van-Kalesi populations in Eastern Turkey. *Croatian Anthropological Society, Zaghreb 34*(4):1257–1262, 2010.

Bayliss, M.: Responses in human cartilage in relation to age. In Russell R, G. G., and Dieppe, P. (Eds.): *Osteoarthritis: Current research and prospects for pharmacological intervention.* London, IBC, 1991.

Beckly, D. E., Anderson, P. W., and Pedegana, L. R.: The radiology of the subtalar joint with special reference to talo-calcaneal coalition. *Clinical Radiology 26*:333–341, 1975.

Beighton, P.: *Inherited disorders of the skeleton.* New York, Churchill Livingstone, 1978.

Belsky, J. L., Hamer, J. S., Hubert, J. E., Insogna, K., and Johns, W.: Torus palatinus: A new anatomical correlation with bone density in postmenopausal women. *The Journal of Endocrinology and Metabolism 88*(5):2081–2086, 2003.

Bender, P.: Genetics of cleft lip and palate. *Journal of Pediatric Nursing 15*(4):242–249, 2000.

Benjamin, M. L., and McGonagle, D.: The anatomical basis for disease localization in seronegative spondyloarthropathy at entheses and related sites. *Journal of Anatomy Nov* (Pt. 5):503–526, 2001.

Benjamin, M. L., and McGonagle, D.: Entheses: Tendon and ligament attachment sites. *Scandinavian Journal of Medical Science and Sports 19*:520–527, 2009.

Benjamin, M. L., Copp, L., and Evans, E. J.: The histology of tendon attachments to bone in man. *Journal of Anatomy 149*:89–100, 1986

Benjamin, M., Kumaim T., Milzm S., Boszczyk, B., Boszczyk, A. and Ralphs, J.: The skeletal attachment of tendon-tendon "entheses." *Comparative Biochemistry and Physiology, Part A;133*:931–945, 2002.

Benjamin, M., Rufai, A., and Ralphs, J. R.: The mechanism of formation of bony spurs (enthesophytes) in the achilles tendon. *Arthritis and Rheumatism 43*(3):576–583, 2000.

Benjamin, M., Toumi, H, Ralphs, J.R., Bydder, G., Best, T. M. and Milz, S.: Where tendons and ligaments meet bone: Attachment sites ('entheses') in relation to exercise and/or mechanical load. *Journal of Anatomy 208*:471–490, 2006.

Bennett, E. H.: On the ossicle occasionally found on the posterior border of the astragalus. *Journal of Anatomy and Physiology 21*(1):59–65, 1886.

Benzel, C. E.: *Biomechanics of spine stabilization.* American Association of Neurological Surgeons, Illinois, 2001.

Berberian, W. S., Hecht, P. J., Wapner, K. L., and Di-Verniero, R.: Morphology of tibiotalar osteophytes in anterior ankle impingement. *Foot and Ankle International 22*(4):313–317, 2001.

Bergman, R. A., Thompson, S. A., Afifi, A. K., and Saadeh, F. A.: *Compendium of human anatomic variation: Text, atlas, and world literature.* Baltimore, Urban and Schwarzenberg, 1988.

Bergsma, D. (Ed.): *Birth defects compendium* (2nd ed.). New York, Alan R. Liss, 1978.

Bernhard, J.: [Seronegative spondyloarthropathies] [article in German]. *Ther Umsch 59*(10): 529–534, 2002.

Berry, A. C.: Factors affecting the incidence of non-metrical skeletal vatiants. *Journal of Anatomy 120*:519–535, 1975.

Berry, A. C., and Berry, R. J.: Epigenetic variation in the human cranium. *Journal of Anatomy 101*(2):361-379, 1967.

Berryman, H. E., and Gunther, W. M.: Keyhole defect production in tubular bone. *Journal of Forensic Sciences 45*(2):483–487, 2000.

Berryman, H. E., and Symes, S. A.: Recognizing gunshot and blunt cranial trauma through fracture interpretation. In Reichs, K. J. (Ed.): *Forensic osteology: Advances in the identification of human remains* (2nd ed.). Springfield, Charles C Thomas, 1998.

Bertaux, T. A.: *L'humerus et le femur* [article in French]. Lile, France, 1891.

Bertram, J. E. A., and Swartz, S. M.: The "law of bone transformation": A case of crying Wolff? *Biological Revue 66*:245–273, 1991.

Bhalaik, V., Chhabra, S., and Walsh, H. P. J.: Bilateral coexistent calcaneonavicular and talocalcaneal tarsal coalition: A case report. *The Journal of Foot and Ankle Surgery 41*(2): 129–134, 2002.

Bilsky, M., and Azeem, S., Multiple myeloma: Primary bone tumor with systemic manifestations. *Neurosurgery Clinics of North America 19*(1):31–40, 2008.

Binford, C. H., and Conner, D. H.: *Pathology of tropical and extraordinary diseases.* Washington, D.C., Armed Forces Institute of Pathology, Von, 1976.

Blackburne, J. S., and Velikas, E. P.: Spondylolisthesis in children and adolescents. *Journal of Bone and Joint Surgery 59-B*(4):490–494, 1977.

Blackwood, H. J. J.: Arthritis of the mandibular joint. *British Dental Journal 115*(8):317, 1963.

Blakey, M. L., Leslie, T. E., and Reidy, J. P.: Frequency and chronological distribution of dental enamel hypoplasia in enslaved African American: A test of the weaning hypothesis. *American Journal of Physical Anthropology 95*:371–383, 1994.

Bland, J. H. *Disorders of the cervical spine: Diagnosis and medical management* (2nd ed.). Philadelphia, W. B. Saunders, 1994.

Bloch, I.: *History of syphilis, system of syphilis.* London, Hodder & Stoughton, 1908.

Bluestone, C. D., and Klein, J. O.: *Otitis media in infants and children.* Philadelphia, Saunders, 1988.

Bohne, W. H. O.: Tarsal coalition. *Current Options in Pediatrics 13*:29–35, 2001.

Boldsen, J. L.: Epidemiological approach to the paleopathological diagnosis of leprosy. *American Journal of Physical Anthropology 115*:380–387, 2001.

Boldsen, J. L.: Leprosy in the early medieval Lauchheim community. *American Journal of Physical Anthropology 135*: 301–310, 2008.

Bolm-Audroff, U.: Intervertebral disc disorders due to lifting and carrying heay weights. *Medical Orthopedic Technology 112*:293–296, 1992.

Borenstein, D. G., Wiesel, S. W., and Boden, S. D.: *Low back pain: Medical diagnosis and comprehensive management* (2nd ed.). W. B. Saunders, Philadelphia, 1995.

Boulle, E. L.: Evolution of two human skeletal markers of the squatting position: A diachronic study from antiquity to the modern age. *American Journal of Physical Anthropology 116*:50–56, 2001.

Bowen, J. R., Kumar, V. P., Joyce, John J. III, and Bowen, J. C.: Osteochondritis dissecans following Perthes' disease. In Burwell, R. B., and Harrison, M. H. M. (Eds.): *Clinical Orthopaedics and Related Research 209*:49–56, 1986.

Boyd, D.C.: Skeletal correlates of human behavior in the Americas. *Journal of Archaeological Method and Theory 2*(2):189–251, 1996.

Boyer, G. S., Templin, D. W., Cornoni-Huntley, J. C., Everett, D. F., Lawrence, R. C., Heyse, S. F., Miller, M. M., and Goring, W. P.: Prevalence of spondylarthropathies in Alaskan Eskimos. *Journal of Rheumatology 21*:2292–2297, 1994.

Boyer, M. I., and Gelberman, R. H.: Operative correction of swan-neck and boutonniere deformities in the rheumatoid hand. *Journal American Academy Orthopedic Surgery 7*(2):92–100, 1999.

Boylston, A.: Evidence for weapon-related trauma in British archaeological samples. In Cox, M., and Mays, S. (Ed.): *Human osteology in archaeological and forensic science.* Cambridge, Cambridge University Press, 2000.

Bradford, D. S.: Spondylolysis and spondylolisthesis. In Chou, S. N., and Seljeskog, E. L. (Eds.): *Spinal deformities and neurological dysfunction.* New York, Raven Press, 1978.

Bradley, J., and Dandy, D. J.: Osteochondritis dissecans and other lesions of the femoral condyles. *Journal of Bone and Joint Surgery 71-B*(3):518–522, 1989.

Brannon, E. W.: Cervical rib syndrome: An analysis of nineteen cases and twenty-four operations. *Journal of Bone and Joint Surgery* 45-A(5):977–998, 1963.

Bridges, P. S.: Spondylolysis and its relationship to degenerative joint disease in the prehistoric Southeastern United States. *American Journal of Physical Anthropology* 79:321–329, 1989.

Bried, M. J., and Galgiani, J. N.: Coccidiodides immitis infections in bones and joints. *Clinical Orthopaedics and Related Research* 211:235–243, 1986.

Bronner, F., Farach-Carson, M. C., and Roach, H. I.: *Bone and development.* London, Springer-Verlag, 2010.

Brook, C. J., Ravikrishnan, K. P., and Weg, J. G.: Pulmonary and articular sporotricosis. *American Review of Respiratory Disorders* 116:141–143, 1977.

Brothwell, D. R.: The real history of syphilis. *Science Journal* 6:27–32, 1970.

_____: Biparietal thinning in early Britian. In Brothwell, D. R., and Sandison, A. T. (Eds.): *Disease in antiquity: A survey of the diseases injuries and surgery of early populations.* Springfield, C. C. Thomas, 1967.

_____: *Digging up bones: The excavation treatment and study of human skeletal remains.* Ithaca: Cornell University Press, 1981.

Brothwell, D. R., and Sandison, A. T. (Eds.): *Diseases in antiquity: A survey of the diseases injuries and surgery of early populations.* Springfield, C. C. Thomas, 1967.

Brower, A. C.: Cortical defect of the humerus at the insertion of the pectoralis major. *American Journal of Roentgenology* 128:677–678, 1977.

Brower, A. C., and Allman, R. M.: The neuropathic joint: A neurovascular bone disorder. *Radiologic Clinics of North America* 19(4):571–580, 1981.

Brown, G. P., Feehery, R. V., Jr., and Grant, S. M.: Case study: The painful os trigonum syndrome. *Journal of Orthopaedics Sports and Physical Therapy* 22(1):22–25, 1995.

Brown, J. C., Klein, E. J., Lewis, C. W., Johnston, B. D., and Cummings, P.: Emergency department analgesia for fracture pain. *Annals of Emergency Medicine* 42(2):197–205, 2003.

Brown, K. R., Pollintine, P., and Adams, M. A.: Biomechanical implications of degenerative joint disease in the apophyseal joints of human thoracic and lumbar vertebrae. *American Journal of Physical Anthropology* 136:318–326, 2008.

Browner, B. D., Jupiter, J. B., Levine, A. M., and Trafton, P. G.: *Skeletal Trauma: Basic science, management and reconstruction* (Vol. 2, 3rd ed.). Philadelphia, Saunders, 2003.

Brukner, P., Bradshaw, C., Kahn, K. M., White, S., and Crossley, K.: Stress fractures: A review of 180 cases. *Clinical Journal of Sport Medicine* 6(2):85–89, 1996.

Bryceson, A., and Pfaltzgraff, R. E.: *Leprosy* (3rd ed.). Edinburgh, Churchill Livingstone, 1990.

Buggenhout, G. van, and Bailleul-Forestier, I.: Mesiodens. *European Journal of Medical Genetics* 51(2):178–181, 2008.

Buikstra, J. E., and Beck, L. A.: (Eds.). *Bioarchaeology: The contextual analysis of human remains.* New York, Elsevier, 2006.

Buikstra, J. E., and Williams S.: *Tuberculosis in the Americas: Current perspectives.* In Ortner, D. J., and Aufderheide, A. C. (Eds.): *Human paleopathology: Current syntheses and future options.* Washington, D.C., Smithsonian Institution Press, pp. 161–172, 1991.

Bullough, P. G.: Ivory exostoses of the skull. *Postgraduate Medical Journal* 41:277–281, 1965.

Bulos, S.: Herniated intervertebral lumbar disc in the teenager. *Journal of Bone and Joint Surgery* 55-B:273–278, 1973.

Bunnell, W. P.: Back pain in children. *Orthopedic Clinics of North America* 13(3):587–604, 1982.

Burgener, F. A., and Kormano, M.: *Differential diagnosis in conventional radiology* (2nd ed.). New York, Georg Thieme Verlag Stuttgart, 1991.

Burke, M. J., Fear, E. C., and Wright, V.: Bone and joint changes in pneumatic drillers. *Annals Rheumatic Diseases* 36:276–279, 1977.

Burman, M. S., and Lapidus, P. W.: Functional disturbances caused by inconstant bones and sesamoids of the foot. *Archives of Surgery* 22:936, 1931.

Burry, H. C.: Sport exercise and arthritis. *British Journal Rheumatology* 25:386–388, 1987.

Busch, M. T., and Morrissy, R. T.: Slipped capital femoral epiphysis. *Orthopaedic Clinics of North America* 18:637–647, 1987.

Bushan, B., Watal, G., Ahmed, A., Saxena, R., Goswami, K., and Pathania, A. G.: Giant ivory osteoma of frontal sinus. *Australasian Radiology* 31(3):306–308, 1987.

Byers, S.: Technical note: Calculation of age at formation of radiopaque transverse lines. *American Journal of Physical Anthropology* 85(3):339–343, 1991.

Cady, L. D.: The incidence of the supracondyloid process in the insane. *American Journal of Physical Anthropology* 5:35–49, 1921.

Caffey, J.: *Pediatric x-ray diagnosis* (6th ed.). Chicago, Year Book Medical Publishers, 1972.

Calce, S. C., and Rogers, T. L.: Evaluation of age estimation technique: Testing traits of the acetabulum to estimate age at death in adult males. *Journal of Forensic Sciences* 56(2):302–311, 2011.

Calin, A.: Ankylosing spondylitis. *Clinics in Rheumatic Diseases* 11(1):41-60, 1985.

Callahan, J. J.: Interesting notes on bipartite patellae. *U. S. Naval Medical Bulletin* 48:229–233, 1948.

Camarda, A. J., Deschamps, C., and Forest, D.: II. Stylohyoid chain ossification: A discussion of etiology. *Oral Surgery Oral Medicine Oral Pathology* 67(5):515–520, 1989.

Campillo, D.: Herniated intervertebral lumbar discs in an individual from the Roman Era, exhumed from the "Quinta de San Rafael" (Tarragona, Spain). *Journal of Paleopathology* 2(2):89–94, 1989.

Canale, G., Scarsi, M., and Mastragostino, S.: Hip deformity and dislocation in spina bifida. *Italian Journal of Orthopaedics and Traumatology* 18(2):155–165, 1992.

Canale, S. T., and Belding, R. H.: Osteochondral lesions of the talus. *Journal of Bone and Joint Surgery* 62-A(1):97–102, 1980.

Canale, S. T., Griffin, D. W., and Hubbard, C. N.: Congenital muscular torticollis. A long term follow-up. *Journal of Bone and Joint Surgery 64*:810–816, 1982.

Capasso, L., Kennedy, K. A. R., and Wilczak, C. A.: *Atlas of occupational markers on human remains.* Journal of Paleontology Monographic Publications 3, Terano, Italy, Edigrafital, S.P.A., 1998.

Caraveo, J., Trowbridge, A. A., Amaral, B. W., Green, J. B., Cain, P. T., and Hurley, D. L.: Bone marrow necrosis associated with a mucor infection. *American Journal of Medicine 62*:404–408, 1977.

Cardoso, F. A.: A portrait of gender in two 19th and 20th century Portuguese populations: A paleopathological perspective, unpublished Ph.D. dissertation, Durham University, 2008.

Cardoso, F. A., and Henderson, C. Y.: Enthesopathy formation in the humerus: Data from known age-at-death and known occupation skeletal collections. *American Journal of Physical Anthropology 141*:550–560, 2010.

Carlson, D., Armelagos, G., and Van Gerven, D.: Factors influencing the etiology of cribra orbitalia in prehistoric Nubia. *Journal of Evolution 3*:405–410, 1974.

Carney, C. N., and Wilson, F. C.: Infections of bones and joints. In Wilson, F.O. (Ed.): *The musculoskeletal system.* Philadelphia, Lippincott, 1975.

Carpintero-Benitez, P., Logrono, C., and Collantes-Estevez, E.: Enthesopathy in leprosy. *Journal of Rheumatology 23*(6):1020–1021, 1996.

Case, D. T., and Heilman, J.: New siding technique for the manual phalanges: A blind test. *International Journal of Osteoarchaeology 16*(4):338–346, 2006.

Casscells, S. W.: The arthroscope in the diagnosis of disorders of the patellofemoral joint. *Clinical Orthopaedics and Related Research 144*:45–50, 1979.

Castilla, E., Paz, J., Mutchinick, O., Munoz, E., Giorgiutti, E., and Gelman, Z.: Polydactyly: A genetic study in South America. *American Journal of Human Genetics 25*:405–412, 1973

Caterall, A.: *Legg-Calve-Perthes disease.* New York: Churchill Livingstone, 1982.

Cave, A. J. E.: The earliet English example of bilateral cervical rib. *British Journal of Surgery 29*(113):47–51, 1941.

————: The nature and morphology of the costoclavicular ligament. *Journal of Anatomy 95*:170–179, 1961.

Cecil, R. C., and Loeb, R. F.: *A textbook of medicine* (8th ed.). Philadelphia, W. B. Saunders, 1951.

Cederlund, C. G., Andren, L., and Olivecrona, H.: Progressive bilateral thinning of the parietal bones. *Skeletal Radiology 8*(1):29–33, 1982.

Ceroni, D., De Coulon, G., Spadola, L., De Rosa, V., and Kaelin, A.: Calcaneous secundarius presenting as calcaneonavicular coalition: A case report. *Journal of Foot and Ankle Surgery 45*(1):25–27, 2006.

Chamay, A., and Tschantz, P.: Mechanical influences in bone remodeling: Experimental research on Wolff's law. *Journal of Biomechanics 5*:173–180, 1972.

Chao, T., Teng, H., Tsu, C., and Chou, W.: Osteomyelitis of multiple lumbar vertebrae associated with infected aortic aneurysm: A case report. *The Kaohsiung Journal of Medical Sciences 19*(9):481–484, 2003.

Chi, J., and Harkness, M.: Elongated stylohyoid process: A report of three cases. *New Zealand Dental Journal 95*(419): 11–13, 1999.

Chimenos-Kustner, E., Batle-Trave, I., Velasquez-Rengijo, S., Garcia-Carabano, T., VinalsIglesias, H., and Rosello-Llabres, X.: Appearance and culture: Oral pathology associated with certain "fashions" (tattos, piercings, etc.). *Medical Oral 8*(3):197–206, 2003.

Chohayeb, A. A., and Volpe, A. R.: Occurrence of torus palatinus and mandibularis among women of different ethnic groups. *American Journal of Dentistry 14*(5):278–280, 2001.

Choi, S., and Harris, L.: Aortic nonanastomotic pseudo-aneurysm eroding lumbar vertebra–A case report. *Vascular Surgery 35*(3):245–250, 2001.

Chopra, S. R. K.: The cranial suture closure in monkeys. *Proceedings of the Zoological Society of London 128*:67, 1957.

Chundru, U., Liebeskind, A., Seidelmann, F., Fogel, J., Franklin, P., and Beltran, J.: Plantar fasciitis and calcaneal spur formation are associated with abductor digiti minimi atrophy on MRI of the foot. *Skeletal Radiology 37*(6):505–510, 2008.

Chung, S. M. K., and Nissenbaum, M. M.: Congenital and developmental defects of the shoulder. *Orthopedic Clinics of North America 6*:381–392, 1975.

Churchill, S. E., and Morris, A. G.: Muscle marking morphology and labour intensity in prehistoric Khoisan foragers. *International Journal of Osteoarchaeology 8*:390–411, 1998.

Cimen, M., and Elden, H.: Numerical variations in human vertebral column: A case report. *Okajimas Folia Anatomy Japan 75*:297–303, 1999.

Claffey, T. J.: Avascular necrosis of the femoral head: An anatomical study. *Journal of Bone and Joint Surgery, 42-B*: 802–809, 1960.

Clanton, T. O., and DeLee, J. C.: Osteochondritis dissecans: History, pathophysiology and current treatment concepts. *Clinical Orthopaedics and Related Research 167*:50– 64, 1982.

Cockburn, T. A.: The origin of the treponematoses. *Bulletin World Health Organization 24*:221–228, 1961.

Cockburn, A., and Cockburn, E.: *Mummies, disease and ancient cultures* (abridged edition). Cambridge, Cambridge University Press, 1980.

Cockshott, P. W.: Anatomical anomalies observed in radiographs of Nigerians–(I) thoracic. *West African Medical Journal 7*(4):179, 1958.

Cohen, M. M., Jr.: History, terminology, and classification of craniosynostosis. In Cohen, M. M., Jr. (Ed.): *Craniosynostosis: Diagnosis, evaluation, and management.* New York, Raven, 1986.

Cohen, M. M., Jr., and MacLean, R. E.: *Craniosynostosis: Diagnosis, evaluation, and management* (2nd ed.). Oxford University Press, 2000.

Coleman, J. R., Jr., and Sykes, J. M.: The embryology, classification, epidemiology, and genetics of facial clefting. *Facial Plastic Surgery Clinics of North America 9*(1):1–13, 2001.

Coley, B.: *Neoplasms of bone and related conditions* (2nd ed.). New York, Paul B. Hoeber, 1950.

Collet, S., Bertrand, B., Cornu, S., Eloy, P., and Rombaux, P.: Is septal deviation a risk factor for chronic sinusitis? Review of literature. *Acta Oto-rhino-laryngologica Belgica* 55(4):299–304, 2001.

Collier, B. D., Johnson, R. P., Carrera, G. F., Meyer, G. F., Schwab, J. P., Flatley, T. J., Isitman, A. T., Hellman, R. S., Zielonka, J. S., and Knobel, J.: Painful spondylolysis or spondylolisthesis studied by radiography and single-photon emission computed tomography. *Radiology 154*: 207–211, 1985.

Colonna, P. C., and Gucker, T.: Blastomycosis of the skeletal system. *Journal of Bone and Joint Surgery 26*:322–328, 1944.

Colquhoun, J.: Butterfly vertebra or sagittal cleft vertebra. *American Journal of Orthopaedic Surgery 10*(2):44–50, 1968.

Congdon, R. T.: Spondylolisthesis and vertebral anomalies in skeletons of American aborigines. *Journal of Bone and Joint Surgery 4-B*:511–524,1931.

Conner, A. N.: The treatment of flexion contractures of the knee in poliomyelitis. *Journal of Bone and Joint Surgery 52-B*(1):138–144, 1970.

Cook, D. C.: Subsistence base and health in prehistoric Illinois Valley: Evidence from the human skeleton. *Medical Anthropology 4*:109–124, 1979.

Cook, D. C., and Buikstra, J. E.: Health and differential survival in prehistoric populations: Prenatal dental defects. *American Journal of Physical Anthropology 51*:649–664, 1979.

Cook, D. C., and Doherty, S. P.: Row, row, row your boat: Activity patterns and skeletal robusticity in a series from Chirikof Island, Alaska. *American Journal of Physical Anthropolology Suppl 32*:53 (abstract), 2001.

Coons, M. S., and Green, S. M.: Boutonniere deformity. *Hand Clinics 11*(3):387–402, 1995.

Cooper, P. D., Stewart, J. H., and McCormick, M. S.: Development and morphology of the sternal foramen. *American Journal of Forensic Medicine and Pathology 9*(4): 342–347, 1988.

Cooper, R., and Misol, S.: Tendon and ligament insertion. *Journal of Bone and Joint Surgery 52A*:1–20, 1970.

Cope, J. M.: *Musculoskeletal attachment site markers and skeletal pathology of the forearm and carpal bones from Tell Abraq, United Arab Emirates, c. 2300 BC.* Unpublished Ph.D. dissertation, University of Massachusetts Amherst, 2007.

Correll, R. W., Jensen, J. L., Taylor, J. B., and Rhyne, R. R.: Mineralization of the stylohyoid-stylomandibular ligament complex. *Oral Surgery Oral Medicine Oral Pathology 48*:286–291, 1979.

Correll, R. W., Jensen, J. L., and Rhyne, R. R.: Lingual cortical mandibular defects: A radiographic incidence study. *Oral Surgery Oral Medicine Oral Pathology 50*(3): 287–291, 1980.

Corruccini, R. S.: An examination of the meaning of cranial discete traits for human skeletal biological studies. *American Journal of Physical Anthropology 40*:425–446, 1974.

Corruccini, R. S., Handler, J. S., and Jacobi, K. P.: Chronological distribution of enamel hypoplasias and weaning in a Caribbean slave population. *Human Biology 57*:699–711, 1985.

Cowell, M. J., and Cowell, H. R.: The incidence of spina bifida occulta in idiopathic scoliosis. *Clinical Orthopaedics and Related Research 118*:16–18, 1976.

Crain, J. B.: *Human paleopathology: A biliographic list.* Sacramento Anthropological Society, Paper 12, Sacramento, 1971.

Crane-Kramer, G. M. M.: *The paleopidemiological examination of treponemal infection and leprosy in Medieval populations from Northern Europe.* Ph.D. dissertation, Department of Archaeology. University of Calgary, 2000.

Crawford, P. J. M., Aldred. M., and Bloch-Zupan, A.: Amelogenesis imperfecta. *Orphanet Journal of Rare Diseases 2*:17:doi:10.1186/1750-1172-2-17, 2007.

Crelin, E. S.: Development of the musculoskeletal system. *Clinical Symposia 33*(1):1–36, 1981.

Crosby, A. W.: The early history of syphilis: A reappraisal. *American Anthropology 71*:218–227, 1969.

Cucina, A.: Brief communication: Diachronic investigation of linear enamel hypoplasia in prehistoric skeletal samples from Trentino, Italy. *American Journal of Physical Anthropology 119*:283–287, 2002.

Cucina, A., and İşcan, M. Y.: Assessment of enamel hypoplasia in a high status burial site. *American Journal of Human Biology 9*:213–222, 1997.

Cummings, S. R., Kelsey, J. L., Nevitt, M. C., and O'Dowd, K. J.: Epidemiology of osteoporosis and osteoporotic fractures. *Epidemiologic Review 7*:178–208, 1985.

Currey, J. D.: *Bones: Structure and mechanics* (2nd ed.). Princeton, Princeton University Press, 2006.

Curtin, A. J.: *On the etiology of occipital lytic lesions in artificially deformed crania.* Poster session presented at the 34th annual meeting of the Paleopathology Association, Philadelphia, March, 2007.

Cybulski, J. S.: Cribra orbitalia, a possible sign of anemia in early historic native populations of the British Columbia Coast. *American Journal of Physical Anthropology 47*(1): 31–39, 1977.

Cyron, B. M., and Hutton, W. C.: The fatigue strength of the lumbar neural arch in spondylolysis. *Journal of Bone and Joint Surgery 60-B*(2):234-238, 1978.

D'Agostino, M.A.: Enthesitis: Best practices and research. *Clinical Rheumatology 20*(3):473–486, 2006.

Dahlberg, A. A.: The dentition of the American Indian. In Laughlin, W. S.: *The physical anthropology of the American Indian.* New York, Viking Fund, pp. 138–76, 1951.

Dahlin, D.: *Bone tumors* (2nd ed.). Springfield, C. C. Thomas, 1967.

Dale, J. M.: *Cribra orbitalia, nutrition and pathogenic stress in prehistoric skeletal remains from the Pender Island Canal sites (DeRt 1, DeRt 2), British Columbia, Canada.* Master's thesis, Simon Fraser University, Canada, 1994.

Dalinka, M. K., Dinnenberg, S., Greendyke, W. H., and Hopkins, R.: Roentgenographic graphic features of osseous coccidioidomycosis and differential diagnosis. *Journal of Bone and Joint Surgery 53-A*:1157–1164, 1971.

Dallman, P., Siimes, R., and Stekel, A.: Iron deficiency in infancy and childhood. *American Journal of Clinical Nutrition 33*:86–118, 1980.

D'Ambrosia, R. D.: Conservative management of metatarsal and heel pain in the adult foot. *Orthopedics 10*(1):137–142, 1987.

D'Ambrosia, R. D., and MacDonald, G. L.: Pitfalls in the diagnosis of Osgood-Schlatter disease. *Clinical Orthopedics 110(Jul-Aug)*:206–209, 1975.

Daniels, E. G., and Nashel, D. J.: Periostitis: A manifestation of venous disease and skeletal hyperostosis. *Journal American Podiatric Association 73*(9):461–464, 1983.

Dar, G., Peleg, S., Masharawi, Y., Steinberg, N., May, H., and Hershkovitz, I.: Demographical aspects of Schmorl nodes: A skeletal study. *Spine 34*(9):E312–315, 2009.

Daveny, J. K., and Ross, M. D.: Cryptococcosis of bone. *Central African Journal of Medicine 15*:78–79, 1969.

David, D. J., Poswillo, D., and Simpson, D.: *The craniosynostoses: Causes, natural history and management.* Berlin, Springer-Verlag, 1982.

David, R., Oria, R. A., Kumar, R., Singleton, E. B., Lindell, M. M., Shirkhoda, A., and Madewell, J. E.: Radiologic features of eosinophilic granuloma of bone. *American Journal of Roentgenology 153*:1021–1026, 1989.

Davis, P. R.: The Thoraco-lumbar mortice joint. *Journal of Anatomy 89*:370–377, 1955.

Day, S. B.: Ossified subperiosteal hematoma. *Journal of American Medical Association 173*:986–990, 1960.

Dean, V. L.: *Effects of cultural deformation and craniosynostosis on cranial venous sinus and middle meningeal vessel pattern expression.* Doctoral dissertation, Indiana University, Bloomington, 1995.

Dee, P. M.: The preauricular sulcus. *Radiology 140*(2):354, 1981.

Deesomchok, U., and Tumrasvin, T.: Clinical comparison of patients with ankylosing spondylitis, Reiter's syndrome and psoriatic arthritis. *Journal Medical Association of Thailand 76*(2):61–70, 1993.

De Graaf, R. J., Matricali, B., and Hamburger, H. L.: Butterfly vertebra. *Clinics Neurology and Neurosurgery 84*(3):163–169, 1982.

Deguine, C., and Pulec, J. L.: Large osteoma of the external auditory canal. *Ear Nose and Throat Journal 80*(1):8, 2001a.

_____: Large nonobstructing exostoses of the external auditory canal. *Ear Nose and Throat Journal 80*(3):134, 2001b.

Delport, E. G., Cucuzzella, T. R., Kim, N., Marley, J., Pruitt, C., and Delport, A. G.: Lumbosacral transitional vertebrae: Incidence in a consecutive patient series. *Pain Physician 9*:53–56, 2006.

Depalma, A. F.: Surgical anatomy of acromioclavicular and sternoclavicular joints. *Surgical Clinics of North America 43(Dec)*:1541–1550, 1963.

De Smet, A. A., Ilahi, O. A., and Graf, B. K.: Untreated osteochondritic dissecans of the femoral condyles: Prediction of patient outcome using radiographic and MR findings. *Skeletal Radiology 26*:463–467, 1997.

Derry, D. E.: Note on the accessory articular facets between the sacrum and ilium, and their significance. *Journal of Anatomy and Physiology 45*:202–210, 1911.

Desai, S. S., Patel, M. R., Michelli, L. J., Silver, J. W., and Lidge, R. T.: Osteochondritis dissecans of the patella. *Journal of Bone and Joint Surgery 69-B*(2):320–325, 1987.

Devoto, F. C.: Shovel-shaped incisors in Pre-Columbian Tastilian Indians. *Journal Dental Research 50*(1):168, 1971.

Devoto, F. C., and Arias, N. H.: Shovel-shaped incisors in early Atacama Indians. *Journal Dental Research 46(6)*: 1478, 1967.

DiBartolomeo, J. R.: Exostoses of the external auditory canal. *Annals Otology Rhinology Laryngology Supplement 88*(6), Part 2, Supplement 61:2–20, 1979.

Dibbets, J. H.: One century of Wolff's Law. In Carson, D. S., and Goldstein, S. A. (Eds.): *Bone dynamics in orthodontic and orthropaedict treatment.* Craniofacial Growth Series: Ann Arbor, Michigan, 1992.

Dickel, D. N., and Doran, G. H.: Severe neural tube defect syndrome from the Early Archaic of Florida. *American Journal of Physical Anthropology 80*:325–334, 1989.

Dickson, R. A.: Conservative treatment for idiopathic scoliosis. *Journal of Bone and Joint Surgery 67-B*:176–181, 1985.

Diekerhof, C. H., Reedt Dortland, R. W., Oner, F. C., and Verbout, A. J.: Severe erosion of lumbar vertebral body because of abdominal aortic false aneurysm: Report of two cases. *Spine 27*(16):E382–384, 2002.

Dieppe, P. A., Bacon, P. A., Bamji, A. N., and Watt, I.: *Atlas of clinical rheumatology.* Philadelphia, Lea & Febiger, 1986.

DiFiori, J. P.: Overuse injuries in children and adolescents. *The Physician and Sports Medicine 27*(1):1-9, 1999. http://www.physsportsmed.com/issues/1999/01_99/Difiori.htm accessed 2005 Apr. 1.

Dingwall, E. J.: *Artificial cranial deformation: A contribution to the study of ethnic mutilations.* London, John Bale, Sons and Danielsson, Ltd., 1931.

Di Maio, V. J. M.: *Gunshot wounds: Practical aspects of firearms, ballistics, and forensic techniques* (2nd ed.). Boca Raton, CRC Press, 1999.

Dingwall, E. J.: *Artificial cranial deformation.* London, John Bale, Sons and Danielson, Ltd., 1931.

Dixon: *Dixon's manual of human osteology.* London, Oxford University Press, 1937. Revised by Jamieson, E. B., 2nd edition.

Dixon, D. S.: Keyhole lesions in gunshot wounds of the skull and direction of fire. *Journal of Forensic Sciences 27*:555–556, 1982.

Dodo, Y.: Aural exostoses in the human skeletal remains excavated in Hokkaido. *Journal of the Anthropological Society, Nippon 80*:11–22, 1972.

Dogra, B. B., Singh, M., and Malik, A.: Congenital proximal radioulnar synostosis. *Indian Journal of Plastic Surgery* [On-line serial], *36*:36–8, 2003 [cited 2011 Apr 8]. Available from: http://ijps.org/text.asp?2003/36/1/36/5778.

Dolan, P., Earley M., and Adams, M. A.: Bending and compressive stresses acting on the lumbar spine during lifting activities. *Journal of Biomechanics 27*:1237–1248, 1994.

Donnelly, L. F., Bisset, G. S., Helms, C. A., and Squire, D. L.: Chronic avulsive injuries of childhood. *Skeletal Radiology 28*:138–144, 1999.

Donoghue, H. D., Hershkovitz, I., Minnikin, D. E., Besra, G. S., Lee, Oona Y-C., Galili, E., Greenblatt, C. L., Lemma, E., Spigelman, M., and Bar-Gal, G. K.: Biomolecular archaeology of ancient tuberculosis: Response to "Deficiencies and challenges in the study of ancient tuberculosis DNA" by Wilbur et al. 2009. www.plosone.org/.../info%3Adoi%2F10.1371%2Fjournal.pone.0003426, 2009.

Douglas, T. E., Jr.: Facial pain from elongated styloid process. *Archives of Otolaryngology 56*:635–638, 1952.

Dugdale, A. E., Lewis, A. N., and Canty, A. A.: The natural history of chronic otitis media. *New England Journal of Medicine* 307:1459–1460, 1982.

Duncan, S. F. M., and Weiland, A. J.: Extraarticular distal radius fractures (Vol. 1, pp. 247–276). In Berger, R. A., and A-P. C. Weiss (Eds.): *Hand surgery.* Philadelphia, Lippincott Williams & Wilkins, 2004.

Dutour, O.: Enthesopathies as indicators of the activities of Neolithic Saharan populations. *American Journal of Physical Anthropology 71*:221–224, 1986.

Dutour, O., Palfi, G., Berato, J., and Brun, J-P. (Eds.): *The origin of syphilis in Europe before or after 1493?* Paris, Editions Errance, 1994.

Dwight, T.: Account of two spines with cervical ribs, one of which cervical ribs, one of which has a vertebra suppressed, and absence of the anterior arch of the atlas. *Journal of Anatomy and Physiology 21*:639–550, 1887.

_____: A bony supracondyloid process in the child. *American Journal of Anatomy 3*:221–228, 1904.

_____: The clinical significance of variations of the wrist and ankle. *Journal of the American Medical Association 47*:252–255, 1906.

Dzurek, W. V., Etter, L. E., Keagy, R. M., and Miller M. W.: Enlarged parietal foramina. A collection of examples from radiologic practices in Pennsylvania. *Medical Radiography and Photography 32*:73–78, 1956.

Eagle, W. W.: Elongated styloid process: Further observations and a new syndrome. *Archives of Otolaryngology 47*:630–640, 1948.

Echols, R. M., Selinger, D. D., Hallowell, C., Goodwin, J. S., Duncan, M. H., and Cushing, A. H.: Rhizopus osteomyelitis: A case report and review. *American Journal of Medicine 66*(1):141–145, 1979.

Edelson, J. G., Nathan, H., and Arensburg, B.: Diastematomyelia-the "double-barrelled" spine. *Journal of Bone and Joint Surgery 69-B*(2):188–189, 1987.

Edelson, J.G., Zuckerman, J., and Hershkovitz, I.: Os Acromiale: Anatomy and surgical implications. *Journal of Bone and Joint Surgery 75-B*:219–273,1993.

Edwards, D. H., and Bentley, G.: Osteochondritis disecans patellae. *Journal of Bone and Joint Surgery, 59-B*(1):58–63, 1977.

Edwards, W. G.: Complications of suppurative otitis media. In Ludman, H. (Ed.): *Mawson's diseases of the ear* (5th ed.). Chicago, Year Book Medical Publishers, Inc., 1988.

Eggen, S., and Natvig, B.: Relationship between torus mandibularis and number of present teeth. *Scandinavian Journal of Dental Research 94*(3):233–240, 1986.

_____: Concurrence of torus mandibularis and torus palatinus. *Scandinavian Journal of Dental Research 102*(1):60–63, 1994.

Egol, K. A., Koval, K. J., and Zuckerman, J. D.: *Handbook of fractures* (4th ed.). Philadelphia, Lippincott & Wilkins, 2010.

Ehara, S., El-Khoury, G. Y., and Bergman, R. A.: The accessory sacroiliac joint: A common anatomic variant. *American Journal of Roentgenology 150*:857–859, 1988.

Eichenholtz, S. N.: *Charcot joints.* Springfield, C. C. Thomas, 1966.

Eisele, S. A., and Sammarco, G. J.: Fatigue fractures of the foot and ankle in the athlete. *Journal of Bone and Joint Surgery 75*(2):290–298, 1993.

Eisenstein, S.: Spondylolysis: A skeletal investigation of two populations. *Journal of Bone and Joint Surgery 60-B*:488–494, 1978.

Elerich, D., and Tyson, R. (Eds.): *Human paleopathology-related subjects: An international bibliography.* San Diego, San Diego Museum of Man, 1997.

Ell, S. R.: Reconstructing the epidemiology of medieval leprosy: Preliminary efforts with regard to Scandinavia. *Perspectives in Biology and Medicine 31*:496–506, 1988.

El Maghraoui, A., Tacache, F., El Khattabi, A., Bezza, A., Abouzahir, A., Ghafir, D., Ohayon, V., and Archane, M. I.: Abdominal aortic aneurysm with lumbar vertebral erosion in Behcet's disease revealed by low back pain: A case report and review of the literature. *Rheumatology (Oxford) 40*(4):472–473, 2001.

El-Najjar, M. Y., and Dawson, G. L.: The effects of artificial cranial deformation on the incidence of wormian bones in the lambdoidal suture. *American Journal of Physical Anthropology 46*:155–160, 1977.

El-Najjar, M. Y., Lozoff, B., and Ryan, D. J.: The paleo-epidemiology of porotic hyperostosis in the American Southwest: Radiological and ecological considerations. *American Journal of Roentgenology Radium Therapy Nuclear Medicine 125*(4):918–924, 1975.

El-Najjar, M., and Robertson, A. L. Jr.: Spongy bones in prehistoric America. *Science 193*(4248):141–143, 1976.

El-Najjar, M. Y., Ryan, D. J., Turner, C. G., and Lozoff, B.: The etiology of porotic hyperostosis among the prehistoric and historic Anasazi Indians of southwestern United States. *American Journal of Physical Anthropology 44*:477–488, 1976.

Enlow, D. H.: *Principles of bone remodeling.* Springfield, C. C. Thomas Publishers, 1963.

Epstein, B. S.: The concurrence of parietal thinness with postmenopausal, senile or idiopathic osteoporosis. *Radiology 60*:29–35, 1953.

Epstein, H. C.: *The spine: A radiological text and atlas* (4th ed.). Philadelphia, Lea and Febiger, 1976.

Erken, E., Ozer, H. T., Gulek, B., and Durgun, B.: The association between cervical rib and sacralization. *Spine 27*(15):1659–1664, 2002.

Ersin, N. K., Candan, U., Alpoz, A. R., and Akay, C.: Mesiodens in primary, mixed and permanent dentitions: A clinical and radiographic study. *Journal of Clinical Pediatric Dentistry 28*(4):295–298, 2004.

Eshed, V., Gopher A., Galli, E., and Hershkovitz, I.: Musculoskeletal stress markers in Natufian hunter gatherers and Neolithic farmers in the Levant: The upper limb. *American Journal of Physical Anthropology 12*:303–315, 2004.

Eshed, V., Latimer, B., Greenwald, C. M., Jellema, L. M., Rothschild, B. M., Wish-Baratz, S., and Hershkovitz, I.: Button osteoma: Its etiology and pathophysiology. *American Journal of Physical Anthropology 118*(3):217–230, 2002.

Esper, J. F.: *Ausfuhrliche Nachricht von neuentdeckten Zoolithen unbekannter vierfussiger Tiere.* Georg Wolfgang Knorrs Seel, Erben, Nurnberg, 1774.

Eubanks, J. D., and Cheruvu, V. K.: Prevalence of sacral spina bifida occulta and its relationship to age, sex, race, and the sacral table angle: An anatomic, osteologic study of three thousand one hundred specimens. *Spine 34*(15):1539–1543, 2009.

Fairgrieve, S. I., and Molto, J. E.: Cribra orbitalia in two temporally disjunct population samples from the Dakhleh Oasis, Egypt. *American Journal of Physical Anthropology 111*(3):319–331, 2000.

Farfan, H. F., Osteria, V., and Lamy C.: The mechanical etiology of spondylolysis and spondylolisthesis. *Clinical Orthopedics 117*:40–55, 1976.

Farkas, A.: Physiological scoliosis. *Journal of Bone and Joint Surgery 23*(3):607, 1941.

Farzan, M., Daneshjou, K., Mortezavi, S. M. J., and Espandar, R.: Congenital radioulnar synostosis: A report of 11 cases and review of literature. *Acta Medica Iranica 40*(2):126–131, 2002.

Fehlandt, A. F., and Micheli, L. J.: Lumbar facet stress fracture in a ballet dancer. *Spine 18*(16):2537-2539, 1993.

Fein, J. M., and Brinker, R. A.: Evolution and significance of giant parietal foramina. *Journal of Neurosurgery 37*: 487–492, 1972.

Feldman, V. B.: Eagle's syndrome: A case of symptomatic calcification of the stylohyoid ligaments. *Journal Canadian Chiropractic Association 47*(1):21–27, 2003.

Felson, D. T., Naimark, A., Anderson, J., Kazis, L., Castelli, W., and Meenan, R. F.: The prevalence of knee osteoarthritis in the elderly: The Framingham osteoarthritis study. *Arthritis and Rheumatism 30*(8):914–918, 1987.

Fenton, J. E., Turner, J., and Fagan, P. A.: A histopathologic review of temporal bone exostoses and osteomata. *Laryngoscope 106*(5 Pt 1):624–628, 1996.

Ferembach, D.: Frequency of spina bifida occulta in prehistoric human skeletons. *Nature 199*:100–101, 1963.

Fernandez, D. L., and Jupiter, J. B.: *Fractures of the distal radius: A practical approach to management* (2nd ed.). New York, Springer-Verlage, 2002.

Ferrario, V. F., Sigurta, D., Daddona, A., Dalloca, L., Miani, A., Tafuro, F., and Sforza, C.: Calcification of the stylohyoid ligament: Incidence and morphoquantitative evaluations. *Oral Surgery Oral Medicine Oral Pathology 69*(4):524–529,1990.

Fiese, M. J.: *Coccidioidomycosis.* Springfield, C. C. Thomas, 1958.

Filipo, R., Rabiani, M., and Barbara, M.: External ear canal exostosis: A physiopathological lesion in aquatic sports. *Journal of Sports Medicine Physical Fitness 22*(3):329–336, 1982.

Fink, I. J., Pastakia, B., and Barranger, J. A.: Enlarged phalangeal nutrient foramina in Gaucher disease and B-thalassemia major. *American Journal of Roentgenology 143*:647-649, 1984.

Fink, A. M., and Maixner, W.: Enlarged parietal foramina: MR imaging features in the fetus and neonate. *American Journal of Neuroradiology 27*:1379–1381, 2006.

Finnegan, M.: Non-metric variation of the infracranial skeleton. *Journal of Anatomy 125*(1):23–37, 1978.

Finnegan, M., and Faust, M. A.: *Bibliography of human and nonhuman non-metric variation.* Research Reports N. 14, Department of Anthropology, University of Massachusetts. Amherst, University of Massachusetts, 1974.

Finnegan, M., and Marcsik, A. M.: Anatomy or pathology: The Stafine defect as seen in archaeological material and modern clinical practice. *Journal of Human Evolution 9*:19–31, 1980.

Fiorato, V., Boylston, A., and Knusel, C.: *Blood red roses: The archaeology of a mass grave from the Battler of Towton, AD 1461.* Oxford, Oxbow Books, 2001.

Fishbien, M. (Ed.): *Birth defects.* Philadelphia, J. B. Lippincott, 1963.

Fitoz, S., Atasoy, C., Yagmurlu, A., and Akyar, S.: Psoas abscess secondary to tuberculous lymphadenopathy: Case report. *Abdominal Imaging 26*(3):323–324, 2001.

Fitton, J. S.: A tooth ablation custom occuring in the Maldives. *British Dental Journal 175*(8):299–300, 1993.

Fiumara, N. J., and Lessell, S.: Manifestation of late congenital syphilis: An analysis of 271 patients. *Archives of Dermatology 102*:78–83, 1970.

———: The stigmata of late congenital syphilis: An analysis of 100 patients. *Sexually Transmitted Diseases 10*(3):126–129, 1983.

Fletcher, H. A., Donoghue, H. D., Holton, J., Pap, I., and Spigelman, M.: Widespread occurrence of mycobacterium tuberculosis DNA from 18th–19th century Hungarians. *American Journal of Physical Anthropology 120*: 144–152, 2003.

Floman, Y., Bloom, R. A., and Robin, G. C.: Spondylolysis in the upper lumbar spine: A study of 32 patients. *Journal of Bone and Joint Surgery 69-B*(4):582, 1987.

Flohr, S., and Schultz, M.: Mastoiditis–Paleopathological evidence of a rarely reported disease. *American Journal of Physical Anthropology 138*:266–273, 2009.

Flourney, L. E., Rogers, N. L., and McCormick, W. F.: The rhomboid fossa of the clavicle as a sex and age estimator. *Journal of Forensic Sciences 45*(1):61–67, 2000.

Fokin, A. A.: Cleft sternum and sternal foramen. *Chest Surgical Clinics of North America 10*(2):261–276, 2000.

Foreman, S. M.: Fractures and dislocations of the cervical spine. In Foreman, S. M., and Croft, A. C. (Eds.): *Whiplash injuries: The cervical acceleration/deceleration syndrome.* Philadelphia, Lippincott, Williams & Wilkins, pp. 271–295, 2001.

Fowler, F. D., Driebe, W. T., and Copeland, M. M.: Supra-condyloid process. *The American Surgeon* 25(6):266–272, 1959.

Fox, R. J., Walji, A. H., Mielke, B., Petruk, K. C., and Aronyke, K. E.: Anatomic details of intradural channels in the parasagittal dura: A possible pathway for flow of cerebrospinal fluid. *Neurosurgery* 39(1): 84–90, 1996.

Franco-Paredes, C., and Blumberg, H. M.: Psoas muscle abscess caused by Mycobacterium tuberculosis and Staphylococcus aureus: Case report and review. *American Journal Medical Science* 321(6):415–417, 2001.

Francois, R. J.: Microradiographic study of the interverte-bral bridges in ankylosing spondylitis and in the normal sacrum. *Annals of Rheumatic Diseases* 24:481–489,1965.

Francois, R. J., Braun, J., and Khan, A. K.: Entheses and enthesitis: A histopathologic review and relevance to spondyloarthrides. *Current Opinion in Rheumatology* 13:255–264, 2001.

Fredrickson, B. E., Baker, D., McHolick, W. J., Yuan, H. A., and Lubicky, J. P.: The natural history of spondylolysis and spondylolisthesis. *Journal of Bone and Joint Surgery* 66-A:699–707, 1984.

Friedman, R. J., and Micheli, L. J.: Acquired spondylolis-thesis following scoliosis surgery. A case report. *Clinical Orthopaedics* 190:132–134, 1984.

Frost, H. M.: Why do long-distance runners not have more bone? A vital biomechanical explanation and an estrogen effect. *Journal of Bone and Mineral Metabolism* 15:9–16, 1997.

Frost, H. M.: Bone's mechanostat: A 2003 update. *Anatomi-cal Record* 275A:1081–1101, 2003.

Frymoyer, J. W: Back pain and sciatica. *New England Journal of Medicine* 318(5):291–300, 1988.

Frymoyer, J. W. (ed.): *The adult spine: Principles and practice.* Philadelphia, Lippincott, 1997.

Gaillard, F.: Ossification of the posterior longitudinal liga-ment. *Radiopaedia.org,* July17, 2008.

Galloway, A. (Ed.): *Broken bones: Anthropological analysis of blunt force trauma.* Springfield, C. C. Thomas, 1999.

Galtés, I., Rodríques-Baeza, A., and Malgosa, A.: Mechanical morphogenesis: A concept applied to the surface of the radius. *The Anatomical Record Part A* 288A:794–805, 2006.

Ganguili, P. K.: *Radiology of bone and joint tuberculosis.* New York, Asia Publishing House, 1963.

Garber, E. A., and Silver, S.: Pedal manifestations of DISH. *Foot and Ankle* 3(1):12–16, 1982.

Garcia-Lechuz, J. M., Julve, R., Alcala, L., Ruiz-Serrano, M. J., and Munoz, P.: Tuberculous spondylodiskitis (Pott's disease): Experience in a general hospital [Article in Spanish, English abstract]. *Enferm Infec Microbiol Clin* 20(1):5–9, 2002.

Gardner, D. L.: *Pathology of the connective tissue diseases.* London: Edward Arnold, 1965.

Garn, S. M.: *The earlier gain and later loss of cortical bone.* Springfield, IL: C. C. Thomas, 1972.

Garn, S. M.: Human growth. *Annual Revue of Anthropology* 9:275–292, 1980.

Garth, W. P., and Van Patten, P. K.: Fractures of the lumbar lamina with epidural hematoma simulating herniation of a disc: A case report. *Journal of Bone and Joint Surgery* 71-A:771–772,1989.

Genant, H. K.: Osteoporosis and bone mineral assessment. In McCarty, D. J. (Ed.): *Arthritis and allied conditions: A textbook of rheumatology* (11th ed.). Philadelphia, Lea and Febiger, 1989.

Genner, B. A.: Fracture of the supracondyloid process. *Journal of Bone and Joint Surgery* 41A:1333–1335, 1959.

Gensburg, R. S., Kawashima, A., and Sandler, C. M.: Scinti-graphic demonstration of lower extremity periostitis secondary to venous insufficiency. *Journal of Nuclear Medicine* 29(7):1279–1282, 1988.

George, G. R.: Bilateral bipartite patellae. *British Journal of Surgery* 22:555, 1935.

Gerber, N., Ambrosini, G. C., Boni, A., Fehr, K., and Wa-genhauser, F. J.: Ankylosing spondylitis (Bechterew) and tissue antigen HLA-B27. II. HLA-b27 negativity in classical clinical ankylosing spondylitis no independent nosological entity (in German). *Zeitschrift fur Rheuma-tologie* 36(7):124–129, 1977.

Gerbino, P. G., and d'Hemecourt, P. A.: Does football cause an increase in degenerative disease of the lumbar spine? *Current Sports Medicine Reports* 1(1):47–51, 2002.

Gerbino, P. G., and Micheli, L. J.: Back injuries in the young athlete. *Clinics in Sports Medicine* 14:571–590, 1995.

Gerschon-Cohen, J., Schraer, H., and Blumberg, N.: Hyper-ostosis frontalis interna among the aged. *American Journal of Roentgenology* 73:396–397, 1955.

Gerscovich, E, O., Greenspan, A., and Szabo, R. M.: Benign clavicular lesions that may mimic malignancy. *Skeletal Radiology* 20:173–180, 1991.

Gerszten, P. C.: An investigation into the practice of cranial deformation among the pre-Columbian peoples of northern Chile. *International Journal of Osteoarchaeology* 3:87–98, 1993.

Gerszten, P. C., Gerszten, E., and Allison, M. J.: Diseases of the skull in pre-Columbian South American mummies. *Neurosurgery* 42(5):1145–1151, 1998.

Gibson, M. J., Szypryt, E. P., Buckley, J. H., Worthington, B. S., and Mulholland, R. C.: Magnetic resonance imaging of adolescent disc herniation. *Journal of Bone and Joint Surgery* 69-B(5):699–703,1987.

Giles, R. G.: A congenital anomaly of the patella. *Texas State Journal of Medicine* 23:731–732, 1928.

Gindhart, P. S.: The frequency of transverse lines in the tibia in relation to childhood illness. *American Journal of Physical Anthropology* 31:17–22, 1969.

Giroux, J. C., and Leclerq, T. A.: Lumbar disc excision in the second decade. *Spine* 7:168–170, 1982.

Gladstone, R. J., and Wakeley, C. P. G.: Cervical ribs and rudimentary first thoracic ribs considered from clinical and etiological standpoints. *Journal of Anatomy* 66:334–370, 1931–32.

Glen-Haduch, E., Szostek, K., and Glab, H.: Cribra or-bitalia and trace element content in human teeth from Neolithic and Early Bronze Age graves in southern

Poland. *American Journal of Physical Anthropology 103*(2): 201–207, 1997.

Godde, K.: An examination of proposed causes of auditory exostoses. *International Journal of Osteoarchaeology 20*(4): 486–490, 2010.

Goff, C. W.: Syphilis. In Brothwell, D., and Sandison, A. (Ed.): *Diseases in antiquity.* Springfield, C. C. Thomas, pp. 279–293, 1967.

Goldsand, G.: Actinomycosis. In Hoeprich, P. D., and Jordan, M. C. (Eds.): *Infectious disease: A modern treatise of infectious processes* (4th ed.). Philadelphia, Lippincott, pp. 666–684, 1989.

Goldsmith, W. M.: The Catlin Mark: The inheritance of an unusual opening in the parietal bones. *Journal of Heredity 13*:69–71, 1922.

Goodman, A. H., Allen, L. H., Hernandez, G. P., Arriola, A., Chavez, A., and Pelto, G. H.: Prevalence and age at development of enamel hypoplasias in Mexican children. *American Journal of Physical Anthropology 72*:7–19, 1987.

Goodman, A. H., and Armelagos, G. J.: Factors affecting the distribution of enamel hypoplasia in human permanent dentition. *American Journal of Physical Anthropology 68*:479–493, 1985.

Goodman, A. H., and Armelagos, G. J.: Childhood stress and decreased longevity in a prehistoric population. *American Anthropologist 90*(4):936–944, 1988.

Goodman, A. H., Armelagos, G. J., and Rose, J. C.: Enamel hypoplasia as indicator of stress in three prehistoric populations from Illinois. *Human Biology 52*:515-528, 1980.

Goodman, A. H., and Clark, G. A.: Harris lines as indicators of stress in prehistoric Illinois populations. In Martin, D. L., and Bumsted, M. P. (Eds.): *Biocultural adaptation: Comprehensive approaches to skeletal analysis.* Research Reports No. 20. Amherst, University of Massachusetts, pp. 35–46, 1981.

Goodman, A. H., and Rose, J. C.: Assessment of systemic physiological perturbations from dental enamel hypoplasias and associated histological structures. *Yearbook of Physical Anthropology 33*:59–110, 1990.

Goodman, A. H., and Rose, J. C.: Dental enamel hypoplasias as indicators of nutritional status. In Kelley, M. A., and Larsen, C. S. (Eds.): *Advances in dental anthropology.* New York, Wiley Liss, pp. 279–293, 1991.

Goodman, A. H., and Song, R. J.: Sources of variation in estimated ages of formation of linear enamel hypoplasias. In Hoppa, R. D., and Fitzgerald, C. M (Eds.): *Human growth in the past.* Cambridge, Cambridge University Press, pp. 210-240, 1999.

Goodman, R. M., and Gorlin, R. J.: *Atlas of the face in genetic disorders.* St. Louis, C. V. Mosby, 1977.

Gorab, G. N., Brahney, C., and Aria, A. A.: Unusual presentation of a Stafne bone cyst. *Oral Surgery Oral Medicine Oral Pathology 61*(3):213–215, 1986.

Gorlin, R. J., Pindborg, J. J., and Cohen, M. M.: *Syndromes of the head and neck.* New York, McGraw-Hill, 1976.

Gorsky, M., Bukai, A., and Shohat, M.: Genetic influence on the prevalence of torus palatines. *American Journal Medical Genetics 75*(2):138–140, 1998.

Gould, A. R., Farman, A. G., and Corbitt, D.: Mutilations of the dentition in Africa: A review with personal observations. *Quintessence International 15*(1):89–94, 1984.

Graham, M. D.: Osteomas and exostoses of the external auditory canal: A clinical histopathologic and scanning electron microscopic study. *Annals Otology Rhinology Laryngology 88*(4 Pt 1):566–572, 1979.

Grainger, R. G., Allison, D. J., Adam, A., and Dixon, A. K. (Eds.): *Diagnostic radiology: A textbook of medical imaging,* (4th ed.). London, Churchill Livingstone, 2001.

Grandmaison, G. L., de la, Brion, F., and Durigon, M.: Frequency of bone lesions: An inadequate criterion for gunshot wound diagnosis in skeletal remains. *Journal of Forensic Sciences 46*(3):593–595, 2000.

Grant, J. C. B.: *A method of anatomy: Descriptive and deductive,* (4th ed.). Baltimore: Williams and Wilkins, 1948.

_____: *Grant's atlas of anatomy* (6th ed.). Baltimore, Williams and Wilkins, 1972.

Grauer, A., and Roberts, C.: Paleoepidemiology, healing and possible treatment of trauma in the medieval cemetery population of St. Helen-on-the-Walls, York England. *American Journal of Physical Anthropology 100*: 531– 544, 1996.

Graves, W. W.: Observations on age changes in the scapula. *American Journal of Physical Anthropology 5*:21–33, 1922.

Gray, H.: *Gray's anatomy.* Pick, T. P., and Howden, R. (Eds.). New York, NY: Bounty Books, 1977.

Gray, H.: *Anatomy of the human body* (25th ed.). Philadelphia, Lea and Febiger, 1948.

Green, W. T. Jr.: Painful bipartite patellae: A report of three cases. *Clinical Orthopaedics and Related Research 110*:197–200, 1975.

Greenfield, G. B.: *Radiology of bone diseases* (2nd ed.). Philadelphia, Lippincott, 1975.

Greer, A. E.: *Disseminating fungal diseases of the lung.* Springfield, C. C. Thomas, 1962.

Gregg, J. B., and Bass, W. M.: Exostoses in the external auditory canals. *Annals of Otology Rhinology and Laryngology 79*:834–839, 1970.

Gregg, J. B., and Gregg, P. S.: *Dry bones: Dakota Territory reflected.* Sioux Falls, Sioux Falls Printing, 1987.

Griffiths, H. J.: *Basic bone pathology.* Norwalk, Appleton-Century-Crofts, 1981.

Grippo, J. O., Simring, M., and Schreiner, S. Attrition, abrasion, corrosion and abfraction revisited. *Journal of the American Dental Association 135*(8):1109–1118, 2004.

Grolleau-Raoux, J-L., Crubezy, D. R., Brugne, J-F., and Saunders, S. R.: Harris lines: A study of age-associated bias in counting and interpretation. *American Journal of Physical Anthropology 103*:209–217, 1997.

Gruber, W.: Ueber einen neuen sekundarrn Tarsalknochen-Calcaneus secundariusmit Bemerkungen uber den Tarsus uberhaupt. *Memoirs de l'Academic des Science de St. Petersbourg, T. 17*, No. 6, 1871.

Grupe, G.: Metastasizing carcinoma in a medieval skeleton: Differential diagnosis and etiolgy. *American Journal of Physical Anthropology 75*:369–374, 1988.

Guidotti, A.: Frequencies of cribra orbitalia in central Italy (19th century) under special consideration of their degrees of expression. *Anthropology Anz, 42*(1):11–16, 1984.

Gulekon, I. N., and Turgut, H. B.: The preauricular sulcus: Its radiologic evidence and prevalence. *Kaiboguku Zasshi, 76*(6):533–535, 2001.

_____: The external occipital protuberance: Can it be used as a criterion in the determination of sex? *Journal of Forensic Sciences 48*(3):513–516, 2003.

Gunzburg, R., and Szpalski M.: *Spondylolysis and spondylolisthesis and degenerative spondylolisthesis.* Philadelphia, Lippincott, Williams & Wilkins, 2006.

Guo, J. J., Luk, K. D. K., Karppinen, J., Yang, H., and Cheung, K.: Prevalence, distribution, and morphology of ossification of the ligamentum flavum: A population study of one thousand seven hundred thirty-six magnetic resonance imaging scans. *Spine 35*(1):51–56, 2010.

Guyuron, B., Uzzo, C. D., and Scull, H.: A practical classification of septonasal deviation and an effective guide to septal surgery. *Plastic Reconstructive Surgery 104*(7):2202–2209, 1999.

Hackett, C. J.: On the origin of the human treponematoses. *Bull WHO 29*:7–41, 1963.

Hackett, C. J.: The human treponematoses. In Brothwell, D. R., and Sandison, A. T. (Eds.): *Diseases in antiquity: A survey of the diseases injuries and surgery of early populations.* Springfield, C. C. Thomas, pp. 152–169, 1967.

Hackett, C. J.: An introduction to diagnostic criteria of syphilis, treponarid and yaws (treponematoses) in dry bones, and some implications. *Virchows Archives Pathology Anatomy Histology 368*(3):229–241, 1975.

_____: Diagnostic criteria of syphilis, yaws and treponarid (treponematoses) and some other diseases in dry bone. *Sitzunosbericht der heidelberger Akademic der Wissenschaft 4,* 1976.

_____:Development of caries sicca in a dry calvaria. *Virchows Archiv. A, Pathological Anatomy and Histology 391*(1): 53–79, 1981.

Hackett, C. J.: Problems in the palaeopathology of the human treponematoses. In Hart, G. D. (Ed.): *Disease in ancient man.* Toronto, Clarke Irwin, pp. 106–128, 1983.

Hagberg, M., and Wegman, D. H.: Prevalence rates and odds ratios of shoulder-neck diseases in different occupational groups. *British Journal of Indian Medicine 44*:602–610, 1987.

Hahn, P. Y., Strobel, J. J., and Hahn, F. J.: Verification of lumbosacral segments on MR images: Identification of transitional vertebrae. *Radiology 182*(2):615–616, 1992.

Haibach, H., Farrell, C., and Gaines, R. W.: Osteoid osteoma of the spine: Surgically correctable cause of painful scoliosis. *Canadian Medical Association Journal 135*(8):895–899, 1986.

Haidar, A., and Kalamchi, S.: Painful dysphagia due to fracture of the styloid process. *Oral Surgery Oral Medicine Oral Pathology 49*(1):5–6, 1980.

Hall, L. T., Esses, S. I., Noble, P. C., and Kamaric, E.: Morphology of the lumbar vertebral endplates. *Spine 23*(14): 1517–1522, 1998.

Hallock, H., and Jones, J. B.: Tuberculosis of the spine. *Journal of Bone and Joint Surgery 36*:219–240, 1954.

Halpern, A. A., and Hewitt, O.: Painful medial bipartite patellae: A case report. *Clinical Orthopaedics and Related Research 34*:180–181, 1978.

Hamilton, W. C. (Ed.): *Traumatic disorders of the ankle.* New York, Springer-Verlag, 1984.

Hamrick, H. W., McPherron, A. C., Lovejoy, C. O., and Hudson, J.: Femoral morphology and cross-sectional geometry of adult myostatin-deficient mice. *Bone 27*(3):343–349, 2000.

Hamrick, M. W., McPherron, A. C., and Lovejoy, C.O.: Bone mineral content and density in the humerus of adult myostatin-deficient mice. *Calcified Tissue International 71*:63–68, 2002.

Handler, J. S., Corruccini, R. S., and Mutaw, R. J.: Tooth mutilation in the Caribbean: Evidence from a slave burial population in Barbados. *Journal of Human Evolution 11*:297–313, 1982.

Hanihara, T., and Ishida, H.: Frequency variations of discrete cranial traits in major human populations. III. Hyperostotic variations. *Journal of Anatomy 199*(Pt 3): 251–272, 2001 (a).

Hanihara, T., and Ishida, H.: Os incae: Variation in frequency in major human population groups. *Journal Anatomy 198*(Pt 2):137–152, 2001 (b).

Hansson, L. G.: Development of a lingual mandibular bone cavity in an 11-year-old boy. *Oral Surgery Oral Medicine Oral Pathology 49*(4):376–378, 1980.

Harbert, J., and Desai, R.: Small calvarial bone scan foci–Normal variations. *Journal Nuclear Medicine 26*(10)1144–1148, 1985.

Harper, K. N., Ocampo, P. S., Steiner, B. M., George, R. W., Silverman, M. S., Bolotin, S., Pillay, A., Saunders, N. J., and Armelagos, G. J. On the origin of the treponematoses: A phylogenetic approach. *PLOS Neglected Tropical Diseases, January 2*(1):e148, 2008 (published online).

Harris, H. A.: The growth of the long bones in childhood: With special reference to certain bony striations of the metaphysic and the role of vitamins. *Archives of Internal Medicine 38*:622–640, 1926.

_____: Lines of arrested growth in long bones in childhood. Correlation of histological and radiographic appearances in clinical and experimental conditions. *British Journal of Radiology 4*:561–588, 1931.

_____: *Bone growth in health and disease: The biological principles underlying the clinical, radiological and histological diagnosis of perversions of growth and disease in the skeleton.* London, Oxford University Press, 1933.

Harris, R. I., and Wiley, J. J.: Acquired spondylolysis as a sequel to spine fusion. *Journal of Bone and Joint Surgery 45-A*:1159–1170, 1963.

Harverson, G., and Warren A. G.: Tarsal bone disintegration in leprosy. *Clinical Radiology 30*(3):317–322, 1979.

Harvey, W., and Noble, H. W: Defects on the lingual surface of the mandible near the angle. *British Journal Oral Surgery 6*:75–83, 1968.

Hatfield, K. D.: The preauricular sulcus. *Australas Radiology* *15*(2):168–169, 1971.

Hatipoğlu, H. G., Çetin, M. A., and Yüksel, E.: Concha bullosa types: Their relationship with sinusitis, osteiomeatal and frontal recess disease. *Diagnostic Intervention Radiology 11*:145-149, 2005.

Hauser, G., and DeStefano, G. F. (Eds.): Introduction. In *Epigenetic variants of the human skull*. Stuttgart, Schweizerbart, 1989.

Havelková, P., and Villotte, S.: Enthesopathies: Test of the reproducibility of the new scoring system based on current medical data. *Slovenská Anthropológia 10*(1):51–57, 2007.

Haverstock, B. D.: Anterior ankle abutment. *Clinics Podiatric Medicine Surgery 18*(3):457–465, 2001.

Hawkey, D. E.: *Use of upper extremity enthesopathies to indicate habitual activity patterns*. Unpublished Master's Thesis, Arizona State University, 1988.

_____: Disability, compassion and the skeletal record: Using musculoskeletal stress markers (MSM) to construct and osteobiography from early New Mexico. *International Journal of Osteoarchaeology 8*(5):326–340, 1998.

Hawkey, D. E., and Merb, C. H.: Activity-induced musculoskeletal stress markers (MSM) and subsistence strategy changes among ancient Hudson Bay Eskimos. *International Journal of Osteoarchaeology 5*:324–338, 1995.

Hawkey, D. E., and Street, S.: Activity-induced stress markers in prehistoric human remains from the eastern Aleutian Islands. *American Journal of Physical Anthropology, Supplement 14*:89 (abstract), 1992.

Hawkins, R. B.: Arthroscopic treatment of sports-related anterior osteophytes in the ankle. *Foot and Ankle 9*(2): 87–90, 1988.

Heithersay, G. S.: Invasive cervical resorption. *Endodontic Topics 7*(1):73–92, 2004.

Heithersay, G. S.: Invasive cervical resorption. In Cohen, S., and Burns, R. C. (Eds.): *Pathways of the pulp* (7th ed.). CH 16, pp. 552–599. St. Louis, Mosby, 1998.

Hellman, M. H.: Charcot joint disease (Charcot joints). In McCarty, D. J. (Ed.): *Arthritis and allied conditions: A textbook of rheumatology* (11th ed.). Philadelphia, Lea and Febiger, 1989.

Hems, T., and Tillman, B: Tendon entheses of the human masticatory muscles. *Anatomy and Embryology 202*:201–208, 2000.

Henderson, C. Y.: When hard work is disease: The interpretation of enthesopathies. In Brickley, M., and Smith, M. (Eds.): *Proceedings of the Eighth Annual Conference of the British Association for Biological Anthropology and Osteoarchaeology* (pp. 17–25). Oxford British Archaeological Reports International Series 1743, 2008.

_____: *Musculo-skeletal stress markers in bioarchaeology: Indicators of activity levels or human variation? A re-analysis and interpretation*. Unpublished Ph.D. dissertation, Durham University, 2009.

Henderson, C. Y., and Gallant, A. J.: Quantitative recording of entheses. *Paleopathology Newsletter 137*:7–12, 2007.

Henderson, M. S.: Cervical Rib: Report of thirty-one cases. *Journal of Bone and Joint Surgery s2-11*:408–430, 1914.

Hengen, O. P.: Cribra orbitalia: pathogenesis and probable etiology. *HOMO 22*:57–75, 1971.

Henrard, J. C., and Bennett, P. H.: Etude epidemiologique de l'hyperostose vertebrale. Enquete dans une population adulte d'Indiens d'Amerique. *Revue du Thumatisme et des MaladiesOsteoarticulaires 40*:681–591, 1973.

Hergan K., Oser, W., and Moriggl, B.: Acetabular ossicles: Normal variant or disease entity? *European Radiology 10*(4):624–628, 2000.

Herring, S. A., and Nilson, K. L.: Introduction to overuse injuries. *Clinical Sports Medicine 6*:225–239, 1987.

Hershkovitz, I., Greenwald, C., Rothschild, B. M., Latimer, B., Dutour, O., Jellema, L. M., and Baratz, S.: Hyperostosis frontalis interna: An anthropological perspective. *American Journal of Physical Anthropology 109*:303–325, 1999.

Hershkovitz, I., Rothschild, B. M., Latimer, B., Dutour, O., Leonetti, G., Greenwald, C. M., Rothschild, C., and Jellema, L. M.: Recognition of sickle cell anemia in skeletal remains of children. *American Journal of Physical Anthropology 104*:213–226, 1997.

Hershkovitz, I., Greenwald, C. M., Jellema, L. M., Wish-Baratz, S., Eshed, V., Dutour, O., and Rothchild, B. M.: Serpens endocranial symmetrica (SES): A new term and a possible clue for identifying intrathoracic disease in skeletal populations. *American Journal of Physical Anthropology 118*(3):210–216, 2002.

Hertslet, L. E., and Keith, A.: Comparison of anomalous parts of two subjects, one with a cervical rib, the other with a rudimentary first rib. *Journal of Anatomy and Physiology 30*:562–567, 1896.

Hertzler, A. E.: Surgical pathology of the diseases of bone. *Hertzler's Monographs on Surgical Pathology*. Chicago, Lakeside, 1931.

Herzog, S., and Fiese, R.: Persistent foramen of Huschke: Possible risk factor for otologic complications after arthroscopy of the temporomandibular joint. *Oral Surgery Oral Medicine Oral Pathology 68*(3):267–270, 1989.

Hess, L.: The metopic suture and the metopic syndrome. *Human Biology 17*:107–136, 1945.

Hida, K., Yano, S., and Iwasaki, Y.: Considerations in the treatment of cervical ossification of the posterior longitudinal ligament. *Clinical Neurosurgery 55*:126–132, 2008.

Hilel, N.: The para-articular processes of the thoracic vertebrae. *Journal of Anatomy 33*:605, 1959.

_____: Osteophytes of the vertebral column: An anatomical study of their development according to age, race, and sex with considerations as to their etiology and significance. *Journal of Bone and Joint Surgery 44-A*:243, 1962.

Hill, C. L., Chaisson, C. E., Skinner, K., Kazis, L., Gale, M. E., and Felson, D. T.: Periarticular lesions detected on magnetic resonance imaging: Prevalence in knees with and without symptoms. *Arthritis and Rheumatism 48*(10):2836–2844, 2003.

Hill, M. C.: Porotic hyperostosis: A question of correlations verses causality. *American Journal of Physical Anthropology 66*:182, 1985.

Hillson, S.: *Teeth.* Cambridge, Cambridge University Press, 2005.

_____: Dental pathology. In Katzenberg, M. A., and Saunders, S. R. (Eds.): *Biological anthropology of the human skeleton.* Hoboken, NJ, Wiley and Sons, 2008.

Hillson, S. W., and Bond, S.: Relationship of enamel hypoplasia to the pattern of tooth crown growth: A discussion. *American Journal of Physical Anthropology 104*:89–103, 1997.

Hillson, S., Grigson, C., and Bond, S.: Dental defects of congenital syphilis. *American Journal of Physical Anthropology 1071*:25–40, 1998.

Hinkes, M. J.: Shovel shaped incisors in human identification. In Gill, G. W., and Rhine, S. (Eds.): *Skeletal attribution of race.* Anthropology Papers No. 4, Maxwell Museum of Anthropology 21–26, 1990.

Hirata, K.: A contribution to the paleopathology of of cribra orbitalia in Japanese. Cribra orbitalia in Edo Japanese. *St. Marianna Medical Journal 16*:6–24, 1988.

Hitchcock, H. H.: Spondylolisthesis: Observations on its development, progression, and genesis. *Journal of Bone and Joint Surgery 22-B*:1, 1940.

Hoeprich, P. D.: Coccidioidomycosis. In Hoeprich, P. D., and Jordan, M. C. (Eds.): *Infectious diseases: A modern treatise of infectious processes* (4th ed.). Philadelphia, Lippincott, pp. 489–502, 1989 (a).

_____: Nonspecific treponematoses. Inn Hoeprich, P. D., and Jordan, M. C. (Eds.): *Infectious diseases: A modern treatise of infectious process* (4th ed.). Philadelphia, Lippincott, pp. 1021–1034, 1989 (b).

Hodge, J. C.: Anterior process fracture of calcaneus secundarius: A case report. *Journal of Emergency Medicine 17*(2):305–309, 1999.

Hodgson, A. R., Wong, W., and Yau, A.: *X-ray appearance of tuberculosis of the spine.* Springfield, C. C. Thomas, 1969.

Hoffman, J. M.: Enlarged parietal foramina–their morphological variation and use in assessing prehistoric biological relationships. In Hoffman, J. M., and Brunker, L. (Eds.): *Studies in California paleopathology.* Contributions of the University of California Archaeological Research Facility, No. 30. Berkeley, University of California, 1976.

Holcomb, R. C.: *Who gave the world syphilis? The Haitian myth.* New York, Froben, 1930.

_____: The antiquity of congenital syphilis. *Medical Lift 42*:275–325, 1935.

Hollender, L.: Enlarged parietal foramina. *Oral Surgery 23*:447–453, 1967.

Holt, C. A.: A re-examination of parturition scars on the human female pelvis. *American Journal of Physical Anthropology 49*(1):91–94, 1978.

Hooton, E. A.: *The Indians of Pecos Pueblo, A study of their skeletal remains.* Papers of the Southwestern Expedition No. 4. New Haven, Yale University Press, 1930.

Hoppenfeld, S.: Back pain. *Pediatric Clinics of North America 24*(4):881–887, 1977.

Horackova, L., and Vargova, L.: Proximal end of femur and problems of its structure. In *New Approaches in Morphology,* 42nd Congress of the Slovak Anatomical Society with International Participation. Košice. ISBN 80-7097-554-7, pp. 35–35, 2004.

Horal, J., Nachemson, A., and Scheller, S.: Clinical and radiological long-term follow-up of vertebral fractures in childern. *Acta Orthopaedica Scandinavica 43*:491–503, 1972.

Horning, G. M., Cohen, M. E., and Neils, T. A.: Buccal alveolar exostosis: Prevalence, characteristics, and evidence for buttressing bone formation. *Journal of Periodontology 71*(6):1032–1042, 2000.

Hoshower, L. M., Buikstra, J. E., Goldstein, P. S., and Webster, A. D.: Artificial cranial deformation at the Omo M10 site: A Tiwanaku complex from the Moquegua Valley, Peru. *Latin American Antiquity 6*:145–164, 1995.

Hough, A. J., and Sokoloff, L.: Pathology of osteoarthritis. In McCarty, D. J. (Ed.): *Arthritis and allied conditions: A textbook of rheumatology* (11th ed.). Philadelphia, Lea and Febiger, 1989.

Houghton, P.: The relationship of the pre-auricular groove of the ilium to pregnancy. *American Journal of Physical Anthropology 41*:381–389, 1974.

_____: The bony imprint of pregnancy. *Bulletin of the New York Academy of Medicine 51*:655–661, 1975.

Hoyte, D. A. N., and Enlow, D. H.: Wolff's law of and problem of muscle attachment on resorptive surfaces of bone. *American Journal of Physical Anthropology 25*:205–214, 1966.

Hrdlička, A.: Artifical deformations of the human skull with special reference to America. In Lehmann-Nitsche, R.: *Actas del XVII Congreso Internacionale de Americanistas. International Society of the Americanists,* pp. 147–149, 1912.

_____: *Anthropological work in Peru in 1913. With notes on the pathology of the ancient Peruvians.* Smithsonian Miscellaneous Collections 61(18):vi–69, 1914.

_____: Shovel-shaped teeth. *American Journal of Physical Anthropology 3*:429–465, 1920.

_____: Incidence of the supercondyloid process in Whites and other races. *American Journal of Physical Anthropology 6*:405–412, 1923.

_____: The principal dimensions, absolute and relative, of the humerus in the white race. *American Journal of Physical Anthropology 16*:431–450, 1932.

_____: Ear exostoses. *Smithsonian Miscellaneous Collections 93*:1–100, 1935.

Hsiao, H. T., Poh, S. B., and Chen., P. M.: Punched-out lesions of the skull in elderly patients with inherited anemia. *European Journal of Haematology 75*(4):273–359, 2005.

Hsu, J. W., Tsai, P. L., Hsiao, T. H., Chang, H. P., Lin, L. M., Liu, K. M., Yu, H. S., and Ferguson, D.: The effect of shovel trait on Carabelli's trait in Taiwan Chinese and Aboriginal populations. *Journal of Forensic Sciences 42*(5):802–806, 1997.

Huang, W. S., Chen, S. L., Chen, S. G., Chiou, T. F., Chen, T. M., Cheng, T. Y., and Wang, H. J.: Polydactyly of the foot: Manifestations and Treatment. *Journal of Medical Science 20*(9):519–530, 2000.

Hudson, E. H.: A footnote on yaws and syphilis: Same or different? *Navy Medical Bulletin 38*:172–176, 1940.

_____: *Non-venereal syphilis: A sociological and medical study of bejel.* Edinburgh, E. & S. Livingstone, 1958.

Hughes, R. J., and Saifuddin, A.: Numbering of lumbosacral transitional vertebrae on MRI: Role of the iliolumbar ligaments. *American Journal of Roentgenology* 187:59–65, 2005.

Hughston, J. C., Hergenroeder, P. T., and Courtenay, B. G.: Osteochondritis dissecans of the femoral condyles. *Journal of Bone and Joint Surgery* 66-A(9):1340–1348, 1984.

Huiskes, R.: If bone is the answer, then what is the question? *Journal of Anatomy* 197:145–156, 2000.

Humphry, G. M.: Depressions in the parietal bones of an Orang and in man–supernumerary molars in Orang. *Journal of Anatomy and Physiology* 8(1):136–141, 1873.

Hunt, D. R., and Bullen, L.: The frequency of os acromiale in the Robert J. Terry Collection. *International Journal of Osteoarchaeology,* 17(3):309–317, 2007.

Hunter, D. J., and Eckstein, F.: Osteoarthritis and exercise. *Journal of Anatomy* 214(2):197–207, 2009.

Hutchinson, D. L., Denise, C. B., Daniel H. J., and Kalmus, G. W.: A reevaluation of the cold water etiology of external auditory exostoses. *American Journal of Physical Anthropology* 103(3):417–422, 1997.

Hutchinson, D. L., and Larsen, C. S.: Determination of stress episode duration from linear enamel hypoplasias: A case study from St. Catherine's Island, Georgia. *Human Biology* 60:93–110, 1988.

Hutter, R., Worcester, J., Francis, K., Foote, F., and Stewart, F.: Benign and malignant giant cell tumors of bone. *Cancer* 15:653–690, 1962.

Ihle, C. L., and Cochran, R. M.: Fracture of the fused os trigonum. *American Journal Sports Medicine* 10(1):47–50, 1982.

İşcan, M. Y., and Kennedy, K. A. R. (Eds.): *Reconstruction of life from the skeleton.* New York, Alan R. Liss, Inc. 1989.

Jackson, R.: Scoliosis in juvenile and adolescent children. *Health Visitor* 61(3):76–77, 1988.

Jaffe, H. L.: *Tumors and tumorous conditions of bones and joints.* Philadelphia, Lea and Febiger, 1958.

_____: Ischemic necrosis of bone. *Medical Radiography and Photography* 86:58–86, 1969.

_____: *Metabolic, degenerative and inflammatory diseases of bone and joints.* Philadelphia, Lea and Febiger, 1975.

Jager, H. J., Gordon-Harris, L., Mehring, U. M., Goetz, G. F, Mathias, K. D.: Degenerative change in the cervical spine and load-carrying on the head. *Skeletal Radiology* 26:475–481, 1997.

Jahss, M. H.: *Disorders of the foot and ankle: Medical and surgical management.* Philadelphia, W. B. Saunders Company, 1991.

Jainkittivong, A., and Langlais, R. P.: Buccal and palatal exostoses: Prevalence and concurrence with tori. *Oral Surgery Oral Medicine Oral Pathology Oral Radiology Endodontics* 90(1):48–53, 2000.

James, R., and Nasmyth-Jones, R.: The occurrence of cervical fractures in victims of judicial hanging. *Forensic Science International* 54:81–91, 1992.

Jarcho, S. (Ed.): *Human paleopathology.* New Haven:Yale University Press, 1966.

Jayson, M. I. V., Dixon, A. J. (Eds): *The lumbar spine and back pain* (4th ed.). New York, Churchill Livingston, 1992.

Jeffery, A. K.: Osteophytes and the osteoarthritic femoral head. *Journal of Bone and Joint Surgery* 57-B(3):314–324, 1975.

Jenkins, D. B.: *Hollingshead's functional anatomy of the limbs and back* (8th ed.). New York, NY: W. B. Saunders Company, 2002.

Jeyasingh, P., Gupta, C. D., Arora, A. K., and Ajmani, M. L.: Incidence of squatting facets on the talus of Indians (Agra region). *Anthropologischer Anzeiger* 37(2):117–122, 1979.

Jit, I., and Kaur, H.: Rhomboid fossa in the clavicles of North Indians. *American Journal of Physical Anthropology* 70:97–103, 1986.

Johansson, L. U.: Bone and related materials. In Hodge, W. M. (Ed.): *In situ archaeological conservation.* Mexico: Instituto Nacional de Antropologia e Historia, 1987.

Johnston, M. C., and Millicovsky, G.: Normal and abnormal development of the lip and palate. *Clinics Plastic Surgery* 12(4):521–532, 1985.

Jones, H. C., and Hedrick, D. W.: Patellar anomalies, roentgenologic and clinical consideration. *Radiology* 38:30–34, 1942.

Jones, R. R., and Martin, D. D.: Blastomycosis of bone. *Surgery* 10:931, 1941.

Julkunen, H., Heinonen, O. P., Knekt, P., and Maatela, J.: The epidemiology of hyperostosis of the spine together with its symptoms and related mortality in general population. *Scandinavian Journal of Rheumatology* 40:581, 1973.

Julkunen, H., Heinonen, O. P., and Pyorala, K.: Hyperostosis of the spine in an adult population: Its relation to hyperglycaemia and obesity. *Annals of Rheumatic Disease* 30:605–612, 1971.

Junge, A., El-Sheik, M., Celik, I., and Gotzen, L.: Pathomorphology, diagnosis and treatment of "hangman's fractures" (article in German). *Unfallchirung* 105(9):775–782, 2002.

Jurmain, R.: *Stories from the skeleton: Behavioral reconstruction in human osteology.* Canada: Gordon and Breach Publishers, 1999.

_____: Paleoepidemiolgical patterns of trauma in a prehistoric population from central California. *American Journal of Physical Anthropology* 115:13–23, 2001.

Jurmain, R. D., and Villotte, S.: Terminology–entheses in medical literature: A brief review. In *Workshop in musculoskeletal stress markers (MSM): Limitations and achievements in the reconstruction of past activity patterns 2010*; http://ww.uc.pt/cia/msm/MSM_terminology3, accessed Sept 25, 2010.

Kaar, S. G., Cooperman, D. R., Blakemore, L, C., Thompson, G. H., Petersilge, C. A., Elder, J. S., and Heiple, K. G.: Association of bladder exstrophy with congenital pathology of the hip and lumbosacral spine: A long-term follow-up study of 13 patients. *Journal of Pediatric Orthopaedics* 22(1):62–66, 2002.

Kahn, M. A.: Ankylosing spondylitis. In Calin, A. (Ed.): *Spondylorthropathies.* Orlando, Grune & Stratton, 1984.

Kajbafzadeh, A. M., Tanhaeivash, R., Elmi, A., Shirazi, M., Talab, S. S., and Shabestari, A.: Three-dimensional anatomy of the pelvic bone in bladder exstrophy: Comparison between patients managed with osteotomy and pubic symphysis internal fixation using metal plates. *Urology 76*(4):934–41; discussion 941, 2010.

Kalantari, B. N., Seeger, L. L., Motamedi, K., and Chow, K.: Accessory ossicles and seasmoid bones: Spectrum of pathology and imaging evaluation. *Applied Radiology 36*(10):28–37, 2007.

Karasick, D., and Schweitzer, M. E.: The os trigonum syndrome: Imaging features. *American Journal of Roentgenology 166*(1):125–129, 1996.

Karpinski, M. R. K., Newton, G., and Henry, A. P. J.: The results and morbidity of varus osteotomy for Perthes disease. In Burwell, R. G., and Harrison, M. H. M. (Eds.): *Clinical Orthopaedics and Related Research 209*:30–40, 1986.

Kasai, Y., Kawakita, E., Sakakibara, T., Akeda, K., and Uchida, A.: Direction of the formation of anterior lumbar vertebral osteophytes. *Musculoskeletal Disorders 10*(4); doi: 10.1186/1471-2474-10-4, 2009.

Kate, B. R.: The incidence and cause of cervical fossa in Indian femora. *Journal of the Anatomical Society of India 12*(2):69, 1963.

Kate, B. R., and Robert, S. L.: Some observations on the upper end of the tibia in squatters. *Journal of Anatomy 99*:137–141, 1965.

Kaufmann, G. A., Sundarum, M., and Mcdonald, D. J.: Magnetic resonance imaging in symptomatic Paget's disease. *Skeletal Radiology 20*:413–418, 1991.

Kawashima, T., and Uhthoff, H. K.: Prenatal development around the sustentaculum tali and its relation to talocalcaneal coalitions. *Journal of Pediatric Orthopaedics 10*:238–243, 1990.

Kay, D. J., Har-El, G., and Lucente, F. E.: A complete stylohyoid bone with a stylohyoid joint. *American Journal of Otolaryngology 22*(5):358–361, 2001.

Kayalioglu, G., Oyar, 0., and Govsa, F.: Nasal cavity and paranasal sinus bony variations: A computed tomographic study. *Rhinology 38*(3):108–113, 2000.

Kaye, M.: *Molecular identification and analysis of treponematosis (syphilis, bejel, yaws, or pinta) in ancient mummified remains from northern Chile and southern Peru.* Doctoral dissertation, University of Alaska, Fairbanks, 2008.

Keats, T. E.: *An atlas of normal roentgen variants that may simulate disease* (34th ed.). Chicago, Year Book Medical, 1988.

Keenleyside, A.: Skeletal evidence of health and disease in pre-contact Alaskan Eskimos and Aleuts. *American Journal of Physical Anthropology 107*(1):51–70, 1998.

Keim, H. A.: Scoliosis. *Clinical Symposia 24*, No. 1, 1972.

_____: Low back pain. *Clinical Symposia 25*, No. 3, 1973.

Kelley, J. O., and Angel, J. L.: Life stresses of slavery. *American Journal of Physical Anthropology 74*:199–211, 1987.

Kelley, M. A.: Parturition and pelvic changes. *American Journal of Physical Anthropology 51*(4):541–546, 1979.

Kelley, M. A., and Eisenberg, L. E.: Blastomycosis and tuberculosis in early American Indians: A biocultural view. *Midcontinental Journal of Archaeology 12*:89–116, 1987.

Kelley, M. A., and El-Najjar, M. V.: Natural variation and differential diagnosis of skeletal changes in tuberculosis. *American Journal of Physical Anthropology 52*:153–167, 1980.

Kelley, M. A., and Micozzi, M. S.: Rib lesions in chronic pulmonary tuberculosis. *American Journal of Physical Anthropology 65*:381–386, 1984.

Kellgren, J. J., and Lawrence, J. S.: Osteoarthritis and disk degeneration in an urban population. *Annals of the Rheumatic Diseases 17*:388–397, 1958.

Kempson, F. C.: Emargination of the patella. *Journal of Anatomy and Physiology 36*:419–420, 1902.

Kennedy, G. E.: The relationship between auditory exostosis and cold water: A latitudinal analysis. *American Journal of Physical Anthropology 71*:401–415, 1986.

Kennedy, K. A. R.: Skeletal markers of occupational stress. In İşcan, W. A., and Krogman, M. A. (Eds.): *Reconstruction of life from the skeleton.* New York, Wiley-Liss. pp. 128–160, 1989.

Kennedy, K. A. R.: Markers of occupational stress: Conspectus and prognosis of research. *International Journal of Osteoarchaeology 8*(5):305–310, 1998.

Kessel, L., and Rang, M.: Supracondylar spur of the humerus. *Journal of Bone and Joint Surgery 48-B*(4):765–769, 1966.

Keur, J. J., Campbell, J. P. S., McCarthy, J. F., and Ralph, W. J.: The clinical significance of the elongated styloid process. *Oral Surgery Oral Medicine Oral Pathology 61*:399–404, 1986.

Khanna, P. C., Thapa, M. M., Iyer, I. S., and Prasad, S. S.: Pictorial essay: The many faces of craniosynostosis. *Indian Journal of Radiology and Imaging 21*(1):49–56, 2011.

Khazhinskaia, V. A., and Ginzburg, M. A.: X-ray anatomic variants of the rhomboid fossa of the clavicle. *Vestnik Rentgenolog Radiologia May–June* (3):332–37, 1975.

Kilgore, L., and Van Gerven, D.: Congenital scoliosis: Possible causes and consequences in a skeleton from Nubia. *International Journal of Osteoarchaeology 20*(6):630–644, 2010.

Kim, N. H., and Suk, K. S.: The role of transitional vertebrae in spondylolysis and spondylolytic spondylolisthesis. *Bulletin Hospital Joint Disease 56*:161–166, 1997.

Kimmerle, E. H., and Baraybar, J. P. (Eds.): *Skeletal trauma: Identification of injuries resulting from human rights abuse and armed conflict.* Boca Raton, CRC Press, 2008.

Klepinger, L. L.: Paleopathologic evidence for the evolution of rheumatoid arthritis. *American Journal of Physical Anthropology 50*:119–122, 1979.

Klepinger, L. L., and Heidingsfelder, J. A.: Probable torticollis revealed in decapitated skull. *Journal of Forensic Sciences 41*(4):693–696, 1996.

Klippel, M., and Feil, A.: Un cas d'absence des vertebres cervicales. *Nouv Iconong Salpetriere 25*:223, 1912.

Klunder, K. B., Rud, B., and Hansen, J.: Osteoarthritis of the hip and knee joint in retired football players. *Acta Orthopaedica Scandinavica 51*:925–927, 1980.

Knight, G. A. M., and Morley, G. H.: Cleft sternum: Case report and brief commentary. *British Journal of Surgery 24*:60–64, 1936–37.

Knowles, A. K.: Acute traumatic lesions. In Hart, G. D. (Ed.): *Disease in ancient man: An international symposium.* Ontario, Irwin, 1983.

Konin, G. P., and Walz, D. M.: Lumbosacral transitional vertebrae: Classification, imaging findings, and clinical relevance. *American Journal of Neuroradiology 31*:1778–1786, 2010.

Kono, S., Hayashi, N., and Kashahara, G.: A study on the etiology of spondylolysis with reference to athletic activities. *Journal of the Japanese Orthopedic Association 49*:125, 1975.

Kortesis, B., Richards, T., David, L., Glazier, S., and Argenta, L.: Surgical management of foramina parietalia permagna. *Journal of Craniofacial Surgery 14*(4):538–544, 2003.

Kose, O.: Fracture of the os trigonum: A case report. *Journal of Orthopaedic Surgery 14*(3):354–356, 2006.

Kostick, E. L.: Facets and imprints on the upper and lower extremities of femora from a Western Nigerian population. *Journal of Anatomy 97*(3):393–402, 1963.

Kraemer, J.: *Intervertebral disc diseases: Causes, diagnosis, treatment, and prophylaxis* (3rd ed.). Stuttgart, Georg Theme Verlag, 2009.

Kramer, S. B., Lee, S. H. S., and Abramson, S. B.: Nonvertebral infections of the musculoskeletal system by Mycobacterium Tuberculosis. In Rom, W. M., Garay, S. M., and Bloom, B. R. (Eds.): *Tuberculosis* (2nd ed.). Philadelphia, Lippincott Williams and Company, 2004.

Krauss, M. J., Morrissey, A. E., Winn, H. N., Amon, E., and Leet, T. L.: Microcephaly: An epidemiologic analysis. *American Journal of Obstetrics and Gynecology 188*(6):1484–1489; discussion 1489–1490, 2003.

Krishnan, J.: Distal radius fractures in adults. *Orthopedics 25*(2):175–179, 2002.

Krogman, W. M: The morphological characteristics of the Australian skull. *Journal of Anatomy 66*(3):399–413, 1932.

_____: The pathologies of pre- and protohistoric man. *Ciba Symposia 2*:432–443, 1940.

Krogman, W. M., and İşcan, M. Y.: The human skeleton in forensic medicine (2nd ed.). Springfield, C. C. Thomas, 1986.

Kromberg, J. G., and Jenkins, T.: Common birth defects in South African blacks. *South African Medical Journal 62*:699–602, 1982.

Kroon, D. F., Lawson, M. L., Derkay, C. S., Hoffmann, K., and McCook, J.: Surfer's ear: External auditory exostoses are more prevalent in cold water surfers. *Otolaryngology Head Neck Surgery 126*(5):499–504, 2002.

Kulkarni, V. N., and Mehta, J. M.: Tarsal disintegration (TD) in leprosy. *Leprosy in India 55*:338–370, 1983.

Kullmann, L., and Wouters, H. W.: Neurofibromatosis, gigantism and subperiosteal haematoma: Report of two children with extensive subperiosteal bone formation. *Journal of Bone and Joint Surgery 54-A*:130–138, 1972.

Kumai, T., and Benjamin, M.: Heel spur formation and the subcalcaneal enthesis of the plantar fascia. *Journal of Rheumatology 29*(9):1957–1964, 2002.

Kurtz, C. A., Humble, B. J., Rodosky, M. W., and Sekiya, J. K: Symptomatic os acromiale. *Journal of the Academy of Orthopaedic Surgeons 14*(1):12–19, 2006.

Kutilek, S., Baxova A., Bayer, M., Leiska, A., and Kozlowski, K.: Foramina parietalia permagna: Report of nine cases in one family. *Journal of Paediatric Child Health 33*(2):168–170, 1997.

Lacout, A., Marsot-Dupucha, K., Smoker, W. R. K., and Lasjaunias, P.: Foramen tympanicum, or foramen of Huschke: Pathologic cases and anatomic CT study. *American Journal of Neuroanatomy 26*:1317–1323, 2005.

Ladisch, S., and Jaffe, E.: The histocytoses. In Pizzo, P. A., and Poplak, D. G. (Eds.): *Principles of pediatric oncology.* Philadelphia, Lippincott, 1989.

Lagia, A., Eliopoulos, C., and Manolis, S.: Thalassemia: macroscopic and radiological study of a case. *International Journal of Osteoarchaeology 17*(3): 269–285, 2007.

Lai, P., and Lovell, N. C.: Skeletal markers of occupational stress in the fur trade: A case from a Hudson Bay Company fur trade post. *International Journal of Osteoarchaeology 2*(3):221–234, 1992.

Lallo, J. W., Armelagos, G. J., and Mensforth, R. P.: The role of diet, disease and physiology in the origin of porotic hyperostosis. *Human Biology 48*(3):471–483, 1977.

Lamb, D. S.: The olecranon perforation. *American Anthropologist April A3*(2):159–174, 1890.

Lambert, P. M.: Rib lesions in a prehistoric Puebloan sample from southwestern Colorado. *American Journal of Physical Anthropology 117*:281–292, 2002.

Lamy, C., Bazergui, A., Kraus, H., and Farfan, H. F.: The strength of the neural arch and the etiology of spondylolysis. *Orthopedic Clinics of North America 6*:215–231, 1975.

Lane, N. E., Bloch, D. A., Jones, H. H., Simpson, U., and Fries, J. F.: Osteoarthritis in the hand: A comparison of handedness and hand use. *Journal of Rheumatology 16*(5):637–642, 1989.

Lang, L., Korogi, Y., Sugahara, T., Ikushima, I., Shigematsu, Y., Takahashi, M., and Provenzale, J. M.: Normal structures in the intracranial dural sinuses: Delineation with 3D contrast-enhanced magnetization prepared rapid acquisition gradient-echo imaging sequence. *American Journal of Neuroradiology 23*:1739–1746, 2002.

Lange, R. H., Lange, T. A., and Rao, B. K.: Correlative radiographic, scintigraphic, and histological evaluation of exostoses. *Journal of Bone and Joint Surgery 66-A*(9):1454–1459, 1984.

Langlais, R. P., Miles, D. A., and Van Dis, M. L.: Elongated and mineralized stylohyoid ligament complex: A proposed classification and report of a case of Eagle's syndrome. *Oral Surgery Oral Medicine Oral Pathology 61*:527–532, 1986.

Langland, O. E., Langlais, R. P., and Preece, J. W.: *Principles of dental imaging.* Philadelphia, Lippincott, 2002.

Lanzkowsky, P.: Radiological features of iron deficiency anemia. *American Journal of Diseases of Children 116*:16–29, 1968.

Larsen, C. S.: *Bioarchaeology: Interpreting behavior from the human skeleton.* London, Cambridge University Press, 1997.

Larsen, W. J.: *Human embryology* (3rd ed.). Philadelphia, Churchill Livingstone, 2001.

Lasater, K., and Groer, M.: Arthritis. In Groer, M. (Ed.): *Advanced pathophysiology: Application to clinical practice.* Philadelphia, J. B. Lippincott Company, pp. 245–265, 2000.

Lastres, J. B., and Cabiese, F.: *La Trepanacion del Craneo en el Antiguo Peru.* Lima: Imprenta de la Universidad Nacional Mayor de San Marcos, 1960.

Latham, R. A., and Burston, W. R.: The postnatal pattern of growth at the sutures of the human skull: A histological survey. *The Dental Practitioner and Dental Record 17*(2):61–67, 1966.

Laughlin, W. S., and Jorgensen, J. B.: Isolate variation in Greenlandic Eskimo crania. *Acta Genetica et Statistica Medica 6*:3–12, 1956.

Laurent, L. E., and Einola, S.: Spondylolisthesis in children and adolescents. *Acta Orthopaedica Scandinavica 31*:45–64, 1961.

Lawrence, J. S.: Rheumatism in cotton cooperative. *British Journal Indian Medicine 18*:270–276, 1961.

_____: Generalized osteoarthritis in a population sample. *American Journal Epidemiology 90*:381–389, 1969.

_____: Relationship of infection to rheumatoid factor in the population. *Annals of Rheumatic Diseases 29*(2):196–197, 1970a.

_____: Paget's disease in population samples. *Annals of Rheumatic Diseases 29*(5):562, 1970b.

Lawson, B. J.: Hutchinson's teeth. *Oral Surgery Oral Medicine Oral Pathology 24*(5):635–636, 1967.

Leach, J. L., Meyer, K., Jones, B. V., and Tomsick, T. A.: Large arachnoid granulations involving the dorsal superior sagittal sinus: Findings on MR imaging and MR venography. *American Journal of Neuroradiology 29*:1335–1339, 2008.

Leak, R. S., Rayan, G. M., and Arthur, R. E.: Longitudinal radiographic analysis of rheumatoid arthritis in the hand and wrist. *Journal of Hand Surgery [Am] 28*(3):427–434, 2003.

Leboeuf, C., Kimber, D., and White, K.: Prevalence of spondylolisthesis, transitional anomalies and low intercrestal line in a chiropractic patient population. *Journal of Manipulative Physiology Therapy 12*(3):200–204, 1989.

Lee, M. C., Kelly, D. M., Sucato, D. J., and Herring, J. A.: Familial bilateral osteochondritis dissecans of the femoral head: A case series. *Journal of Bone and Joint Surgery 91*:2700–2707, 2009.

Lee, T. C., and Taylor, D.: Bone remodeling: Should we cry Wolff? *Irish Journal of Medical Sciences 168*(2):102–105, 1999.

Lee, T. M., and Mehlman, C. T.: Hyphenated history: Park-Harris growth arrest lines. *American Journal of Orthopedics 32*(8):408–411, 2003.

Leestma, J. E.: *Forensic neuropathology* (2nd ed.). Boca Raton, CRC Press, 2009.

Le Gros Clark, W. E.: On the Pacchionian bodies. *Journal of Anatomy 55*:40–48, 1920.

Lehman, T. J.: *Enthesitis, arthritis and heel pain. Journal of the American Podiatric Medical Association 89*(1):18–19, 1999.

Lepow, G. M., and Cafiero, J. M.: Heterotopic bone formation: A case report. *Journal of Foot Surgery 19*(2):63–65, 1980.

Lester, C. W., and Shapiro, H. L.: Vertebral arch defects in the lumbar vertebrae of pre-historic American Eskimos: A study of skeletons in the American Museum of Natural History, chiefly from Point Hope, Alaska. *American Journal of Physical Anthropology 28*:43–47, 1968.

Lestini, W. F., and Wiesel, S. W.: The pathogenesis of cervical spondylosis. *Clinical Orthopaedics and Related Research 239*:69–93, 1989.

Letts, M., Smallman, T., Afanasiev, R., and Gouw, G.: Fracture of the pars inter-articularis in adolescent athletes: A clinical-biomechanical analysis. *Journal of Pediatric Orthopedics 6*:40–46, 1986.

Leung, J. S. M., Mok, C. K., Leong, J. C. Y., and Chan, W. C.: Syphilitic aortic aneurysm with spinal erosion: Treatment by aneurysm replacement and anterior spinal fusion. *Journal of Bone and Joint Surgery 59-B*(1):89–92, 1977.

Levine, S. M., Lambiase, R. E., and Petchprapa, C. N.: Cortical lesions of the tibia: Characteristic appearances at conventional radiography. *Radiographics 23*(1):157–177, 2003.

Liberson, F.: Os acromiale–A contested anomaly. *Journal of Bone and Joint Surgery 19*:683–689, 1937.

Libson, E., Bloom, R. A., and Dinari, G.: Symptomatic and asymptomatic spondylolysis and spondylolisthesis in young adults. *International Orthopedics 6*:259–261, 1982.

Lichenstein, L.: Histiocytosis X. Integration of eosinophilic granuloma of bone, letterer-Siwe disease, and Schuller-Christian disease as related manifestations of a single nosologic entity. *Archives of Pathology 56*:84, 1953.

_____: *Diseases of bone and joints.* Saint Louis, Mosby, 1970.

Lidov, M., and Som, P. M.: Inflammatory disease involving a concha bullosa (enlarged pneumatized middle nasal turbinate): MR and CT appearance. *American Journal Neuroradiology 11*(5):999–1001, 1990.

Lieberman, D. E., Pearson, O. M., Polk, J. D., Demes, B., and Crompton, A. W.: Optimization of bone growth and remodeling in response to loading in tapered mammalian limbs. *Journal of Experimental Biology 206*:3215–3128, 2003.

Lieverse, A. R., Bazaliiskii, V. I., Gouriunova, O. I., and Weber, A. W.: Upper limb musculoskeletal stress markers among Middle Holocene foragers of Siberia's Cis-Baikal region. *American Journal of Physical Anthropology 138*:458–472, 2009.

Lima, Mauricio D. L. P.: *Contribucicao ao Estudo do Os Trigonum Tarsi.* San Paulo, These Inaugural, 1928.

Limson, M.: Metopism as found in Filipino skulls. *American Journal of Physical Anthropology 7*(3):317–324, 1924.

Lindahl, S., Nyman, R. S., Brismar, J., Hugosson, C., and Lundstedt, C.: Imaging of tuberculosis. IV. Spinal manifestation in 63 patients. *Acta Radiologica 37*:506–511, 1996.

Lindau, T. R., Aspenberg, P., Arner, M., Redlundh-Johnell, I., and Hagberg, L.: Fractures of the distal forearm in young adults. An epidemiologic description of 341 patients. *Acta Orthopaedica Scandinavica 70*(2):124–128, 1999.

Lindblom, K.: Bachache and its relation to ruptures of the intravertebral disks. *Radiology 57*:710, 1951.

Lisowski, F. P.: Prehistoric and early historic trepanation. In Brothwell, D. R., and Sandison, A. T. (Eds.): *Diseases in antiquity: A survey of diseases injuries and surgery of early populations.* Springfield, C. C. Thomas, 1967.

Littleton, J., and Townsend, G. C.: Linear enamel hypoplasis and historical change in a central Australian community. *Australilan Dental Journal 50*(2):101–107, 2005.

Livingstone, F.: On the origin of syphilis: An alternative hypothesis. *Current Anthropology 32*:587–590, 1991.

Llguy, M., Llguy, D., Guler, N., and Bayirli, G.: Incidence of the type and calcification patterns in patients with elongated styloid process. *Journal of International Medical Research 33*:96–102, 2005.

Lodge, T.: Thinning of the parietal bones in early Egyptian populations and its aetiology in the light of modern observations. In Brothwell, D. R., and Sandison, A. T. (Eds.): *Diseases in antiquity: A survey of diseases, injuries, and surgery of early populations.* Springfield, C. C. Thomas, 1967.

Lombardi, C. M., Silhanek, A. D., and Connolly, F. G.: Modified arthroscopic excision of the symptomatic os trigonum and release of the flexor hallucis longus tendon: Operative technique and case study. *Journal of Foot and Ankle Surgery 38*(5):347–351, 1999.

Long, B. W., and Rafert, J. A.: *Orthopaedic radiography.* Philadelphia, W. B. Saunders Company, 1995.

Longia, G. S., Agarwal, A. K., Thomas, R. J., Jain, P. N., and Saxena, S. K.: Metrical study of rhomboid fossa of clavicle. *Anthropologischer Anzeiger 40*(2):111–115, 1982.

Longridge, N. S.: Exostosis of the external auditory canal: A technical note. *Otology Neurology 23*(3):260–261, 2002.

Lorena, S. C. M., Oliviera, D. T., and Odell, E. W.: Multiple dental anomalies in the maxillary incisor region. *Journal of Oral Science 45*(1):47–50, 2003.

Lovell, N. C.: Analysis and interpretation of skeletal trauma. In Katzenberg, M. A., and Saunders, S. R. (Eds.): *Biological anthropology of the human skeleton.* Hoboken, NJ, Wiley and Sons, 2008.

Lowe, J., Libson, J., Ziv, I., Nyska, M., Floman, Y., Bloom, R. A., and Robin, G. C.: Spondylolysis in the upper lumbar spine: A study of 32 patients. *Journal of Bone and Joint Surgery 39-B*(69):582–586, 1987.

Lu, H., Gu, G., and Zhu, S.: Heel pain and calcaneal spurs (article in Chinese, English abstract). *Zhonghua Wai Ke Za Zhi 34*(5):294–296, 1996.

Lucas, M. F.: (1) Two cases of cervical ribs. (2) An anomalous arrangement of the vagi. *Journal of Anatomy and Physiology 69*:336–342, 1915.

Lukacs, J., and Rodriguez Martin, C.: Lingual cortical mandibular defects (Stafne's defect): An anthropological approach based on prehistoric skeletons from the Canary Islands. *International Journal of Osteoarchaeology 12*:112–126, 2002.

Lundy, J. K.: A report on the use of Fully's anatomical method to estimate stature in military skeletal remains. *Journal of Forensic Sciences 33*(2):534, 1988.

Luong, A. A., and Salonen, D. C.: Imaging of the seronegative spondyloarthropathies. *Current Rheumatology Preparation 2*(4):288–296, 2000.

Ly, J. Q., Sanders, T. G., Mulloy, J. P., Soares, G. M., Beall, D. P., Parsons, T. W., and Slabaugh, M. A.: Osseous change adjacent to soft-tissue hemangiomas of the extremities: Correlation with lesion size and proximity to bone. *American Journal Roentgenology 180*(6):1695–1700, 2003.

Maas, M., Slim, E. J., Heoksma, A. F., van der Kleij, A. J., Akkerman, E. M., den Heeten, G. J., and Faber, W. R.: MR imaging of neuropathic feet in leprosy patients with suspected osteomyelitis. *International Journal Leprosy Other Mycobacterial Disease 70*(2):97–103, 2002.

MacAusland, W. R., and Mayo, R. A.: *Orthopedics: A concise guide to clinical practices.* Boston, Little, Brown, 1965.

MacInnis, E. L., Hardie, J., Baig, M., and al-Sanea, R. A.: Gigantiform Torus palatinus: Review of the literature and report of a case. *International Dental Journal 48*(1): 40–43, 1998.

Madewell, J. E., Ragsdale, B. D., and Sweet, D. E.: Radiologic and pathologic analysis of solitary bone lesions. Part I: Internal margins. *Radiologic Clinics of North America 19*(4):715–748, 1981a.

_____: Radiologic and pathologic analysis of solitary bone lesions. Part II: Periosteal reactions. *Radiologic Clinics of North America 19*(4):749–783, 1981b.

Madkour, M. M. (Ed.): *Tuberculosis.* New York, Springer-Verlag, 2004.

Magee, D. J.: *Orthopedic physical assessment* (4th ed.). Philadelphia: Saunders, 2002.

Mahato, N. K.: Association of rudimentary sacral zygapophyseal facets and accessory and ligamentous articulations: Implications for load transmission at the L5–S1 junction. *Clinical Anatomy 23*(6):707–711, 2010.

Manchester, K.: *The archaeology of disease.* Bradford, University of Bradford, 1983.

_____: Bone changes in leprosy: Pathogenesis and paleopathology. Paper presented at the Annual meeting of the American Association of Physical Anthropologists, San Diego, 1989.

Mankin, H. J.: The structure, chemistry and metabolism of articular cartilage. *Bulletin of Rheumatic Diseases 17*(7): 447–452, 1967.

_____: The effect of aging on articular cartilage. *Bulletin of the New York Academy of Medicine 44*(5):545–552, 1968.

_____: The reaction of articular cartilage to injury and osteoarthritis (second of two parts). *New England Journal of Medicine 291*:1335–40, 1974a.

_____: Rickets, osteomalacia, and renal osteodystrophy, part I. *Journal of Bone and Joint Surgery 56-A*:101, 1974b.

_____: Rickets, osteomalacia, and renal osteodystrophy, part II. *Journal of Bone and Joint Surgery 56-B*:352–386, 1974c.

_____: The response of articular cartilage to mechanical injury. *Journal of Bone and Joint Surgery 64*(3):460–466, 1982.

_____: Rickets, osteomalacia, and renal osteodystrophy. An update. *Orthopedic Clinics of North America 21*(1):81–96, 1990.

Mann, R. W.: Calcaneus secundarius: Description of a common accessory ossicle. *Journal of American Podiatric Medical Association 79*(8):363–366, 1989.

_____: The calcaneus secundarius: Frequency and description of an accessory ossicle in six samples. *American Journal of Physical Anthropology 87*:17–25, 1990.

_____: Enlarged parietal foramina and craniosynostosis in an American Indian child. *American Journal of Roentgenology 154*:658, 1990.

_____: A method for siding and sequencing human ribs. *Journal of Forensic Sciences 38*(1):151–155, 1993.

_____: *Stafne's defect in the human mandible.* Doctoral dissertation, University of Hawaii, Manoa, 2001.

_____: A Method for articulating and displaying the human spine. *Journal of Forensic Sciences 54*:1229–1230, 2009.

Mann, R. W., and Hunt, D. R.: *Photographic regional atlas of bone disease: A guide to pathological and normal variation in the human skeleton* (2nd ed.). Springfield, C. C. Thomas, 2005.

Mann, R. W., and Murphy, S. P.: Skeletal indicators of stress in 28 American soldiers who died in the Battle of Fort Erie, War of 1812. *Northeastern Anthropology Association 26*, 1989.

_____: *Regional atlas of bone disease: A guide to pathologic and normal variation in the human skeleton.* Springfield, C. C. Thomas, 1990.

Mann, R. W., and Owsley, D. W.: Anatomy of uncorrected talipes equinovarus in a fifteenth century American Indian. *Journal of the American Podiatric Medical Association 79*(9):436–440, 1989.

Mann, R. W., and Owsley, D. W: Os trigonum: Variation of a common accessory ossicle of the talus. *Journal of the American Podiatric Medical Association 80*(11):623–625, 1990.

Mann, R. W., and Owsley, D. W.: Human osteology: Key to the sequence of events in a post mortem shooting. *Journal of Forensic Sciences 37*(5):1386–1392, 1992.

Mann, R. W., and Tsaknis, P. J.: Cortical defects in the mandibular sulcus. *Oral Surgery Oral Medicine Oral Pathology 71*:514–516, 1991.

Mann, R. W., and Verano, J. W: Congenital spinal anomalies in a prehistoric adult female from Peru. *Paleopathology Newsletter 72*, 1990.

Mann, R. W., Meadows, L., Bass, W. M., and Watters, D. R.: Description of skeletal remains from a black slave cemetery from Montserrat, West Indies. *Annals of Carnegie Museum 56*:319–336, 1987.

Mann, R. W., Owsley, D. W., and Sledzik, P. S.: Seronegative spondyloarthropathy of the foot. *Journal of the American Podiatric Medical Association 80*(7):360–363, 1990.

Mann, R. W., Sledzik, P. S., Owsley, D. W., and Droulette, M. R.: Radiographic examination of Chinese foot binding. *Journal of the American Podiatric Medical Association 80*(8):405–409, 1990.

Manzanares, M. C., Goret-Nicaise, M., and Dhem, A.: Metopic sutrual closure in the human skull. *Journal of Anatomy 161*:203–215, 1988.

Marden, K., and Ortner, D. J.: A case of treponematosis from pre-Columbian Chaco Canyon, New Mexico. *International Journal of Osteoarchaeology 21*(1):19–31, 2011.

Margetts, E. L.: Trepanation of the skull by the Medicine-men of primative cultures, with particular reference to present day native east African practice. In Brothwell, D. R., and Sandison, A. T. (Eds.): *Diseases in antiquity: A survey of the diseases, injuries, and surgery of early populations.* Springfield, C. C. Thomas, pp. 673–701, 1967.

Marino, R., and Gonzales-Portillo, M. Preconquest Peruvian neurosurgeons: A study of Inca and Pre-Columbian trephination and the art of medicine in ancient Peru. *Neurology 47*(4):940–955, 2000.

Marjolein, C. H., van der Meulen, and Huiskes, R. : Why mechanobiology? A survey article. *Journal of Biomechanics 35*:401–414, 2002.

Mariotti, V., Facchini, F., and Belcastro, M. G.: Enthesopathies–Proposal of a standardized scoring method and applications. *Collegium Anthropologicum 28*:145–159 2004.

Mariotti, V., Facchini, F., and Belcastro, M. G.: The study of entheses: Proposal of a standardized scoring method for twenty-three entheses of the postcranial skeleton. *Collegium Anthropologicum 31*:191–313, 2007.

Markowitz, H. A., and Gerry, R. G.: Temporomandibular joint disease. *Oral Surgery, Oral Medicine, Oral Pathology 3*:75, 1950.

Marks, M. K., and Hamilton, M. D.: Metastatic carcinoma: Palaeopathology and differential diagnosis. *International Journal of Osteoarchaeology 17*(3):217–234, 2007.

Marotta, J. J., and Micheli, L. J.: Os trigonum impingement in dancers. *American Journal of Sports Medicine 20*(5):533–536, 1992.

Marquis, J. W., Bruwer, A. J., and Kieth, H. M.: Supracondyloid process of the humerus. *Mayo Clinic Proceedings 32*(24, Nov. 27):691–697, 1957.

Martin, D. S., and Smith D. T.: Blastomycosis. *American Revue of Tuberculosis 39*:275–304, 1939.

Masciocchi, C., Catalucci, A., and Barile, A.: Ankle impingement syndromes. *European Journal of Radiology 27* Suppl 1:S70–73, 1998.

Masharawi, Y. M., Peleg, S., Albert, H. B., Dar, G., Steingberg, N., Medlej, B., Abbas, J., Salame, K., Mirovski, Y., Peled, N., and Hershkovitz, I.: Facet asymmetry in normal vertebral growth: Characterization and etiologic theory of scoliosis. *Spine 33*(8):898–902, 2008.

Masi, A. T., and Medsger, T. A.: Epidemiology of the rheumatic diseases. In McCarty, D. J. (Ed.): *Arthritis and allied conditions: A textbook of rheumatology* (11th ed.). Philadelphia, Lea and Febiger, 1989.

Massada, J. L.: Ankle overuse in soccer players: Morphological adaptation of the talus in the anterior impingement. *Journal of Sports Medicine and Physical Fitness 31*(3): 447–451, 1991.

Matsumura, G., Uchiumi, T., Kida, K., Ichikawa, R., and Kodama, G.: Developmental studies on the interparietal part of the human occipital squama. *Journal of Anatomy 182*(Pt. 2):197–204, 1993.

Mays, S. A.: *The archaeology of human bones*. London, Routledge Heritage, 1998.

_____: Effect of age and occupation on cortical bone in a group of 18th–19th century British men. *American Journal of Physical Anthropology 116*:34–44, 2001.

_____: Septal aperture of the humerus in a mediaeval human skeletal population. *American Journal of Physical Anthropology 136*(4):432–440, 2008

Mays, S., Fysh, E., and Taylor, G. M.: Investigation of the link between visceral surface rib lesions and tuberculosis in a medieval skeletal series from England using ancient DNA. *American Journal of Physical Anthropology 119*(1): 27–36, 2002.

Mays, S., Crane-Kramer, G., and Bayliss, A.: Two probable cases of treponemal disease of medieval date from England. *American Journal of Physical Anthropology 20*:133–143, 2003.

Mays, S., Strouhal, E., Vyhnanek, L., and Nemeckova, A.: A case of metastatic carcinoma of medieval date from Wharram Percy, England. *Journal of Paleopathology 8*:33–42, 1996.

Mays, S., Taylor, G. M., Legge, A. J., Young, D. B., and Turner-Walker, G.: Paleopathological and biomolecular study of tuberculosis in a medieval skeletal collection from England. *American Journal of Physical Anthropology 114*:298–311, 2001.

McCarty, D. J. (Ed.): *Arthritis and allied conditions: A textbook of rheumatology* (11th ed.). Philadelphia, Lea and Febiger, 1989.

McCormick, W. F.: Mineralization of the costal cartilages as an indicator of age: Preliminary observations. *Journal of Forensic Sciences 25*:736–741, 1980.

_____: Sternal foramina in man. *American Journal of Forensic Medicine and Pathology 2*:249–252, 1981.

McCormick, W. F., and Stewart, J. H.: Ossification patterns of costal cartilages as an indicator of sex. *Archives of Pathology and Laboratory Medicine 107*:206–210, 1985

McCormick, W. F., Stewart, J. H., and Langford, L. A.: Sex determination from chest plate roentgenograms. *American Journal of Physical Anthropology 68*(2):173–195, 1985.

McCort, J. J., and Mindelzun, R. E.: *Trauma radiology*. New York, Churchill Livingstone, 1990.

McDougall, A.: The os trigonum. *Journal of Bone and Joint Surgery 37-B*:257–265, 1955.

McKee, B. W., Alexander, W. J., and Dinbar, J. S.: Spondylolysis and spondylolistesis in children. *Journal of the Canadian Association of Radiology 22*:100, 1971.

McKern, T. W., and Stewart, T. D.: *Skeletal age changes in young American males: Analyzed from the standpoint of age identification*. Technical Report EP-45, Quartermaster Research and Development Center, Massachusetts, 1957.

McKusick, V. A.: *Human genetics* (2nd ed.). Englewood Cliffs, Prentice Hall, 1969.

Meisel, A. D., and Bullough, P. G.: *Atlas of osteoarthritis*. New York, Lea and Febiger, 1984.

Mellado, J. M., Ramos, A., Salvado, E., Camins, A., Danus, M., and Sauri, A.: Accessory ossicles and accessory bones of the foot: Imaging findings, clinical significance and differential diagnosis. *European Radiology 13*(4): L164–177, 2003.

Mensforth, R. P., Lovejoy, C. O., Lallo, J. W., and Armelagos, G. J.: The role of constitutional factors, diet, and infectious disease in the etiology of porotic hyperostosis and periosteal reactions in prehistoric infants and children. *Medical Anthropology 2*, 1-59, 1978.

Merbs, C. F.: *Patterns of activity-induced pathology in a Canadian Inuit population*. Archaeological Survey of Canada Paper, Mercury Series 119. Ottawa, National Museum of Man, 1983.

_____: Incomplete spondylolysis and healing: A study of ancient Canadian Eskimo skeletons. *Spine 20*:2328–2334, 1995.

_____: Spondylolysis of the sacrum in Alaskan and Canadian Inuit skeletons. *American Journal of Physical Anthropology 101*(3):357–367, 1996.

_____: Degenerative spondylolisthesis in ancient and historic skeletons from New Mexico Pueblo sites. *American Journal of Physical Anthropology 116*:285–295, 2001.

_____: Asymmetrical spondylolysis. *American Journal of Physical Anthropology 119*:156–174, 2002.

Merbs, C. F., and Euler, R. C.: Atlanto-occipital fusion and spondylolisthesis in an Anasazi skeleton from Bright Angle Ruin, Grand Canyon National Park, Arizona. *American Journal of Physical Anthropology 67*:381–391, 1985.

Meschan, I.: *Roentgen signs in diagnostic imaging* (2nd ed.). Philadelphia: W. B. Saunders, 1984.

Meyer, A. W: The "cervical fossa" or Allen. *American Journal of Physical Anthropology 7*(2):257–269, 1924.

Micheli, L. J., Slater, J. A., Woods, E., and Gerbino, P. G.: Patella alta and the adolescent growth spurt. *Clinical Orthopedics 213*:159–162, 1986.

Micozzi, M. S., and Kelley, M. A.: Evidence for precolumbian tuberculosis at the Point of Pines Site, Arizona: Skeletal pathology in the sacro-iliac region. In Merbs, C. F., and Miller, R. J. (Eds.): *Health and disease in the prehistoric Southwest*. Anthropological Research Papers No. 34. Tempe, Arizona State University Press, 1985.

Miedzybrodzka, Z.: Congenital talipes equinovarus (clubfoot): A disorder of the foot but not the hand. *Journal of Anatomy 202*(1):37–42, 2003.

Mihran, O., and Tachdjian, M. D.: *Atlas of pediatric orthopaedic surgery*. Philadelphia, W. B. Saunders, 1994.

Mii, S., Mori, A., Yamaoka, T., and Sakata, H.: Penetration by a huge abdominal aortic aneurysm into the lumbar vertebrae: Report of a case. *Surgery Today 29*(12):1299–1300, 1999.

Miki, T., Tamura, T., Senzoku, F., Kotani, H., Hara, T., and Masuda, T.: Congenital laminar defect of the upper lumbar spine associated with pars defect. A report of eleven cases. *Spine 16*:353–355, 1991.

Miles, J. S.: *Orthopedic problems of the Wetherill Mesa populations.* Washington, D.C., Publications in Archaeology 7G. National Park Service, 1975.

Ming-Tzu, P. A.: Septal apertures in the Chinese. *American Journal of Physical Anthropology 20*(2):165–170, 1935.

Minowa, K., Kobayashi, I., Matsuda, A., Ohmori, K., Kurokawa, Y., Inoue, N., Totsuka, Y., and Nakamura, M.: Static bone cavity in the condylar neck and mandibular notch of the mandible. *Australian Dental Journal 54*:49–53, 2009.

Mintz, G., and Fraga, A.: Severe osteoarthritis of the elbow in foundry workers. *Archives Environment Health 27*:78–80, 1973.

Mirra, J. M.: Pathogenesis of Paget's disease based on viral etiology. *Clinical Orthopaedics and Related Research 217*:162–170, 1987.

Mirra, J. M., Brien, E. W., and Tehranzadeh, J.: Paget's disease of bone: Review with emphasis on radiologic features, part I. *Skeletal Radiology 24*:163–171, 1995 (a).

_____: Paget's disease of bone: Review with emphasis on radiologic features, part II. *Skeletal Radiology 24*:173–184, 1995 (b).

Mittler, D. M., and Van Gerven, D. P.: Developmental, diachronic, and demographic analysis of cribra orbitalia in the medieval Christian populations of Kulubnarti. *American Journal of Physical Anthropology 93*(3): 287–297, 1994.

Moe, J. H., Winter, R. B., Bradford, D. S., and Lonstein, J. E.: *Scoliosis and other spinal deformities.* Philadelphia, Saunders, 1978.

Møller-Christensen, V.: *Ten lepers from Naestved in Denmark: A study of skeletons from a medieval Danish leper hospital.* Copenhagen, Danish Science Press, 1953.

_____: *Bone changes in leprosy.* Copenhagen: Munksgaard, 1961.

_____: Evidence of leprosy in earlier peoples. In Brothwell, D. R., and Sandison, A. T. (Eds.): *Diseases in antiquity: A survey of the diseases, injuries, and surgery of early populations.* Springfield, C. C. Thomas, pp. 295–306, 1967.

_____: Changes in the anterior nasal spine and alveolar process of maxilla in leprosy: A clinical examination. *International Journal of Leprosy 42*:431–435, 1974.

_____: *Leprosy changes in the skull.* Odense, Odense University Press, 1978.

_____: Leprosy and tuberculosis. In Hart, G. D. (Ed.): *Disease in ancient man: An international symposium.* Ontario, Irwin, pp. 129–138, 1983.

Møller-Christensen, V., and Sandison, A. T.: Ursa orbitae (cribra orbitalia) in the collection of crania in the anatomy department of the University of Glascow. *Pathologia et Microbiologia 26*:175–183, 1963.

Molleson T., Andrews, P., Boz, B., Sofaer Derevenski, J., and Pearson, J.: *Human remains up to 1998.* Catalhoyuk 1998 Archive Report (catal.arch.cam.ac.uk/index.html).

Molnar, P.: Tracing prehistoric activities: Musculoskeletal stress marker analysis of a stone-age population on island of Gotland in the Baltic Sea. *American Journal of Physical Anthropololgy 129*:12–23, 2006.

_____: Patterns of physical activity and material culture on Gotland Sweden, during the Middle Neolithic. *International Journal of Osteoarchaeology 20*:1–14, 2010.

Molto, J. E.: *Biological relationships of Southern Ontario Woodland peoples: The evidence of discontinuous cranial morphology.* National Museum of Man (Mercury series), Archeological Survey of Canada, Paper 117, 1983.

Monsour, P. A., and Young, W. G.: Variability of the styloid process and stylohyoid ligament in panoramic radiographs. *Oral Surgery, Oral Medicine, Oral Pathology 61*: 522–526, 1986.

Moore, K. L.: *The developing human: Clinically oriented embryology.* London, Saunders, 1982.

Moore, K. L., and Persaud, T. V. N.: *Before we are born: Essentials of embryology and birth defects* (6th ed.). Philadelphia, Saunders, 2003.

Moore, R. M., and Green, N. E.: Blastomycosis of bone: A report of six cases. *Journal of Bone and Joint Surgery 64-A* (7):1097–1101, 1982.

Moore, S.: *Hyperostosis cranii.* Springfield, C. C. Thomas, 1955.

Moore, S. L., and Rafii, M.: Imaging of musculoskeletal and spinal tuberculosis. *Radiologic Clinics of North American 39*:329–342, 2001.

Moore, T. E., Kathol, M. H., El-Khoury, G. Y., Walker, C. W., Gendall, P. W., and Whitten, C. G.: Unusual radiological features in Paget's disease of bone. *Skeletal Radiology 23*:257–260, 1994.

Moore, T. E., King, A. R., Kathol, M. H., El-Khoury, G. Y., Palmer, R., and Downey, P. R.: Sarcoma in Paget disease of bone: Clinical, radiologic and pathologic features in 22 cases. *American Journal of Radiology 156*: 1199–1203, 1990.

Morrissy, R. T., Weinstein, S. L., and Kida, B.: *Atlas of pediatric prthopaedic surgery.* Philadelphia, J. B. Lippincott, 2000.

Morse, D.: Prehistoric tuberculosis in America. *American Review of Respiratory Disease 83*:489–504, 1961.

_____: *Ancient disease in the Midwest.* Illinois State Museum Reports of Investigations, No. 15, 1969.

Morton, N. E.: Birth defects in racial crosses. In *Congenital Malformation: Proceedings of the Third International Conference.* New York, Excerpta Medica, 1970.

Mosely, J. E.: The paleopathologic riddle of "symmetrical osteoporosis." *American Journal of Roentgenology 95*(1): 135–142, 1965.

Moseley, J. E.: *Bone changes in hematologic disorders.* New York, Grune and Stratton, 1963.

Moskowitz, R. W., Howell, D. S., Goldberg, V. M., and Mankin, H. J.: *Osteoarthritis: Diagnosis and management.* Philadelphia, Saunders, 1984.

Moss, M. L.: The pathogenesis of artificial cranial deformation. *American Journal of Physical Anthropology 16*:269–286, 1958.

Muckley, T., Schutz, T., Kirschner, M., Potulski, M., Hofmann, G., and Buhren, V.: Psoas abscess: The spine as a primary source of infection. *Spine 28*(6):E106–113, 2003.

Mudge, M. K., Wood, V. E., and Frykman, G. K.: Rotator cuff tears associated with os acromiale. *Journal of Bone and Joint Surgery 66-A*(3):427–429, 1984.

Mulhern, D. M., Wilczak, C. A. and Dudar, J. C.: Brief communication: Unusual finding at Pueblo Bonito: Multiple cases of hyperostosis frontalis interna. *American Journal of Physical Anthropology 130*(4):480–484, 2006.

Muller, F., O'Rahilly, R., and Benson, D. R.: The early origin of vertebral anomalies, as illustrated by a "butterfly vertebra." *Journal of Anatomy 149*:157–169, 1986.

Murphy, J., and Gooding, C. A.: Evolution of persistently enlarged parietal foramina. *Radiology 97*:391–392, 1970.

Murphy, S. P., and Mann, R. W.: Cortical defects of the proximal humerus: An indicator of physical stress. *American Journal of Physical Anthropology 81*(2):273, 1990.

Murray, J. F., Merriweather, A. M., and Freedman, M. L.: Endemic syphilis in the Bakwena Reserve of the Bechuanaland protectorate. *Bulletin of World Health Organization 15*:975–1039, 1956.

Murray-Leslie, D. F., Lintott, D. J., and Wright, V.: The knees and ankles in sport and veteran military parachutists. *Annals Rheumatic Disease 36*:327–337, 1977.

Myers, W. J.: Anterior ankle impingement exostoses. *Journal of American Podiatric Medical Association 77*(7):347–350, 1987.

Naffsiger, H. C., Inman, V., and Saunders, J.: Lesions of intervertebral disc and ligamenta flava: Clinical and anatomical studies. *Surgery, Gynecology and Obstetrics 66*:288, 1938.

Nagy, B. L. B.: Age, activity, musculoskeletal stress markers. *American Journal of Physical Anthropology, Supplement 26*:168–169 [abstract], 1998.

Nagy, B. L. B., and Hawkey, D. E.: Musculoskeletal stress markers as indicators of sexual division of labor: Multivariate analyses. *American Jouornal of Physical Anthropololgy, Supplement 20*:158 [abstract], 1995.

Nathan, H.: Spondylolysis: Its anatomy and mechanism of development. *Journal of Bone and Joint Surgery 41-A*:303–320, 1959.

Nakashima, T., Hojo, T., Suzuki, K., and Ijichi, M.: Symphalangism (two phalanges) in the digits of the Japanese foot. *Annals of Anatomy 177*(3):275–278, 1995.

Nelsen, K., Tayles, N., and Domett, K.: Missing lateral incisors in Iron Age South-East Asians as possible indicators of dental agenesis. *Archives Oral Biology 46*(10):963–971, 2001.

Neugebauer, F. L.: The classic: A new contribution to the history and etiology of spondylolisthesis. In Urist, M. R. (editor-in-Chief): *Clinical orthopaedics and related research* (No. 107). Philadephia, J. B. Lippincott Company, pp. 4–22, 1976.

Neumann, G. K.: Evidence for the antiquity of scalping from Central Illinois. *American Antiquity 5*:287–289, 1940.

Newman, P. H.. and Stone K. H.: The etiology of spondylolysis. *Journal of Bone and Joint Surgery 45-B*(1):39–59, 1963.

Nielsen, S. P., Xie, X., and Barenholdt, O.: Geometric properties of distal radius and pathogenesis of Colles fracture: A peripheral quantitative computed tomography study. *Journal of Clinical Densitometry 4*(3):209–219, 2001.

Niepl, G. A., and Sit'aj, S.: Enthesopathy. *Journal of Clinical Rheumatology 5*:857–872, 1979.

Niswander, J. D., Barrow, M. V., and Bingle, G. J.: Congenital malformations in the American Indian. *Social Biology 22*(3):203–215, 1975.

Norman, A.: Roentgenologic diagnosis. In Moskowitz, R. W., Howell, D. S., Goldberg, V. M., and Mankin, H. J. (Eds.): *Osteoarthritis: Diagnosis and management.* Philadelphia, Saunders, 1984.

Norman, A. P.: *Congenital abnormalities in infancy.* Philadelphia, F.A. Davis, 1963.

Nowak, O., and Piontek, J.: The frequency of appearance of transverse (Harris) lines in the tibia in relationship to age at death. *Annals of Human Biology 29*(3):314–325, 2002 (a).

_____: Does the occurrence of Harris lines affect the morphology of human long bones? *Homo 52*(3):254–276, 2002 (b).

Nugent, C. A., Gall, E. P., and Pitt, M. J.: Osteoporosis, osteomalacia, rickets and Paget's disease. *Primary Care 11*(2):353–368, 1984.

Nyland, H., and Krogness, K. G.: The normal calvaria indices of the human skull. 1. Absolute measurements. 2. Proportional measurements. *Pediatric Radiology 7*(1):1–3, 1978.

O'Beirne, J. G., and Horgan, J. G.: Stress fracture of the lamina associated with unilateral spondylolysis. *Spine 13*:220–222, 1988.

Oettenking, B.: Anomalous patellae. *Anatomical Record 23*:269–279, 1922.

Ogden, A. R., Pinhasi, R., and White, W. J.: Gross enamel hypoplasia in molars from subadults in a 16th–18th century London graveyard. *American Journal of Physical Anthropology 133*:957–966, 2007.

Ogden, J. A., McCarthy, S. M., and Jokl, P.: The painful bipartite patella. Journal of Pediatric *Orthopedics 2*(3):263–269, 1982.

Ohmori, K. Ishida, Y., Takatsu, T., Inque, H., and Suzuki, K.: Vertebral slip in lumbar spondylolysis and spondylolisthesis: Long-term follow-up of 22 adult patients. *Journal of Bone and Joint Surgery 77-B*(5), 771– 773, 1995.

Okumura, M. M., Boyadjian, C. H. C., and Eggers, S.: Auditory exostoses as an aquatic activity marker: A comparison of coastal and inland skeletal remains from tropical and subtropical regions of Brazil. *American Journal of Physical Anthropology 132*:558–567, 2007.

Olivieri, I., Barozzi, L., and Padula, A.: Enthesopathy: Clinical manifestations, imaging and treatment. *Bailliere's Clinical Rheumatology 12*(4):665–681, 1998.

Oloff, L. M., Schulhofer, S. D., and Bocko, A. P.: Subtalar joint arthroscopy for sinus tarsi syndrome: A review of 29 cases. *Journal of Foot and Ankle Surgery 40*(3):152–157, 2001.

Olmsted, W. W.: Some skeletogenic lesions with common calvarial manifestations. *Radiologic Clinics of North America 19*:703–713, 1981.

Omey, M. L., Micheli, L. J., and Gerbino, P. G.: Idiopathic scoliosis and spondylolysis in the female athlete: Tips for treatment. *Clinical Orthopedics 372*:74–84, 2000.

Omnell, K. A., Gandhi, C., and Omnell, M. L.: Ossification of the human stylohyoid ligament: A longitudinal study. *Oral Surgery Oral Medicine Oral Pathology Oral Radiology Endodontics 85*(2):226–232, 1998.

O'Neal, M. L., Dwornik, J. J., Ganey, T. M., and Ogden, J. A.: Postnatal development of the human sternum. *Journal of Pediatric Orthopedics, 18*(3):398–405, 1998.

O'Neill, D. B., and Micheli, L. J.: Overuse injuries in the young athlete. *Clinical Sports Medicine 7*(3):591–610, 1988.

O'Rahilly, R. M., and Muller, F.: *Human embryology and teratology* (2nd ed.). New York, Wiley-Liss, 1996.

O'Rahilly, R. M., and Twohig, M.: Foramina parietalia permagna. *American Journal of Roentgenology 67*:551–561, 1952.

Ortner, D. J.: Descriptions and classifications of degenerative bone changes in the distal joint surfaces of the humerus. *American Journal of Physical Anthropology 28*:139–155, 1968.

_____: Observations on the pathogenesis of skeletal disease in leprosy. In Roberts, C. A., Lewis, M. E. and Manchester, K. (Eds.): *The past and present of leprosy: Archaeological, historical, paleopathological and clinical approaches.* Oxford, Archaeopress, pp. 73–80, 2002.

_____: *Identification of pathological conditions in human skeletal remains.* Washington, D.C., Smithsonian Institution Press, 2003.

_____: Skeletal manifistations of leprosy. In Magilton, J., Lee, F., and Boylston, A.: *Lepers outside the gate. Chichester Excavations v. 10, CBA Research Report 158.* Council for British Archaeology, York: Council for British Archaeology, pp. 198–207, 2008.

_____: Differential diagnosis of skeletal lesions in infectious disease. In Pinhasi, R., and Mays, S. (Eds.): *Advances in human paleopathology.* Sussex, John Wiley & Sons, Ltd., 2008.

Ortner, D. J., and Aufderheide, A. C (Eds.): *Human paleopathology: Current syntheses and future options.* Washington, D. C., Smithsonian Institution Press, 1991.

Ortner, D. J., Butler, W., Cafarella, J., and Milligan, L.: Evidence of probable scurvy in subadults from archeological sites in North America. *American Journal of Physical Anthropology 114*(4):343–351, 2001.

Ortner, D. J., and Putschar, W. G. J.: *Identification of pathological conditions in human skeletal remains.* Washington, D.C., Smithsonian Institution Press, 1985.

Ortner, D. J., and Turner-Walker, G.: The biology of skeletal tissues. In Ortner, D. J.: *Identification of pathological conditions in human skeletal remains* (2nd ed.). Amsterdam, Academic Press, 2003: pp. 11–35.

Outerbridge, R. E.: Further studies on the etiology of chondromalacia patellae. *The Journal of Bone and Joint Surgery 46-B*(2):179–190, 1964.

Owsley, D. W., and Mann, R. W.: An Amercian Indian skeleton with clubfoot from the Cabin Burial Site (A1184), Hemphill County, Texas. *Plains Anthropologist 35*:93–101, 1990.

Owsley, D. W., Orser, C. E., Mann, R. W., Moore-Jansen, P. H., and Montgomery, R. L.: Demography and pathology of an urban slave population from New Orleans. *American Journal of Physical Anthropology 74*:185– 197, 1987.

Oygucu, I. H., Kurt, M. A., Ikiz, I., Erem, T., and Davies, D. C.: Squatting facets of the talus and extensions of the trochlear surface of the talus in late Byzantine males. *Journal of Anatomy 192*(Pt. 2):287–291, 1998.

Ozbek, M.: Cranial deformation in a subadult sample from degirmentepe (Chalcolithic, Turkey). *American Journal of Physical Anthropology 115*:238–244, 2001.

Ozek, M. M., Cinalli, G., and Maixner, W. J. (Eds.): *Spina bifida: Management and outcome.* Milan, Springer-Verlag Italia, 2008.

Paget, J.: On the production of some of the loose bodies in the joints. *St. Bartholomew's Hospital Report 6*:1, 1870.

_____: On a form of chronic inflammation of bones (osteitis deformans). *Transactions of the Medical Chirurgical Society 60*:37–64, 1877.

Pálfi, G., Dutour, O., Dedk, J., and Hutas, I. (Eds.): *Tuberculosis: Past and present.* Tuberculosis Foundation, Budapest, Szeged and Golden Book, 1999.

Palkovich, A. M.: Endemic disease patterns in paleopathology: Porotic hyperostosis. *American Journal of Physical Anthropology 74*:526–537, 1987.

Palmer, P. E. S., and Reeder, M. M.: *The imaging of tropical diseases: With epidemiological, pathological and clincal correlation* (Vol. 2.). Berlin, Springer-Verlag, 2001.

Pandey, S. K., and Singh, S.: Study of squatting facet/extension of the talus in both sexes. *Medical Science and Law 30*(2):159–164, 1990.

Pang, D., and Lin, A.: Symptomatic large parietal foramina. *Neurosurgery 11*:33–37, 1982.

Panush, R. S., Schmidt, C., Caldwell, J. R., Edwards, N. L., Longley, S., Yonker, R., Webster, E., Nauman, J., and Pettersson, H.: Is running associated with degenerative joint disease? *Journal of American Medical Association 255*:1152–1154, 1986.

Papadakis, M., Papadokostakis, G., Kampanis, N., Sapkas, G., Papadakis, S. A., and Katonis, P.: The association of spinal osteoarthritis with lumbar lordosis. *BMC Musculoskeletal Disorders 11*:1, doi: 10.1186/1471-2474-11-1, 2010.

Papadopoulos, C. C.: Temporal variation and sex differences in the incidence of cranial porotic hyperostosis in Peru. *Paleopathology Newsletter 9*:11–14, 1977.

Paparella, M. M., Goycoolea, M. V., and Meyerhoff, W. L.: Inner ear pathology and otitis media: A review. *Annals of Otology Rhinology and Laryngology 89*(68):249–253, 1980.

Paraskevas, G., Natsis, K., Spanidou, S., Tzaveas, A., Kitsoulis, P., Raikos, A., Papaziogas, B., and Anastasopoulos, N.: Excavated-type of rhomboid fossa of the clavicle: A radiological study. *Folia Morphol (Warsz) 68*(3):163–6, 2009.

Parashuram, R.: *Morphometrical study of sacral hiatus using isolated dry human sacra.* Doctoral dissertation, Bangalore Medical College and Research Institute, Bangalore, 2008.

Park, D. W., Sohn, J. W., Kim, E-H., Cho, D-Il., Lee, J-H., Kim, K-T., Ha, K-Y., Jeon, C-H., Shim, D-M., Lee, J-S., Lee, J-B., Chun, B. C., and Kim, M. J.: Outcome and management of spinal tuberculosis according to the severity of disease: A retrospective study of 137 adult patients at Korean teaching hospitals. *Spine 32*(4):E130–E135, doi: 10.1097/01.brs.0000255216.54085.21, 2007.

Park, E. A.: The imprinting of nutritional disturbances on the growing bone. *Pediatrics 33*:815–862, 1964.

Park, J. G., Lee, J. K., and Phelps, C.T.: Os acromiale associated with rotator cuff impingement: MR Imaging of the shoulder. *Musculoskeletal Radiology 193*:255–257, 1994.

Parkes, J. C. II, Hamilton, W. G., Patterson, A. J., and Rawles, J. G. Jr.: The anterior impingement syndrome of the ankle. *Journal of Trauma 20*(10):895–898, 1980.

Parkinson, C. E.: The supracondyloid process. *Radiology 62*:556–558, 1954.

Parsons, F. G.: On the proportions and characteristics of the modern English clavicle. *Journal of Anatomy 51*:71, 1916.

Patankar, T., Krishnan, A., Patkar, D., Kale, H., Prasad, S., Shah, J., and Castillo, M.: Imaging in isolated sacral tuberculosis: A review of 15 cases. *Skeletal Radiology 29*:392–396, 2000.

Paterson, D. E.: Bones changes in leprosy. *Leprosy in India 28*:128, 1965.

Paterson, D. E., and Job, C. K.: Bone changes and absorption in leprosy. In Cochrane, R. G., and Davey, T. F. (Eds.): *Leprosy in theory and practice.* Bristol, John Wright and Sons, 1964.

Patil, K. M., and Jacob, S.: Mechanics of tarsal disintegration and plantar ulcers in leprosy by stress analysis in three dimensional foot models. *Indian Journal of Leprosy 72*(1):69–86, 2000.

Patni, V. M., Gadewar, D. R., and Pillai, K. G.: Ossification of stylohyoid ligament with pseudojoint formation–A case report. *Journal of Indian Dental Association 58*(6):227–231, 1986.

Pavithran, K.: Saber tibiae in lepromatous leprosy. *International Journal of Leprosy and Other Mycobaterial Disorders 58*(2):385–387, 1990.

Pearson, O. M., and Lieberman, D. E.: The aging of Wolff's "Law": Ontogeny and responses of mechanical loading in cortical bone. *Yearbook of Physical Anthropology 47*:63–99, 2004.

Pecina, M. M., and Bojanic, I.: *Overuse injuries of the musculoskeletal system* (2nd ed.). Boca Raton, CRC Press, 2004.

Pedersen, A. K., and Hagen, R.: Spondylolysis and spondylolisthesis. *Journal of Bone and Joint Surgery 70-A*(1):15–24, 1988.

Peng, B., Wu, W., Hou, S., Shang, W., Wang, X., and Yang, Y. The pathogenesis of Schmorl's nodes. *Journal of Bone and Joint Surgery 85(B)*:879–882, 2003.

Pepper, O. H. P., and Pendergrass, E. P.: Heredity occurrence of enlarged parietal foramina. *American Journal of Roentgenology 35*(1):1–8, 1936.

Percy, E. C., and Mann, D. L.: Tarsal coalition: A review of the literature and presentation of 13 cases. *Foot and Ankle 9*(1):40–44, 1988.

Perou, M.: *Cranial hyperostosis.* Springfield, C. C. Thomas, 1964.

Peterson, J.: The Natufian hunting conundrum: Spears, atlatls, or bows? Musculoskeletal and armature evidence. *International Journal of Osteoarchaeology 8*:378–389, 1998.

Peterson, J.: *Sexual revolutions: Gender and labor at the dawn of agriculture.* Altamira Press, Walnut Creek, CA, 2002.

Petra, H., and Villotte, S.: Enthesopathies: Test of reproducibility of new scoring system based on current medical data. *Slovenská anthropológia 10*(1):51–57, 2007.

Pfeiffer, S.: Age changes in the external dimensions of adult bone. *American Journal of Physical Anthropology 52*:529–532, 1980.

_____: Morbidity and mortality in the Uxbridge Ossuary. *Canadian Review of Physical Anthropology 4*(2):23–31, 1985.

_____: Rib lesions and New World tuberculosis. *International Journal of Osteoarchaeology 1*:191–198. 1991.

Pfirrmann, C. W. A., and Resnick, D.: Schmorl nodes of the thoracic and lumbar spine: Radiographic-pathologic study of prevalence, characterization, and correlation with degenerative changes of 1,650 spinal levels in 100 cadavers. *Radiology 219*(2):368–374, 2001.

Philipsen, H. P., Takata, T., Reichart, P. A., Sato, S., and Suei, Y.: Lingual and buccal mandibular bone depressions: A review based on 583 cases from a world-wide literature survey, including 69 new cases from Japan. *DentoMaxilloFacial Radiology 31*:281–290, 2002.

Piek, J., Lidke, G., and Terberger, T.: Ancient trephinations in neolithic people–Evidence for stone age neurosurgery? *Nature Proceedings hdl*: 10101/npre.2008. 1615. 1: Posted 21 Feb 2008 (accessed online 22 March 2011).

Pikula, J. R.: Supracondyloid process of the humerus: a case report. *Journal of Canadian Chiropractic Association 38*(4): 211–215, 1994.

Pindborg, J. J.: *Pathology of the dental hard tissue.* Philadelphia, Saunders, 1970.

Pineda, C., Mansilla-Lory, J., Martínez-Lavín, M., Leboreiro, I., Izaguirre, A., and Pijoan, C.: Rheumatic diseases in the ancient Americas: The skeletal manifestations of treponematoses. *Journal of Clinical Rheumatology 15*(6): 280–283, 2009.

Pinhasi, R., and Mays, S. (Eds.): *Advances in human paleopathology.* Sussex, John Wiley and Sons, Ltd., 2008.

Piotrowski, B. T., Gillette, W. B., and Hancock, E. B.: Examining the prevalence and characteristics of abfraction-like cervical lesions in a population of U.S. veterans. *Journal of American Dental Association 132*(12): 1694–1701, 2001.

Pitt, M. J.: Rachitic and osteomalacic syndromes. *Radiologic Clinics of North America 19*(4):581–599, 1981.

Pitt, M. J.: Rickets and osteomalacia are still around. *Radiologic Clinics of North America 29*(1):97–118, 1991.

Plotz G. M., Rrymka, M., Knoch, M. V., Markova, B., and Ulrich, H. W.: Spontaneous fusion of "os acetabuli" after triple pelvic osteotomy. *Archives Orthopaedics Trauma Surgery 122*(9–10):526–529, 2002.

Poirier, P.: *Traite' d'anatomie humaine 1*:515. Paris, Poireir & Charpy, 1911.

Pommerville, J. C.: *Alcamo's fundamentals of microbiology.* Sudbury, Jones and Bartlett Publishers, LLC, 2010.

Ponec, D. J., and Resnick, D.: On the etiology and pathogenesis of porotic hyperostosis of the skull. *Investigative Radiology 19*(4):313–317, 1984.

Ponseti, I. V.: Growth and development of the acetabulum in the normal child: Anatomical, histological, and roentgenographic studies. *Journal of Bone and Joint Surgery 60:* 575–585, 1978.

_____: Treatment of congenital clubfoot. *Journal of Orthopaedics Sports Physical Therapy 20*(1):1, 1994.

Ponseti, I. V., El-Khoury, G. Y., Ippolito, E., and Weinstein, S. L.: A radiographic study of skeletal deformities in treated clubfeet. *Clinical Orthopaedics and Related Research 160*:30–42, 1981.

Popowsky, J.: Bifurcated extremities of the ribs. *Anatomischen Anzeiger 75*:284, 1918.

Porter, R. W., and Park, W.: Unilateral spondylolysis. *Journal of Bone and Joint Surgery 64-B*(3):344–348, 1982.

Posch, T. J., and Puckett, M. L.: Marrow MR signal abnormality associated with bilateral avulsive cortical irregularities in a gymnast. *Skeletal Radiology 27*:511–514, 1998.

Postacchini, F., and Rauschning, W.: Anatomy, pp. 17–55. In Postacchini, F. (Ed.): *Lumbar disc herniation.* New York, Springer-Verlag, 1999.

Poswall, B. D.: Coccidioidomycosis and North American blastomycosis: Differential diagnosis of bone lesions in pre-Columbian American Indians. *American Journal of Physical Anthropology 44*:199–200, 1976.

Powell, M. L., and Cook, D. C.: Treponematosis: inquiries into the nature of a protean disease. In Powell, M. L., and Cook, D. C. (Eds.): *The myth of syphilis: The natural history of treponematosis in North America.* Gainesville, University Press of Florida, pp. 9–62, 2005.

Poznanski, A. K.: *The hand and radiologic diagnosis.* Philadelphia, Saunders, 1974.

Prado, F. B., de Mello Santos, L. S., Caria, P. H. F., Kawagucki, J. T., Preza, A. d'O. G., Daruge, Jr. E., da Silva, E. F., and Daruge, E.: Incidence of clavicular rhomboid fossa (impression for costoclavicular ligament) in the Brazilian population: Forensic application. *Journal Forensic Odontostomatologia 27*(1):12–16, 2009.

Prasad, K. C., Kamath, M. P., Reddy, K. J., Raju, K., and Agarwal, S.: Elongated styloid process (Eagle's syndrome): A clinical study. *Journal of Oral Maxillofacial Surgery 60*(2):171–175, 2002.

Procknow, J. J., and Loosli, C. G.: Treatment of the deep mycosis. *American Medical Association Archives of Internal Medicine 101*:765–802, 1958.

Prokopec, M., Simpson, D., Morris, L., and Pretty, G.: Craniosynostosis in a prehistoric aboriginal skull: A case report. *OSSA 9/11*:111–118, 1982–1984.

Prusick, V. R., Samberg, L. C., and Wesolowski, D. P.: Klippel-Feil syndrome associated with spinal stenosis: A case report. *Journal of Bone and Joint Surgery 67-A*:161–164, 1985.

Pulec, J. L., and Deguine, C.: Osteoma of the external auditory canal. *Ear Nose and Throat Journal 79*(12):908, 2000.

_____: Nonobstructing exostoses of the auditory canal. *Ear Nose and Throat Journal 80*(2):66, 2001a.

_____: Exostoses of the external auditory canal. *Ear Nose and Throat Journal 80*(4):190, 2001b.

Puranen, J., Ala-Ketola, L., Peltokallio, P., and Saarela, J.: Running and primary osteoarthritis of the hip. *British Medical Journal 2*:424–425, 1975.

Purves, R. K., and Wedin, P. H.: Familial incidence of cervical ribs. *Journal of Thoracic Surgery 79*:952–956, 1950.

Pusey, W. A.: The beginning of syphilis. *Journal of American Medical Association 44*:1961–1964, 1915.

Pynn, B. R., Kurys-Kos, N. S., Walker, D. A., and Mayhall, J. T.: Tori mandibularis: A case report and review of the literature. *Journal Canadian Dental Association 61*(12): 1957–1058, 1063–1066, 1995.

Quatrehomme, G., and İşcan, Y. M.: Characteristics of gunshot wounds in the skull. *Journal of Forensic Sciences 44*(3):568–576, 1999.

Queneau, P., Gabbai, A., Perpoint, B., Salque, J. R., Laurent, H., Decousus, H., and Boucheron, S.: Acro-osteolysis in leprosy. Apropos of 19 personal cases (Article in French). *Rev Rhum Mal Osteoartic 49*(2):111–119, 1982.

Quesada-Gómez, C., Valmaseda-Castellón, E., Berini-Aytés, L., Gay-Escoda, C.: Stafne bone cavity: A retrospective study of 11 cases. *Med Oral Patol Oral Cir Bucal 11:* E277–280, 2006.

Radin, E. L., Parker, H. C., and Paul I. L.: Pattern of degenerative arthritis. Preferential involvement of distal finger joints. *Lancet 1*:377–379, 1971.

Ragsdale, B. D., Madewell, J. E., and Sweet, D. E.: Radiologic and pathologic analysis of solitary bone lesions. Part II: Periosteal reactions. *Radiologic Clinics of North America 19*:749–783, 1981.

Raina, S., Kaushal, S. S., Gupta, D., Goyal, A., and Sood, V.: Charcot's knee joints. *Journal of Physicians India 55*:786, 2007.

Raisz, L. G.: Osteoporosis. *Journal of the American Geriatrics Society 30*(2):127–137, 1982.

_____: Physiology and pathophysiology of bone remodeling. *Clinical Chemistry 45*(8):1353–8, 1999.

Rankine, J. J., and Dickson, R. A. Unilateral spondylolysis and the presence of facet joint tropism. *Spine 35*(21): E111–E114, 2010.

Rasore-Quartino, A., Vignola, G., and Camera, G.: Hereditary enlarged parietal foramina (foramina parietalia permagna): Prenatal diagnosis, evolution and family study. *Pathologica 77*(1050):449–455, 1985.

Rasool, M. N., and Govender, S.: The skeletal manifestations of congenital syphilis. A review of 197 cases. *Journal of Bone and Joint Surgery 71-B*(5):752–755, 1989.

Rateitschak, K. H., and Wolf, H. F. (Eds.).: *Color atlas of dental medicine: TMJ disorders and orofacial pain: The role of dentistry in multidisciplinary diagnostic approach.* Stuttgart, Georg Thieme Verlag, 2002.

Rau, R. K., and Sivasubrahmanian, D.: Supracondyloid process. *Journal of Anatomy 65*:392–394, 1931.

Ravichandran, G.: Upper lumbar spondylolysis. *International Orthopedics 5*:31–35, 1981.

Rees, J. S.: The biomechanics of abfraction. Proceedings of the Institute of Mechanical Engineers, Part H. *Journal of Engineering in Medicine 220*(1):69–80, 2006.

Reeves, R. J., and Pederson, R: Fungous infection of bones. *Radiology 62*:55, 1954.

Reichart, P.: Facial and oral manifestations in leprosy. An evaluation of seventy cases. *Oral Surgery Oral Medicine Oral Pathology 31*(3):385–399, 1976.

Reid, D. J., and Dean, M. C.: Brief communication: Timing of linear enamel hypoplasia on human anterior teeth. *American Journal of Physical Anthropology 113*:135–139, 2000.

Resnick, D.: *Diagnosis of bone and joint disorders* (4th ed.). Oxford, UK: Elsevier Science, 2002.

Resnick, D., and Greenway, G.: Distal femoral cortical defects, irregularities, and excavations. *Radiology 143*(2): 345–354, 1982.

Resnick, D., and Niwayama, G.: Radiographic and pathologic features of spinal involvement in diffuse idiopathic skeletal hyperostosis (DISH). *Radiology 119*(3):559–368, 1976.

_____: Intravertebral disk herniations: cartilaginous (Schmorl's) nodes. *Radiology 126*(1):57–65, 1978.

_____: *Diagnosis of bone and joint disorders.* Philadelphia, Saunders, 1981.

_____: Anatomy of individual joints. In Resnick, D., and Niwayama, G. (Eds.): *Diagnosis of bone and joint disorders.* Philadelphia, Saunders, 1981.

_____: "Entheses and enthesopathy." anatomical, pathological and radiological correlation. *Radiology 146*:1–9, 1983.

_____: *Diagnosis of bone and joint disorders* (2nd ed.). Philadelphia, Saunders, 1988.

Resnick, D., Niwayama, G., and Goergen, T. G.: Comparison of radiographic abnormalities of the sacroiliac joint in degenerative disease and ankylosing spondylitis. *American Journal of Roentgenology 128*(2):189–196, 1977.

Restrepo, S., Palacios, E., and Rojas, R.: Eagle's syndrome. *Ear Nose Throat Journal 81*(10):700–701, 2002.

Rhine, S.: Non-metric skull racing. In Gill, G. W., and Rhine, S. (Eds.): *Skeletal attribution of race.* Maxwell Museum of Anthropology, Anthropological Papers No. 4:9–21, 1990.

Rhode, M. P.: *Habitual subsistence practices among prehistoric Andean populations: Fishers and farmers.* Unpublished Ph.D. dissertation, University of Missouri-Columbia, 2006.

Rhode, M. P., and Arriaza, B. T.: Influence of cranial deformation on facial morphology among prehistorc South Central Andean populations. *American Journal of Physical Anthropology 130*:462–470, 2006.

Richards, L. C.: Temporomandibular joint morphology in two Australian aboriginal populations. *Journal Dental Research 66*:1602–1607, 1987.

_____: Degenerative changes in the temporomandibular joint in two Australian aboriginal populations. *Journal Dental Research 67*:1529–1533, 1988.

Richards, L. C., and Brown, T.: Dental attrition and degenerative arthritis of the temporomandibular joint. *Journal Oral Rehabilitation 8*:293–307, 1981.

Richardson, E. G.: Tarsal coalition. In Myerson, M. S.: *Foot and ankle disorders* (Vol. 1). Philadelphia, Saunders, pp. 729–748, 2000.

Richardson, M. L.: Chinese foot binding: Radiographic findings and case report. *Radiology Case Reports 4*(1):270, (doi: 10.2484/rcr.v4i1.270, 2009).

Riepert, T., Drechsler, T., Urban, R., Schild, H., and Mattern, R.: The incidence, age dependence and sex distribution of the calcaneal spur. An analysis of its x-ray morphology in 1027 patients of the central European population (article in German; abstract in English). *Fortschritte auf dem Gebiete der Rontgenstrahlen und der Neuen Bildgebenden Verfahren (Stuttgart) 162*(6):502–505, 1995.

Rifkinson-Mann, S.: Cranial surgery in ancient Peru. *Neurosurgery 23*(4):411–416, 1988.

Rippon, J. W.: *Medical mycology: The pathogenic fungi and the pathogenic actinomycetes* (3rd ed.). Philadelphia, Saunders, 1988.

Riseborough, E. J.: Scoliosis in adults. *Current Practice in Orthopaedic Surgery 7*:36-55, 1977.

Ritschl, P., Hajek, P. C., and Pechmann, U.: Fibrous metaphyseal defects. Magnetic resonance imaging appearances. *Skeletal Radiology 18*:253–259, 1989.

Ritschl, P., Karnel, F., and Hajek, P.: Fibrous metaphyseal defects–determination of their origin and natural history using a radiomorphological study. *Skeletal Radiology 17*:8–15, 1988.

Robb, J. E.: Skeletal signs of activity in the Italian metal ages: Methodological and interpretive notes. *Human Evolution 9*(3):215–229, 1994.

Robb, J. E.: The interpretation of skeletal muscle sites: A statistical approach. *International Journal of Osteoarchaeology 8*:354–377, 1998.

Robb, J., Bigazzi, R., Lazzarini, L., Scarsini, C., and Sonego, F.: Social "status" and biological "status": A comparison of grave goods and skeletal indicators from Pontecagnano. *American Journal of Physical Anthropology, 115*(3): 213–222, 2001.

Robbins, S. L.: *Pathology* (3rd ed.). Philadelphia, Saunders, 1968.

Robinson, H. B. G., and Miller, A. S.: *Color atlas of pathology* (4th ed.). Philadelphia, J. B. Lippincott, 1983.

Roberts, C. A., Boylston, A., Buckely, L., Chamberlain, A. C., and Murphy, E. M.: Rib lesions and tuberculosis: The palaeopathological evidence. *Tubercle Lung Disease 79*:55–60, 1996.

Roberts, C., Lucy, D., and Manchester, K.: Inflammatory lesions of ribs: An analysis of the Terry Collection. *American Journal of Physical Anthropology 95*:169–182, 1994.

Roche, M. B.: The pathology of neural-arch defects. *Journal of Bone and Joint Surgery, 31-A*:529–537, 1949.

Roche, M. B., and Rowe, G. G.: Anomalous centers of ossification for inferior articular processes of the lumbar vertebrae. *Anatomical Record 109*(2):253–259, 1951.

_____: The incidence of separate neural arch and coincident bone variation. *Anatomical Record 109*:233–252, 1953.

Rockwood, C. A.: *Fractures in children* (4th ed.). Philadelphia, Saunders, 1989.

Rockwood, C. A., and Green, D. P. (Eds.): *Fractures in adults.* Philadelphia, Lippincott, Vols. 1–3, 1975.

Rockwood, C. A., and Matsen, F. A. III.: *The shoulder* (2nd ed., Vols. 1 and 2). Philadelphia, Saunders, 1998.

Roesler H.: The history of some fundamental concepts in bone biomechanics. *Journal of Biomechanics 20*(11, 12): 1025–1034, 1987.

Rogers, J., Shepstone, L., and Dieppe, P.: Bone formers: Osteophyte and enthesophyte formation are positively associated. *Annals Rhuematic Disease 56*(2):85–90, 1997.

Rogers, L. F.: *Radiology of skeletal trauma.* New York, Churchill Livingstone, Vols. 1–2, 1982.

Rogers, N. L., Flournoy, L. E., and McCormick, W F.: The rhomboid fossa of the clavicle as a sex and age estimator. *Journal of Forensic Science 45*(1):61–67, 2000.

Rogers, J., and Waldron, T.: The paleopathology of enthesopathy. In Capasso, L. (Ed.): Advances in paleopathology. *Proceedings of the VIIth European Meeting of the Paleopathology Association, Lyon, September 1988*, pp. 169–173, 1989.

Rogers, J., and Waldron, T.: *A field guide to joint disease in archaeology.* New York, Wiley, 1995.

Rogers, S., and Merbs, C. F.: *Trephined skulls* (42 slide set). San Diego Museum of Man, 1980.

Rom, W. N., Garay, S. M., and Bloom, B. R.: *Tuberculosis* (2nd ed.). Philadelphia, Lippincott, Williams and Wilkins, 2004.

Rosenberg, N. J., Bargar, W. L., and Friedman, B.: The incidence of spondylolysis and spondylolisthesis in nonambulatory patients. *Spine 6*:35–38, 1981.

Roser, L. A., and Clawson, D. K.: Football injuries in the very young athlete. *Clinical Orthopedics and Related Research 69*:219–223, 1970.

Ross, A. H.: Caliber estimation from cranial entrance defect measurements. *Journal of Forensic Sciences 41*(4):629–633, 1996.

Rothschild, B.: Porotic hyperostosis as a marker of health and nutritional conditions. *American Journal of Human Biology 13*(6):709–717, 2001.

Rothschild, B. M.: History of syphilis. *Clinical Infectious Diseases 40*:1454–1463, 2005.

Rothchild, B., and Rothchild, C.: Treponemal disease revisited: Skeletal discriminators of yaws, bejel and venereal syphilis. *Clinical Infectious Diseases 20*(5):1402–8, 1995.

Roux, W.: *Der zuchtende Kampf der Teile, oder die "Teilauslese" im Organismus (Teorie der 'funktionellen Antassung).* Leipzig: Wilhelm Engelmann, 1881.

_____: Beitrage zur morphologie der funktionellen anspassung. *Archives of Anatomie Physiologie Abt. 9*:120–158, 1885.

Rowe, G. G., and Roche, M. B.: The etiology of separate neural arch. *Journal of Bone and Joint Surgery 35-A*:102–110, 1953.

Rowe, L., and Steiman, I: Anterolisthesis in the cervical spine – spondylolysis. *Journal of Manipulative Physiological Therapeutics 10*(1):11–20, 1987.

Rudolph, A. H.: Syphilis. In Hoeprich, P. D., and Jordan, M. C. (Eds.): *Infectious diseases: A modern treatise of infectious processes* (4th ed.). Philadelphia, Lippincott, pp. 666–684, 1989.

Ruff, C. B., Holt, B., and Trinkaus, E.: Who's afraid of the Big Bad Wolff?: "Wolff's Law" and bone functional adaptation. *American Journal of Physical Anthropology 129*:484–498, 2006.

Ruff, C. B., Walker, A., and Trinkaus, E.: Postcranial robusticity in Homo III: ontogeny. *American Journal of Physical Anthropology 93*:35–54, 1994.

Ruge, D., and Wiltse, L. L.: *Spinal disorder: Diagnosis and treatment.* Philadelphia, Lea and Febiger, 1977.

Russell, R. G. G.: Paget's disease. In Nordin, B. E. O. (Eds.).: *Metabolic bone and stone disease.* London, Churchill Livingstone, pp. 190–233, 1984.

Ryan, D. E.: Painful temporomandibular joint. In McCarty, D. J. (Ed.): *Arthritis and allied conditions: A textbook of rheumatology* (11th ed.). Philadelphia, Lea and Febiger, 1989.

Sager, P.: *Spondylosis cervicalis: A pathological and osteoarchaeological study.* Copenhagen, Munksgaard, 1969.

Saifuddin, A., White J., Tucker, S., and Taylor, A. B.: Orientation of lumbar pars defects: Implications for radiological detection and surgical management. *Journal of Bone and Joint Surgery 80-B*(2):208–211, 1998.

Saini, T. S., Kharat, D. U., and Mokeen, S.: Prevalence of shovel-shaped incisors in Saudi Arabian dental patients. *Oral Surgery Oral Medicine Oral Pathology 70*(4):540–544, 1990.

Salter, R. B.: *Textbook of disorders and injuries of the musculoskeletal system.* Baltimore, Williams and Wilkins, 1970.

Saluja, P. G.: The incidence of spinal bifida occulta in a historic and a modern London population. *Journal of Anatomy 158*:91–93, 1988.

Saluja, P. G., Fitzpatrick, F., Bruce, M., and Cross, J.: Schmorl's nodes (intervertebral ruinations of intervertebral disc tissue) in two historic British populations. *Journal of Anatomy 145*:87–96, 1986.

Salvadei, L., Ricci, F., and Manzi, G.: Porotic hyerostosis as a marker of health and nutritional conditions during childhood: Studies at the transition between Imperial Rome and the Early Middle Ages. *American Journal of Human Biology 13*(6):709–717, 2001.

Sambrook, P. N.: The skeleton in rheumatoid arthritis: Common mechanisms for bone erosion and osteoporosis? *Journal of Rheumatology 27*(11):2541–2542, 2000.

Sammarco, G. J. and Cooper, P. S.: *Foot and ankle manual* (2nd ed.). Philadelphia, Williams and Wilkins, 1998.

Sammarco, V. J.: Os Acromiale: Frequency, anatomy and clinical implications. *Journal of Bone and Joint Surgery 82-A*:394–400, 2000.

Sanan, A., and Haines, S. J.: Repairing holes in the head: A history of cranioplasty. *Neurosurgery 40*(3):588–603, 1997.

Sankar, W. N., Wills, B. P. D., Dormans, J. P., and Drummond, D. S.: Os odontoideum revisited: The case for a multifactorial etiology. *Spine 31*(9):979–984, 2006.

Santos, A. L., and Roberts, C. A.: Anatomy of a serial killer: Differential diagnosis based on rib lesions from the Coimbra Identified Skeletal Collection. *American Journal of Physical Anthropology [Supplement] 32*:130 [abstract], 2001.

Sapp, J. P., Eversole, L. R., and Wysocki, G. W.: *Contemporary oral and maxillofacial pathology.* St. Louis, Mosby, 1997.

Sarnat, B. G., and Schour, I.: Enamel hypoplasia (chronic enamel aplasia) in relation to systemic disease: A chronologic, morphologic and etiologic classification. *Journal of American Dental Association 28*:1989–2000, 1941.

Sarrafian, S. K.: *Anatomy of the foot and ankle: Descriptive, topographic, functional* (2nd ed.)). Philadelphia,Lippincott, 1993.

Sauer, N. J.: The timing of injuries and manner of death: Distinguishing among antemortem, perimortem and postmortem trauma. In Reichs, K. (Ed.): *Forensic osteology: Advances in the identification of human remains* (2nd ed.). Springfield, C. C. Thomas, pp. 321–333, 1998.

Saunders, S. R.: *The development and distribution of discontinuous morphological variation of the human infracranial skeleton.* Ottawa, National Museum of Man Mercury Series, 1978.

Saunders, S. R., and Mayhall, J. T.: Developmental patterns of human dental morphological traits. *Archives of Oral Biology 27*(1):45–49, 1982.

Savini, R., Martucci, E., Prosperi, P., Gusella, A., and Di Silverstre, M.: Osteoid osteoma of the spine. *Journal Ortopedic Traumatologia 14*(2):233–238, 1988.

Schaeffer, J. P. (Ed.): *Morris' human anatomy: A complete systematic treatise* (19th ed.). Philadelphia, Blakiston, 1942.

Schajowicz, F.: *Tumors and tumorlike lesions of bones and joints.* New York, Springer-Verlag, 1981.

Schajowicz, F., Sainz, M. C., and Slullitel, J. A.: Juxta-articular bone cysts (intra-osseous ganglia). *Journal of Bone and Joint Surgery 61-B*(1):107–116, 1979.

Schendel, S. A., Tessier, P., and Tulasne, J.-F.: Facial clefting disorders and craniofacial synostoses: Skeletal considerations. In Turvey, T. A., Vig, K. W. L., and Fonseca, R. J. (Eds.): *Facial clefts and craniosynostosis: Principles and management.* Philadelphia, Saunders, 1996.

Scheuer, L., and Black, S.: *Developmental juvenile osteology.* New York, Academic Press, 2000.

Schijman, E.: Artificial cranial deformation in newborns in the pre-Columbian Andes. *Childs Nervous System 21*:945–950, 2005.

Sciubba, J. J., Regezi, J. A., and Rogers, R. S., III: *PDQ oral disease: Diagnosis and treatment.* Hamilton, BC Decker, Inc., 2002.

Schlesinger, I., and Waugh, T.: Slipped capital femoral epiphysis, unsolved adolescent hip disorder. *Orthopaedic Review 16*:33-48, 1987.

Schmorl, G., and Junghanns, H.: Die gesunde and dranke Wirbelsaule in Rontgenbild (article in German). *Fortschr. Rontgenstr.*, Supplement 43, Leipzig, 1932.

_____: *The human spine in health and disease* (2nd ed.). (American). New York, Grune and Stratton, 1971a.

_____: Chapter IV, Variations and malformations of the spine: and Chapter XIII, Vertebral slipping and displacement. In Besemann, E. F. (Ed.): *The human spine in health and disease* (Second American edition). New York, Grune and Stratton, 1971b.

Schoenecker, P. L.: Slipped capital femoral epiphysis. *Orthopaedic Review 74*:289, 1985.

_____: Legg-Calve-Perthes disease. *Orthopaedic Review 75*(9):561–574, 1986.

Schoeninger, M. J.: *Dietary reconstruction at Chalcatzinoo, a formative period site in Morelos, Mexico.* Technical report No. 9, Museum of Anthropology. Ann Arbor, University of Michigan, 1979.

Schultz A. H.: The fontanella metopica and its remnants in the adult skull. *American Journal of Anatomy 23*:259–271, 1918.

Schultz, M.: Diseases of the ear region in early and prehistoric populations. *Journal of Human Evolution 8*(6):575–580, 1979.

Schultz, M.: Microscopic investigation of excavated skeletal remains: A contribution to pale opathology and forensic medicine. In Haglund, W. D., and Sorg, M. H.: *Forensic taphonomy: The postmortem fate of human remains.* Boca Raton, CRC Press, pp. 201–222, 1997.

Schultz, M.: Paleohistopathology of bone: A new approach to the study of ancient diseases. *American Journal of Physical Anthropology 33*:106–147, 2001.

Schultz, M.: Light microscopic analysis of skeletal paleopathology. In Ortner, D. J.: *Identification of pathological conditions in human skeletal remains* (2nd ed.). Amsterdam, Academic Press, pp. 73–107, 2003.

Schwarz, E.: A typical disease of the upper femoral epiphysis. In Burwell, R. B., and Harrison, M. H. M. (Eds.): *Clinical Orthopaedics and Related Research 209*:5, 1986.

Scott, G. R. C., and Turner, C. G. II: *The anthropology of modern human teeth: Dental morphology and its variation in recent human populations.* Cambridge, Cambridge University Press, 1997.

Scutellari, P. N., Orzincolo, C., and Castaldi, G.: Association between diffuse idiopathic skeletal hyperostosis and multiple myeloma. *Skeletal Radiology 24*:489–492, 1995.

Seah, Y. H.: Torus palatinus and torus mandibularis: A review of the literature. *Australian Dental Journal 40*(5):318–321, 1995.

Sease, C.: Benzotriazole: A review for conservators. *Studies in Conservation 23*:76–85, 1987.

Sefcakova, A., Strouhal, E., Nemeckova, A., Thurzo, M., and Stassikova-Stukovska, D.: Case of metastatic carcinoma from end of the 8th–Early 9th Century Slovakia. *American Journal of Physical Anthropology 776*:216–229, 2001.

Seitsalo, S., Osterman, K., Poussa, M., and Laurent, L. E.: Spondylolisthesis in children under 12 years of age: Long-term results of 56 patients treated conservatively or operatively. *Journal of Pediatric Orthopaedics 8*:516–521, 1988.

Sekiguchi, M., Yabuki, S., Koichiro, S., and Shinichi, K.: An anatomic study of the sacral hiatus: A basis for successful caudal epidural block. *Clinical Journal of Pain 20*(1): 51–54, 2004.

Seligsohn, R., Rippon, J. W., and Lerner, S. A.: Aspergillus terreus osteomyelitis. *Archives of Internal Medicine 137*: 918–920, 1977.

Sella, E. J., and Barrette, C.: Staging of Charcot neuroarthropathy along the medial column of the foot in the diabetic patient. *The Journal of Foot and Ankle Surgery 38*(1): 34–40, 1999.

Semrad, W., Vanek, I., Taborsky, J., and Urban, T.: Aneurysm of the descending aorta causing destruction of vertebral bodies. *Sbornik Lekarsky* (Praha) *101*(3):273–279, 2000.

Shaaban, M. M.: Trephination in ancient Egypt and the report of a new case from Dakleh Oasis. *OSSA 9/11*: 135, 1982–1984.

Shah, D. S., Sanghavi, S. J., Chawda, J. D., and Shah, R. M.: Prevalence of torus palatinus and torus mandibularis in 1000 patients. *Indian Journal Dental Research 3*(4):107–110, 1992.

Shaibani, A., Workman, R., and Rothschild, B. M.: The significance of enthesopathy as a skeletal phenomenon. *Clinical and Experimental Rheumatology 11*(4):399–403, 1993.

Shands, A. R.: *Handbook of orthopedic surgery.* St. Louis, Mosby, 1951.

Sharma, J. C.: Dental morphology and odontometry of the Tibetan immigrants. *American Journal of Physical Anthropology 61*(4):495–505, 1983.

Sharma, P. D., and Dawkins, R. S.: Patent foramen of Huschke and spontaneous salivary fistula. *Journal Laryngology Otology 98*(1):83–85, 1984.

Sharon, R., Weinberg, H., and Husseini, N.: An unusually high incidence of homozygous MM in ankylosing spondylitis. *Journal of Bone and Joint Surgery 67-B*(1):122–123, 1985.

Shauffer, I. A., and Collins, W. V.: The deep clavicular rhomboid fossa. Clinical significance and incidence in 10,000 routine chest photofluorograms. *Journal of American Medical Association 195*:778–779, 1966.

Shaw, C. N., and Stock, J. T.: Intensity, repetitiveness, and directionality of habitual adolescent mobility patterns influence the tibial diaphyseal morphology of athletes. *American Journal of Physical Anthropology 140*:140–159, 2009a.

———: Habitual throwing and swimming corresponds with upper limb diaphyseal strength and shape in modern human athletes. *American Journal of Physical Anthropology 140*:160–172, 2009b.

She, R., and Szakacs, J.: Hyperostosis frontalis interna: Case report and review of literature. *Annals of Clinical and Laboratory Science 34*:206–208, 2004.

Sheehy, J. J.: Diffuse exostoses and osteomata of the external auditory canal: A report of 100 operations. *Otolaryngology Head Neck Surgery 90*:337–342, 1982.

Shepherd, F. J.: Symmetrical depressions on the exterior surface of the parietal bones (with notes of three cases). *Journal of Anatomy and Physiology 27*(4):501–504, 1893.

Shields, E. D., and Mann, R. W.: Salivary glands and human selection: A hypothesis. *Journal of Craniofacial Genetics and Developmental Biology 16*(2):126–136, 1996.

Shipman, P., Walker, A., and Bichell, D.: *The human skeleton.* Cambridge, Harvard Press, 1985.

Shimizu, M., Osa, N., Okamura, K., and Yoshiura, K.: CT analysis of the Stafne's bone defects of the mandible. *DentoMaxilloFacial Radiology 35*:95–102, 2006.

Shiota, K., and Matsungago, E.: A genetic and epidemiologic study of polydactyly in human embryos in Japan. *Journal of Human Genetics 23*(2):173–192, 1978

Shore, L. R.: A report of the nature of certain bony spurs arising from the dorsal arches of the thoracic vertebrae. *Journal of Anatomy 65*:379, 1931.

Sikanjic, P. R., and Vlak, D.: Elongated styloid process in late medieval skeletons from Uzdolje-Grablje, Croatia. *International Journal of Osteoarchaeology 20*:248–252, 2010.

Sillen, A.: Strontium and diet at Hayonim Cave, Israel. Ph.D. dissertation. University of Pennsylvania, Philadelphia, 1981.

Silverman, F. N., and Kuhn, J. P.: *Caffey's pediatric x-ray diagnosis: An integrated imaging approach* (Vols. 1–2). St. Louis, Mosby, 1993.

Simmons, E. H., and Jackson, R. P.: The management of nerve root entrapment syndromes associated with the collapsing scoliosis of idiopathic lumbar and thoracolumbar curves. *Spine 4*(6):533–541, 1979.

Simpson, W. M., and McIntosh, C. A.: Actinomycosis of the vertebrae (actinomycotic Pott's disease): Report of four cases. *Archives of Surgery 14*:1166, 1927.

Singer, F. R.: *Paget's disease of bone.* New York, Plenum, 1977.

Singhal, S., and Rao, V.: Supratrochlear foramen of the humerus. *Anatomical Service International 82*:105–107, 2007.

Sisk, T. D.: Fractures of lower extremity. In Crenshaw, A. H. (Ed.): *Campbell's operative orthopaedics* (7th ed., Volume 3). St. Louis, C. V. Mosby, 1987.

Sisman, Y., Etöz, O. A., Mavili, E., Sahman, H., and Tarim Ertas, E.: Anterior Stafne bone defect mimicking a residual cyst: A case report. *Dentomaxillofacial Radiology 39*:124–126, 2010.

Sivers, J. E., and Johnson, G. K.: Diagnosis of Eagle's syndrome. *Oral Surgery Oral Medicine Oral Pathology 59*:575–577, 1985.

Sjovold, T.: A report on the heritability of some cranial measurements and non-metric traits. In Van Vark, G. N., and Howells, W. W. (Eds.): *Multivariate statistics in physical anthropology.* Dordrecht, D. Reidel, pp. 223–246, 1984.

Slatkin, M. B.: MRI of the shoulder (2nd ed.). Philadelphia, Lippincott Williams & Wilkins, 2003.

Smillie, I. S.: *Osteochondritis dessicans.* Edinburgh, Livingstone, 1960.

———: *Injuries of the knee joint* (3rd ed.). Baltimore, Williams and Wilkins, 1962.

Smith, G. E.: *The archaeological survey of Nubia: Report on the human remains.* Cairo, National Printing Department, 1910.

Smith, O. C., Berryman, H. E., and Lahren, C. H.: Cranial fracture patterns and estimate of direction from low velocity gunshot wounds. *Journal of Forensic Sciences 32*:1416–1421, 1987.

Smith, O. C., Berryman, H. E., Symes, S. A., Francisco, J. T., and Hnilica, V.: Atypical gunshot exit defects to the cranial vault. *Journal of Forensic Sciences 38*(2):339–343, 1993.

Smith, J. T., Skinner, S. R., and Shonnard, N. H.: Persistent synchondrosis of the second cervical vertebra simulating a hangman's fracture in a child. *Journal of Bone and Joint Surgery 75*(8):1228–1230, 1993.

Snapper, I.: *Bone diseases in medical practice.* New York, Grune and Stratton, 1957.

Snodgrass, J. J., and Galloway, A.: Utility of dorsal pits and pubic tubercle height in parity assessment. *Journal of Forensic Sciences 48*(6):1226–1130, 2003.

Snow, C. E.: *Early Hawaiians: An initial study of skeletal remains from Mokapu, Oahu.* Lexington, University Press of Kentucky, 1974.

Solomon, L. B., Ruhli, F. J., Taylor, J., Ferris, L., Pope, R., and Henneberg, M.: A dissection and computer tomograph study of tarsal coalitions in 100 cadaver feet. *Journal of Orthopaedic Research 21*: 352–358, 2003.

Spitz, D. J., and Newberg, A. H.: Imaging of stress fractures in the athlete. *Radiologic Clinics of North America 40*:313–331, 2002.

Spitz, W. U.: Injury by gunfire: Gunshot wounds. In Spitz, W. U., and Fisher, R. S. (Eds.): *Medicolegal investigation of death: Guidelines for the application of pathology to crime investigation.* Springfield, C. C. Thomas, 1980.

Spjut, H. J., Dorfman, H. D., Fechner, R. E., and Ackerman, L. V.: *Tumors of bone and cartilage.* Washington, D. C., Armed Forces Institute of Pathology, Vol. 1, 1971.

Sponseller, P. D., Bisson, L. J., Gearhart, J. R., Jeffs, R. D., Magid, D., and Fishman, E.: The anatomy of the pelvis in the exstrophy complex. *Journal of Bone and Joint Surgery 77-A*(2):177–189, 1995.

Sponseller, P. D., Jani, M. M., Jeffs, R. D., and Gearhart, J. P.: Anterior innominate osteotomy in repair of bladder exstrophy. *Journal of Bone and Joint Surgery 83-A*(2):184–193, 2001.

Spring, D. B., Lovejoy, C. O., Bender, G. N., and Duerr, M.: The radiographic preauricular groove: Its non-relationship to past parity. *American Journal of Physical Anthropology 79*:247–252, 1989.

Sreedharan, S., and Li, Y. H.: Diffuse idiopathic skeletal hyperostosis with cervical spinal cord injury: A report of 3 cases and a literature review. *Annals Academy of Medicine 34*(3):257–261, 2005.

Stafne, E. C.: Bone cavities situated near the angle of the mandible. *Journal of American Dental Association 29*:1969, 1942.

Stallworthy, J. A.: A case of enlarged parietal foramina associated with metopism and irregular synostosis of the coronal suture. *Journal of Anatomy 67*:168–174, 1932.

Stanitski, C. L.: Anterior knee pain syndromes in the adolescent. *Journal of Bone and Joint Surgery 75-A*(9):1407–1416, 1993.

Steele, D. G., and Bramblett, C. A.: *The anatomy and biology of the human skeleton.* College Station, Texas A&M University Press, 1988.

Steen, S. L., and Lane, R. W.: Evaluation of habitual activities among two Alaskan Eskimo populations based on musculoskeletal stress markers. *International Journal of Osteoarchaeology 8*(5):341–353, 1998.

Steinbock, T. R.: *Paleopathological diagnosis and interpretation.* Springfield, C. C. Thomas, 1976.

Steinke, H., Hammer, N., Slowik, V., Stadler, J., Josten, C., Böhme, J., and Spanel-Borowski, K.: Novel insights into the sacroiliac joint ligaments. *Spine 35*(3):257–263, 2010.

Stewart, T. D.: Incidence of separate neural arch in the lumbar vertebrae of Eskimos. *American Journal of Physical Anthropology 16*:51–62, 1931.

Stewart, T. D.: The circular type of cranial deformity in the United States. *American Journal of Physical Anthropology 28*:343–351, 1941.

_____: The age incidence of neural-arch defects in Alaskan natives, considered from the standpoint of etiology. *Journal of Bone and Joint Surgery 35-A*:937–950, 1953.

_____: Examination of the possibility that certain skeletal characters predispose to defects in the lumbar neural arches. *Clinical Orthopaedics and Related Research 8*:44–60, 1956.

_____: Distortion of the pubic symphyseal surface in females and its effect on age determination. *American Journal of Physical Anthropology 15*:9–18, 1957.

_____: Stone age skull surgery. Annual report of the Smithsonian Institution. Washington, D.C., Smithsonian Press, pp. 469–491, 1957.

_____: The rate of development of vertebral osteoarthritis in American Whites and its significance in skeletal age identification. *The Leech 28*:144–151, 1958.

_____: Identification of the scars of parturition in the skeletal remains of females. In Stewart, T.D. (Ed.): *Personal identification in mass disasters.* Washington, D.C., Smithsonian Press, 1970.

_____: *The people of America.* New York, Charles Scribner, 1973.

_____: Cranial dysraphism mistaken for trephination. *American Journal of Physical Anthropology 42*(3):435–437, 1975.

_____: Are supra-inion depressions evidence of prophylactic trephination? *Bulletin of the History of Medicine 50*:414–434, 1976.

_____: *Essentials of forensic anthropology.* Springfield, C. C. Thomas, 1979.

_____: Scaphocephaly in blacks: A variant form of pathologic head deformity. *Bulletin et Memoirs de la society d'Anthropologia de Paris*, t. 9, serie 13:267–279, 1982.

Stewart, T. D., and Groome, J. R.: The African custom of tooth mutilation in America. *American Journal of Physical Anthropology 28*(1):31–42, 1968.

Stewart, T. D., and Spoehr, A.: Evidence on the paleopathology of yaws. *Bulletin of the History of Medicine 26*:538–553, 1952.

Stibbe, S. P.: Anatomical note. Skull showing perforations of the parietal bone, or enlarged parietal foramina. *Journal of Anatomy 63*:277, 1929.

Stirland, A. J.: A Possible correlation between os acromiale and occupation in the burials from Mary Rose. *Proceedings of the 5th European Meeting, Sienna, Paleopathology Association*, pp. 327–334, 1984.

_____: Pre-Columbian treponematosis in Mediaeval Britain. *International Journal of Osteoarchaeology 1:39-47*, 1991 (a).

_____: Diagnosis of occupationally related paleopathology: Can it be done? In Ortner, D. J., and Aufderheide, A. C. (Eds.): *Human paleopathology: Current syntheses and future Options*. Washington, D.C., Smithsonian Institution Press, pp. 40–47, 1991b.

_____: Asymmetry and activity-related changes in the male humerus. *International Journal of Osteoarchaeology 3*:105–113, 1993.

_____: Patterns of trauma in a unique medieval parish cemetery. *International Journal of Osteoarchaeology 6*(1): 92–100, 1996

_____: Musculoskeletal evidence for activity: Problems of evaluation. *International Journal of Osteoarchaeology 8*: 354–362, 1998.

_____: *Human bones in archaeology*. Buckinghamshire, Shire Publications, 1999.

Stoller, S. M., Hekmat, F., and Kleiger, B.: A comparative study of the frequency of anterior impingement exostoses of the ankle in dancers and nondancers. *Foot and Ankle 4*(4):201–203, 1984.

Strassberg, M.: The epidemiology of anencephalus and spinal bifida: A review. Part I: Introduction, embryology, classification and epidemiological terms. *Spina Bifida Therapy 4*(2):53, 1982.

Struthers, J.: Supra-condyloid process in man. *The Lancet*, Feb:1, 1873.

Stuart-Macadam, P. L.: Porotic hyperostosis: Representative of childhood condition. *American Journal of Physical Anthropology 66*:391–398, 1985.

_____: A radiographic study of porotic hyperostosis. *American Journal of Physical Anthropology 74*(4):511–520, 1987a.

_____: Porotic hyperostosis: New evidence to support the anemia theory. *American Journal of Physical Anthropology 74*(4):521–526, 1987b.

_____: Porotic hyperostosis: Relationship between orbital and vault lesions. *American Journal of Physical Anthropology 80*:187–193, 1989.

_____: Porotic hyperostosis: A new perspective. *American Journal of Physical Anthropology 87*(1):39–47, 1992.

Suchey, J. M., Wiseley, D. V., Green, R. F., and Noguchi, T. T.: Analysis of dorsal pitting in the os pubis in an extensive sample of modern American females. *American Journal of Physical Anthropology 51*:517–539, 1979.

Suchey, J. M., Wisely, D. V., and Katz, D. Evaluation of the Todd and McKern-Stewart methods of aging the male os pubis. In Reichs, K. (Ed.): *Forensic osteology: Advances in the identification of human remains*. Springfield, C. C. Thomas, 1986.

Suzuki, T.: Paleopathological study on osseous syphilis in skulls of the Ainu remains. *OSSA 9/11*:153–167, 1982–1984.

_____: Paleopathological study on a case of osteosarcoma. *American Journal of Physical Anthropology 74*:309, 1987.

Swank, S. M., and Barnes, R. A.: Osteoid osteoma in a vertebral body: Case report. *Spine 12*(6):602, 1987.

Sweet, P. A. S., Buonocore, M. G., and Buck, I. F.: Prehispanic Indian dentistry. *Dental Radiography and Photography 36*:3, 1963.

Symes, S. A., Williams, J. A., Murray, E. A., Hoffman, J. M., Holland, T. D., Saul, J. M., Saul, F. P., and Pope, E. J. Taphonomic context of sharp-force trauma in suspected cases of mutilation and dismemberment. In Haglund, W. D., and Sorg, M. H. (Eds.): *Advances in forensic taphonomy: Method, theory, and archaeological perspectives*, pp. 404-434, Boca Raton, CRC Press, 2002.

Symington, J.: On separate acromion process. *Journal of Anatomy and Physiology 34*:287, 1900.

Symmers, W.: A skull with enormous parietal foramina. *Journal of Anatomy and Physiology 29*:329–330, 1895.

Tague, R. G.: Morphology of the pubis and preauricular area in relation to parity and age at death in Macaca mulatta. *American Journal of Physical Anthropology 82*(4): 517–525, 1990.

Takada, Y., Ishikura, R., Ando, K., Morikawa, T., and Nakao, N.: Imaging findings of elongated styloid process syndrome (Eagle's syndrome): Report of two cases (article in Japanese). *Nippon Igaku Hoshasen Gakkai Zasshi 63*(1):56–58, 2003.

Tanamas, S. K., Wluka, A. E., Pelletier, J-P., Martel-Pelletier, J., Abram, F., Wang, Y., and Cicuttini, F. M.: The association between subchondral bone cysts and tibial cartilage volume and risk of joint replacement in people with knee osteoarthritis: A longitudinal study. *Arthritis Research and Therapy 12*:R58 (doi:10.1186/ar2971), 2010.

Taylor, J. A. M., and Resnick, D.: *Skeletal imaging: Atlas of the spine and extremeties*. Philadelphia, Saunders, 2000.

Teele, D. W., Klein, J. O., and Rosner, B. A.: Otitis media with effusion during the first three years of life and development of speech and language. *Pediatrics 74*:282–287, 1984.

Tehranzadeh, J., Andrews, C., and Wong, E.: Lumbar spine imaging normal variants, imaging pitfalls, and artifacts. *Radiologic Clinics of North America 38*:1207–1253, 2000.

Terry, R. J.: A study of the supracondyloid process in the living. *American Journal of Physical Anthropology 4*:129, 1921.

_____: New data on the incidence of the supracondyloid variation. *American Journal of Physical Anthropology 9*:265–270, 1926.

_____: On the racial distribution of the supracondyloid variation. *American Journal of Physical Anthropology 14*: 459–462, 1930.

_____: Osteology. In Jackson, C. M. (Ed.): *Morris' human anatomy*. Philadelphia, Blakiston's, 1933.

Tessier, P.: Anatomical classification of facial, cranio-facial and laterofacial clefts. *Journal of Maxillofacial Surgery 4*:69–92, 1976.

Testut, L.: *Traite' d' anatomie humaine*. Paris, Poireir & Charphy, 1911.

Thawley, S. E., Panje, W. R., Batsakis, J. G., and Lindberg, R. D.: *Comprehensive management of head and neck tumors* (Vol. 2). Philadelphia, Saunders, 1987.

Thieme, F. P.: *Lumbar breakdown caused by erect posture in man: With emphasis on spondylolisthesis and herniated intervertebral discs*. Anthropological Papers, Museum of Anthropology. Ann Arbor, University of Michigan Press, No. 4, 1950.

Thijn, C. J. P., and Steensma, J. Y.: *Tuberculosis of the skeleton. Focus on radiology*. New York, Springer-Verlag, 1990.

Thomas, C. L. (Ed.): *Taber's cyclopedic medical dictionary* (15th ed.). Philadelphia, Davis, 1985.

Thomas, J. L., Christensen, C., Kravitz, S. R., Mendicino, R. W., Schuberth, J. M., Vanore, J. V., Weil, L. S., Zlotoff, H. J., and Couture, S. D.: The diagnosis and treatment of heel pain. *Journal of Foot and Ankle Surgery* 40(5):329–340, 2001.

Thot, B., Revel, S., Mohandas, R., Rao, A. V., and Kumar, A.: Eagle's syndrome. Anatomy of the styloid process. *Indian Journal Dental Research* 11(2):65–70, 2000.

Tini, P. G., Wieser, C., and Zinn, W. M.: The transitional vertebra of the lumbosacral spine: Its radiological classification, incidence, prevalence, and clinical significance. *Rheumatology Rehabilitation* 16:180–85, 1977.

Tkocz, I., and Bierring, F.: A medieval case of metastasizing carcinoma with osteosclerotic bone lesions. *American Journal of Physical Anthropology* 65:373–380, 1984.

Tod, P. A., and Yelland, J. D. N.: Craniostenosis. *Clinical Radiology* 22:472–486, 1971.

Todd, T. W.: "Cervical" rib: Factors controlling its presence and its size. Its bearing on the morphology and development of the shoulder. *Journal of Anatomy* 46:244–288, 1912.

Todd, T. W., and McCally, W. C.: Defects of the patellar border. *Annals of Surgery* 74:775–782, 1921.

Tol, J. L., Slim, E., van Soest, A. J., and van Dijk, C. N.: The relationship of the kicking action in soccer and anterior ankle impingement syndrome: A biomechanical analysis. *American Journal of Sports Medicine* 30(1):45–50, 2002.

Tolman, D. E., and Stafne, E. C.: Developmental bone defects of the mandible. *Oral Surgery, Oral Medicine, Oral Pathology* 24(4):488–490, 1967.

Toohey, J. S.: Skeletal presentation of congenital syphilis: Case report and review of the literature. *Journal of Pediatric Orthopaedics* 5(1):104–106, 1985.

Torgersen, J. H.: A radiological study of the metopic suture. *Acta Radiologica* 33:1–11, 1950.

_____: The developmental genetics and evolutionary meaning of the metopic suture. *American Journal of Physical Anthropology* 9:98–102, 193–210, 1951.

Trinkhaus, E.: Femoral neck-shaft angles of Qafzeh-Skhul early modern humans, and activity levels among immature Near Eastern Middle Paleolithic hominids. *Journal of Human Evolution* 25:393–416, 1993.

Trinkhaus, E., Churchill, S. E., and Ruff, C.B.: Postcranial robusticity in Homo II: Bilateral asymmetry and bone plasticity. *American Journal of Physical Anthropology* 93:1–34, 1994a.

_____: Postcranial robusticity in Homo III: Ontogeny. *American Journal of Physical Anthropology* 93:35–54, 1994b.

Trope, M.: Root resorption of dental and traumatic origin: Classification based etiology. *Practical Periodontics and Aesthetic Dentistry* 10(4):515–522, 1998.

Trotter, M.: Septal apertures in the humerus of American white and Negro. *American Journal of Physical Anthropology* 19:213–227, 1934.

_____: Accessory sacroiliac articulations. *American Journal of Physical Anthropology* 22:247–261, 1937.

_____: A common anatomic variation in the sacro-iliac region. *Journal of Bone and Joint Surgery* 22:283–299, 1940.

Troup, J. D. G.: Mechanical factors in spondylolisthesis and spondylolysis. *Clinical Orthopaedics and Related Research* 117:59–67, 1976.

Trueta, J.: *Studies of the development and decay of the human frame*. Philadelphia, Saunders, 1968.

Tsai, S. J., and King, N. M.: A catalogue of anomalies and traits of the permanent dentition of southern Chinese. *Journal Clinical Pediatric Dentistry* 22(3):185–194, 1998.

Tuli, S. M.: *Tuberculosis of the spine*. National Library of Medicine. Preur Printing Press, 1975.

Turk, J. L.: Syphilitic caries of the skull–the changing face of medicine. *Journal Royal Society Medicine*, 88(3):146–148, 1995.

Turkoglu, K., and Orhan, K.: Stafne cavity in the anterior mandible. *Journal of Craniofacial Surgery* 21(6):1769–1775, 2010.

Turlik, M. A.: Seronegative arthritis as a cause of heel pain. *Clinics in Podiatric Medicine and Surgery* 7(2):369–375, 1990.

Turner, M. M.: On exostoses within the external auditory meatus. *Journal of Anatomy and Physiology* 13(2):200–203, 1879.

Turner, W.: On supernumerary cervical ribs. *Journal of Anatomy and Physiology* 4(1):130–139, 1869.

_____: Cervical ribs, and the so-called bicipital ribs in man, in relation to corresponding structures in the Cetacea. *Journal of Anatomy and Physiology* 17(3):384–400, 1883.

Turvey, T. A., Vig, K. W L., and Fonseca, R. J.: *Facial clefts and craniosynostosis: Principles and management*. Philadelphia, Saunders, 1996.

Twomey, L. T., and Taylor, J. R.: Age changes in lumbar vertebrae and intervertebral discs. *Clinical Orthopaedics and Related Research* 224:97–104, 1987.

Tyson, R., and Dyer, E. S. (Eds.): *Catalogue of the Hrdlička paleopathology collection*. San Diego, San Diego Museum of Man, 1980.

Ubelaker, D. H.: *Human skeletal remains: Excavation, analysis, interpretation* (3rd ed.). Washington, Taraxacum, 1999.

_____: The development of American paleopathology. In Spencer, F. (Ed.): *A History of American Physical Anthropology 1930-1980*. New York, Academic, 1982.

Uemura, S., Fujishita, M., and Fuchihata, H.: Radiographic interpretation of so-called developmental defect of mandible. *Oral Surgery Oral Medicine Oral Pathology* 41(1):120–128, 1976.

Umans, J.: Ankle inpingement syndromes. *Seminar Musculoskeletal Radiology 6*(2):133–139, 2003.

Utsinger, P. D.: Diffuse idiopathic skeletal hyperostosis (DISH, ankylosing hyperostosis). In Moskowitz, R. W., Howell, D. S., Goldberg, V. M., and Mankin, J. J. (Eds.): *Osteoarthritis: Diagnosis and management.* Philadelphia, Saunders, 1984.

———: Diffuse idiopathic skeletal hyperostosis. *Clinics in Rheumatic Diseases 11*(2):325–351, 1985.

Van Dijk, C. N., Wessel, R. N., Tol, J. L., and Maas, M.: Oblique radiograph for the detection of bone spurs in anterior ankle impingement. *Skeletal Radiology 31*(4): 214–221, 2002.

Van Gilse, P.: Des observations ulterieures sur las genese des exostoses du conduit externe pas l'irrigation d'eau froide. *Acta Otolaryngology 26*:343–352, 1938.

van Saase, J. L., van Romunde, L. K., Cats, A., Vandenbroucke, J. P., and Valkenburg, H. A.: Epidemiology of osteoarthritis: Zoetermeer survey. Comparison of radiological osteoarthritis in a Dutch population with that in 10 other populations. *Annals of the Rheumatic Diseases 48*(4):271–280, 1989.

Verano, J. W., and Andrushko, V. A.: Cranioplasty in ancient Peru: A critical review of the evidence, and a unique case from the Cuzco area. *International Journal of Osteoarchaeology 20*(3):269–279, 2010.

Vercellotti, G., Caramella, D., Formicola, V., Fornaciari, G., and Larsen, C. S.: Potic hyperostosis in a Late Upper Palaeolithic skeleton (Villabruna 1, Italy). *International Journal of Osteoarchaeology 20*(3):358–368, 2010.

Versfeld, G. A., and Solomon, A.: A diagnostic approach to tuberculosis of bones and joints. *Journal of Bone and Joint Surgery 64-B*(4):446–447, 1982.

Velasco-Suarex, M., Bautista Martinez, J., Garcia Oliveros, R., and Weinstein, P. R.: Archaeological origins of cranial surgery: Trephination in Mexico. *Neurosurgery 31*(2): 313–318, 1992.

Vidic, B.: Incidence of torus palatinus in Yugoslav skulls. *Journal of Dental Research 45*:1511–1515, 1966.

Villotte, S., Castex, D., Couallier, V., Dutour, O., Knusel, C. J., and Henry-Gambier, D.: Enthesopathies as occupational stress markers: Evidence from the upper limb. *American Journal of Physical Anthropology 142*:224–234, 2010.

Vincelette, P., Laurin, C. A., and Levesque, H. P.: The footballer's ankle and foot. *Canadian Medical Association Journal 107*(9):872–874, 1972.

Vodanovic, M., Slaus, M., Galic, I., Marotti, M., and Brkic, H.: Stafne's defects in two mandibles from archaeological sites in Croatia. *International Journal of Osteoarchaeology 21*:119–126, 2011.

Waddington, M. M.: *Atlas of the human skull.* Vermont, Academy Books, 1981.

Wagner, A. L., Murtagh, F. R., Arrington, J. A., and Stallworth, D.: Relationship of Schmorl's nodes to vertebral body endplate fractures and acute endplate disk extrusions. *American Journal of Neuroradiology 21*:276–281, 2000.

Waldron, T.: Unilateral spondylolysis. *International Journal of Osteoarchaeology 2*:177–181, 1992.

———: A case-referent study of spondylolysis and spina bifida and transitional vertebrae in human skeletal remains. *International Journal of Osteoarchaeology 3*:55–57, 1993.

———: *Counting the dead: The epidemiology of skeletal populations.* New York: John Wiley, 1994.

———: *Paleopathology.* Cambridge, Cambridge University Press, 2008.

Walker, P. L.: Porotic hyperostosis in a marine-dependent California Indian population. *American Journal of Physical Anthropology 69*:345–354, 1986.

———: Cranial injuries as evidence of violence in prehistoric southern California. *American Journal of Physical Anthropology 80*:313–323, 1989.

———: A bioarchaeological perspective on the history of violence. *Annual Review of Anthropology 30*:573–596, 2011.

Walker, P. L., Bathurst, R. R., Richman, R., Gjerdrum, T., and Andrushko, V. A.: The causes of porotic hyperostosis and cribra orbitalia: A reappraisal of the iron-deficiency-anemia hypothesis. *American Journal of Physical Anthropology 139*:109–125, 2009.

Wang, R. G., Bingham, B., Hawke, M., Kwok, P., and Li, J. R.: Persistence of the foramen of Huschke in the adult: An osteological study. *Journal of Otolaryngology 20*(4): 251–253, 1991.

Wang, X., Rosenberg, Z. S., Mechlin, M. B., and Schweitzer, M. E.: Normal variants and diseases of the peroneal tendons and superior peroneal retinaculum: MR imaging features. *Radiographics 25*:587–602, 2005.

Wang,, Z., Parent, S., Mac-Thiong, J-M., Petit, Y., and Labelle, H.: Influence of sacral morphology in developmental spondylolisthesis. *Spine 33*(20):2185–2191, 2008.

Warwick, D., Prothero, D., Field, J., and Bannister, G.: Radiological measurement of radial shortening in Colles' fracture. *The Journal of Hand Surgery: Journal of the British Society for Surgery of the Hand 18*(1):50–52, 1993.

Wastie, M. L.: Radiological changes in serial x-rays of the foot and tarsus in leprosy. *Clinical Radiology 26*(2):285–292, 1975.

Watt, I., and Dieppe, P.: Osteoarthritis revisited. *Skeletal Radiology 19*:1–3, 1990.

Weaver, J. K.: Bipartite patella as a cause of disability in the athlete. *American Journal of Sports Medicine 5*:137–143, 1977.

Webb, S. G.: *Paleopathology of aboriginal Australians. Health and disease across a hunter-gatherer continent.* Cambridge, Cambridge University Press, 1995.

———: Two possible cases of trephination from Australia. *American Journal of Physical Anthropology 75*(4):541–548, 1988.

Weinfeld, R. M., Olson, P. N., Maki, D. D., and Griffiths, H. J.: The prevalence of diffuse idiopathic skeletal hyperostosis (DISH) in two large American Midwest metropolitan hospital populations. *Skeletal Radiology 26*:222–225, 1997.

Weinstein, P. R., Ehni, G., and Wilson, C. B.: *Lumbar spondylosis: Diagnosis, management and surgical treatment.* Chicago, Year Book Medical Publishers, 1977.

Weiss, E.: Understanding muscle markers: Aggregation and construct validity. *American Journal of Physical Anthropology 121*:230–240, 2003a.

————: Effects of rowing on humeral strength. *American Journal of Physical Anthropology 121*:293–303, 2003b.

————: Understanding muscle markers: Lower limbs. *American Journal of Physical Anthropology 125*:232–238, 2004.

————: Muscle markers revisited: Activity pattern reconstruction with controls in a central Californian Amerind population. *American Journal of Physical Anthropology 133*: 931–940, 2007.

————: Sex differences in humeral bilateral asymmetry in two hunter-gatherer populations: California Amerinds and British Columbia Amerinds. *American Journal of Physical Anthropology 140*:19–24, 2009.

Weiss, E., Corona, L., and Schultz, B.: Sex differences in musculoskeletal stress markers: Problems with activity pattern reconstructions. *International Journal of Osteoarchaeology*, online 28 Jun 2010. (Wiley.com/doi/10.1002/oa.1183), 2010.

Welcker, H.: Cribra orbitalia. Ein ethnologish-diagnostisches merkmal am schadel mebruer menschenrassen. *Archives Anthropologie 17*:1–18, 1888.

Wells, C.: *Bones, bodies, and disease.* London: Thames and Hudson, 1964.

————: Pseudopathology. In Brothwell, D. R., and Sandison, A. T. (Eds.): *Diseases in antiquity. A survey of the diseases, injuries, and surgery of early populations.* Springfield, C. C. Thomas, pp. 5–19, 1967.

————: Osteochondritis dissecans in ancient British skeletal material. *Medical History 18*:365–369, 1974.

————: Ancient lesions of the hip joint. *Medical and Biological Illustration 26*:171–177, 1976.

Wells, C., and Woodhouse, N.: Paget's disease in an Anglo-Saxon. *Medical History 19*:396–400, 1975.

West, N. F.: The aetiology of ankylosing spondylitis. *American Rheumatological Disease 8*:143–148, 1949.

Wheat, L. J.: Histoplasmosis. In Hoeprich, P. D., and Jordan, M. C. (Eds.): *Infectious diseases: A modern treatise of infectious processes* (4th ed.). Philadelphia, Lippincott, pp. 481–488, 1989.

White, S. C., and Pharoah, M. J.: *Oral radiology: Principles and interpretation* (4th ed.). St. Louis, Mosby, 2000.

White, T. D., and Folkens, P. A.: *Human osteology.* New York, Academic Press, 1991.

Whyte, M. P.: Paget's disease of bone. *New England Journal of Medicine 355*:593–600, 2006.

Wilbur, A. K., Bouwman, A. S., Stone, A. C., Roberts, C. A., Pfistera, L. A., Buikstra, J. E., and Brown, T. A.: Deficiencies and challenges in the study of ancient tuberculosis DNA. *Journal of Archaeological Science 36*(9): 1990–1997, 2009.

Wilczak, C. A.: Consideration of sexual dimorphism, age, and asymmetry in quantitative measurements of muscle insertion sites. *International Journal of Osteoarchaeology 8*(5):311–325, 1998.

————: New directions in the analysis of musculoskeletal stress markers. Paper presented at the Workshop in Musculoskeletal Stress Markers (MSM): Limitations and Achievements in the Reconstruction of past Activity Patterns. University of Coimbra, July 2–3, 2009.

Wilczak, C. A., Kennedy, K. A. R., and Mostly, M. O. S.: Technical aspects of identification of skeletal markers of occupational stress. In Reichs, K. J. (ed.): *Forensic osteology: Advances in the identification of human remains.* Springfield, C. C. Thomas, pp. 461–490, 1998.

Wilczak C. A., and Ousley S. D.: Test of the relationship between sutural ossicles and cultural cranial deformation: Results from Hawikuh, New Mexico. *American Journal of Physical Anthropology. 139*(4):483–93, 2009.

Wilkins, K. E.: The uniqueness of the young athlete: Musculoskeletal injuries. *American Journal of Sports Medicine 8*(5):377–382, 1980.

Williams, H. U.: The origin and antiquity of syphilis: The evidence from diseased bones. *Archives of Pathology 13*:779–814, 931–983, 1932.

Williams, P. L., and Warwick, R. (Eds.): *Gray's anatomy* (36th ed.). Edinburgh, Churchill Livingstone, 1980.

Williams, T. G.: Hangman's fracture. *Journal of Bone and Joint Surgery 57-B*(1):82–88, 1975.

Willis, T. A.: Bachache from vertebral anomaly. *Surgery, Gynecology and Obstetrics May*:658–665, 1924.

————: The separate neural arch. *Journal of Bone and Joint Surgery 13-A*:709, 1931.

Wilson, A. K.: Roentgenological findings in bilateral symmetrical thinness of the parietal bones (senile atrophy). *American Journal of Roentgenology 51*:685, 1944.

Wilson, F. C.: Fractures and dislocations of the ankle. Rockwood, C. A., and Green, D. P. (Eds.): *Fractures* (Vol. 2). Philadelphia, J. B. Lippincott Company, pp. 1361–1399, 1975.

Wilson, G. E.: *Fractures and their complications.* Toronto, Macmillan Company of Canada Limited, 1930.

Wiltse, L. L.: The etiology of spondylolisthesis. *American Journal of Orthopedics 44A*:539–560, 1962.

Wiltse, L. L., Widell, E. H., Jr., and Jackson, D. W: Fatigue fracture: The basic lesion in isthmic spondylolisthesis. *Journal of Bone and Joint Surgery 57-A*:17–21, 1975.

Wiltse, L. L., Newman, P. H., and Macnab, I.: Classification of spondylolysis and spondylolisthesis. In Urist, M. R (Editor-in Chief): *Clinical Orthopaedics and Related Research* (No. 107). Philadelphia, J. B. Lippincott Company, pp. 23–30, 1976.

Witt, C. M.: The supracondyloid process of the humerus. *Journal of the Missouri Medical Association 47*:445–446, 1950.

Wolff, J. (translated by Scheck, M.): The classic: concerning the interrelationship between form and function of the individual part of the organism. *Clinical Orthopaedics 223*:2–11, 1988.

Wong, B. J., Cervantes, W., Doyle, K. J., Karamzadeh, A. M., Boys, P., Brauel, G., and Mushtaq, E.: Prevalence of external auditory canal exostoses in surfers. *Archives Otolaryngology Head Neck Surgery 125*(9):969–972, 1999.

Wright, W.: Case of accessory patellae in the human subject, with remarks on emargination of the patella. *Journal of Anatomy and Physiology 38*(1):65–67, 1903.

Wright, V.: Osteoarthritis-epidemiology. Presented at the Conference of International Symposium on Epidemiology of Osteoarthritis. Paris, June 30, 1980.

Wroble, R. R., and Weinstein, S. L.: Histocytosis X with scoliosis and osteolysis. *Journal of Pediatrics and Orthopedics 8*(2):213–218, 1988.

Wuyts, W., Cleiren, E., Homfray, T., Rasore-Quartino, A., Vanhoenacker, F., and Van Hul W: The ALX4 homeobox gene is mutated in patients with ossification defects of the skull (foramina parietalia permagna, OMIM 168500). *Journal of Medical Genetics 37*(12):916–920, 2000.

Wyman, J.: *Observations on crania.* Boston, A. A. Kingman, 1868.

Wynne-Davies, R.: Family studies and etiology of club foot. *Journal of Medical Genetics 2*:227–232, 1965.

Wynne-Davies, R., and Scott, J. H.: Inheritance and spondylolisthesis: A radiographic family survey. *Journal of Bone and Joint Surgery 67-B*(3):301–305, 1979.

Yamazaki, T., Maruoka, S., Takahashi, S., Saito, H., Takase, K., Nakamura, M, and Sakamoto, K.: MR findings of avulsive cortical irregularity of the distal femur. *Skeletal Radiology 24*(1):43–45, 1995.

Yazici, M., Kandemir, U., Atilla, B., and Eryilmaz, M.: Rotational profile of lower extremities in bladder exstrophy patients with unapproximated pelvis: A clinical and radiologic study in childern older than 7 years. *Journal of Pediatric Orthopaedics 19*(4):531–535, 1999.

Yekeler, E., Tunaci, M., Tunaci, A., Dursun, M., Acunas, G.: Frequency of sternal variations and anomalies evaluated by MDCT. *American Journal of Roentgenology 186*: 956–960, 2006.

Yochum, T. R., and Rowe, L. J.: *Essentials of skeletal radiology* (2nd ed.). Baltimore, Williams & Wilkins, 1996.

Youmans, G. G. P., Paterson, P. Y., and Sommers, H. M.: *The biologic and clinical basis of infectious disease.* Philadelphia, Saunders, 1980.

Younes, M., Ben Ayeche, M. L., Bejia, I., Ben Hamida, R., Dahmene, J., and Moula, T.: Tubercular abscess of the psoas without associated spinal involvement. A case report (Article in French). *Revue de Medecine Interne* (Paris) *23*(6):549–553, 2002.

Yu, W., Feng, F., Dion, E., Yang, H., Jiang, M., and Genant, H. K.: Comparison of radiography, computed tomography and magnetic resonance imaging in the detection of sacroiliitis accompanying ankylosing spondylitis. *Skeletal Radiology 27*:311–320, 1998.

Zaaijer, T.: Untersuchungen uber die form des beckens javanischen Frauen, Naturrk. Verhandel. Holland Maatsch. *Wentesch Haarlem 24*:1, 1866.

Zabek, M.: Familial incidence of foramina parietalia permagna. *Neurochirurgia* (Stuttgart), *30*(1): 25–27, 1987.

Zaino, E.: *Symmetrical osteoporosis, a sign of severe anemia in the prehistoric Pueblo Indians of the Southwest.* In Wade, W. (Ed.): Miscellaneous papers in paleopathology. Museum of Northern Arizona Technical Series No. 7, 1967.

Zaino, D. E., and Zaino, E. C.: Cribra orbitalia in the aborigines of Hawaii and Australia. *American Journal of Physical Anthropology 42*(1):91–93, 1975.

Zaki, M. E., Sarry El-Din, A. M., Soliman, M. A-T., Mahmoud, N. H., and Basha, W. A. B.: Limb amputation in ancient Egyptians from Old Kingdom. *Journal of Applied Sciences Research 6*(8):913–917, 2010.

Zeppa, M. A., Laorr, A., Greenspan, A., McGahan, J. P., and Steinbach, L. S.: Skeletal coccidioidomycosis: Imaging findings in 19 patients. *Skeletal Radiology 25*: 337–343, 1996.

Zhang, Y., Jun, J., Hiroaki, I., and Katsuya, N.: Footballer's ankle: A case report. *Chinese Medical Journal* (Beijing) *115*(6):942–943, 2002.

Zimmerman, M., and Kelley, M.: *Atlas of human paleopathology.* New York, Praeger, 1982.

Zink, A., Haas, C. J., Reischl, U., Szeimies, U., and Nerlich, A. G.: Molecular analysis of skeletal tuberculosis in an ancient Egyptian population. *Journal of Medical Microbiology 50*:355–366, 2001.

Zinreich, S., Albayram, S., Benson, M., and Oliversio, P.: The ostiomeatal complex and functional endoscopic surgery. In Som P. (Ed): *Head and neck imaging* (4th ed.). St. Louis, Mosby, pp. 149–173, 2003.

Zumwalt, A.: A new method for quantifying the complexity of muscle attachment sites. *The Anatomical Record Part B: The New Anatomist 286B*:21–28, 2005.

_____: The effect of endurance exercise on the morphology of muscle attachment sites. *Journal of Experimental Biology 209*:444–454, 2006.

INDEX

403

416 *Photographic Regional Atlas of Bone Disease*